Oxidative Stress in Applied Basic Research and Clinical Practice

Editor-in-Chief
Donald Armstrong

For other titles published in this series, go to
http://www.springer.com/series/8145

Note from the Editor-in-Chief

All books in this series illustrate point-of-care testing and critically evaluate the potential of antioxidant supplementation in various medical disorders associated with oxidative stress. Future volumes will be updated as warranted by emerging new technology, or from studies reporting clinical trials.

Donald Armstrong
Editor-in-Chief

Lester Mandelker · Peter Vajdovich
Editors

Studies on Veterinary Medicine

Humana Press

Editors
Lester Mandelker
Fellow, American Academy of Veterinary Pharmacology and Therapeutics
Community Veterinary Hospital
Largo, FL 33770, USA
lestervet2@aol.com

Peter Vajdovich
Faculty of Veterinary Sciences
Department of Internal Medicine and Clinics
Szent Istvan University
Istvan u. 2, 1143, Budapest, Hungary
Vajdovich.Peter@aotk.szie.hu

ISBN 978-1-61779-070-6 e-ISBN 978-1-61779-071-3
DOI 10.1007/978-1-61779-071-3
Springer New York Dordrecht Heidelberg London

Library of Congress Control Number: 2011921347

Printed on acid-free paper

Humana Press is part of Springer Science+Business Media (www.springer.com)

Preface

The production of free radicals (ROS) is an unavoidable consequence of life in an aerobic environment. Free radicals produced from the metabolic activities of oxygen attack biological membranes and lipoproteins via oxidation in a process called lipid perioxidation. This attack damages cells and lipids often in a chain reaction with carbon-based molecules such as polyunsaturated fatty acids (PUFA) in a reaction with molecular oxygen. This creates oxidative stress and damage to tissues.

Free radicals also damage chromosomal DNA. It is more likely that damage to DNA occurs from external sources rather than mitochondrial-produced free radicals. Synthetic compounds, pollutants, radiation, xenobiotics (drugs), and food components make up the most likely factors that damage DNA. Proteins are another target of free radicals as proteins and their amino acids can be modified and degraded through free radical mediated reactions. Oxidized proteins then become the target of specialized proteases that turn them into less biologically active smaller peptides and amino acids.

Growing evidence supports a major role of oxidative stress in aging and disease. It has been almost a half century since Denham Harman first proposed the free radical theory in relation to disease. This hypothesis simply states that oxidative stress is the most important determining factor for aging and age-related diseases. Evidence continues to suggest that oxidative stress limits the chronological lifespan which is consistently shortened when there is a reduction of antioxidant enzymes. Furthermore, there is also ample evidence to indicate that reducing oxidative stress is both important and necessary for an extended lifespan. However, even though there appear to be beneficial effects of antioxidant treatment against pathological disease, major preventative clinical trials of dietary antioxidants have failed to prove benefits in increasing longevity. It may be that oxidative stress is more like an active bystander instead of an active component in increasing longevity. No assessment of the free radical theory of aging and pathogenesis of age-related diseases would be complete without an up-to-date account of the major impact oxidative free radicals have in the pathology of disease.

There are various disorders with clear signs of oxidative damage (i.e., paracetamol (Tylenol®) toxicosis, hypoxia reperfusion injury, etc.), and there are some in which oxidative stress is a side effect (i.e., diabetes mellitus, inflammation, liver failure, etc.). Our opinion is that the appropriate way to treat a disorder is multimodular and antioxidant therapy is one important member among the options. In cases of clear oxidative damages we still have to use other treatment therapies (e.g., fluid therapy, antimicrobial therapy, etc.) with antioxidants. There are also diseases in which oxidative stress is a secondary problem, such as diabetes mellitus where the primary treatment is insulin, but antioxidant therapy appears beneficial as well in improving insulin sensitivity.

The purpose of this book is to inform clinicians, students, and others of the vast effects of free radical damage on various cells, tissues, and organs and in different species of animals. In addition, the effects of oxidative stress are analyzed in aging and various disease states such as diabetes, cognitive dysfunction, and heart disease. Each author presents his or her interpretation of the effects of oxidative damage in disease and in several species and the challenges in controlling oxidative damage with antioxidant therapies. We have compiled ideas and scientific information from scientists, veterinarians, and the medical community from around the world. This is surely a universal effort to promote further understanding of oxidative stress and its effects on various animals and organ systems. We would like to commend the authors for their vision and undertaking such a daunting task.

Largo, FL — Lester Mandelker
Budapest, Hungary — Peter Vajdovich

Contents

Contributors

Krisztián Balogh, Ph.D.
Research Group of Animal Breeding and Hygiene, University of Kaposvár, H-7400 Kaposvár, Guba S. u. 40, Hungary
balogh.krisztian@mkk.szie.hu

Wendy Baltzer, D.V.M., Ph.D.
Small Animal Surgery, College of Veterinary Medicine, Oregon State University, Corvallis, OR 97331-4801, USA
wendy.baltzer@oregonstate.edu

Pietro Celi, D.V.M., Ph.D.
University of Sydney, C04 MC Franklin Lab, PMB 3, 425 Werombi Rd, Camden, NSW 2570, Australia
pietro.celi@sydney.edu.au

Stefano Comazzi, D.V.M., Ph.D., dipl. ECVCP diabetes
University of Milan, Department of Veterinary Pathology, Hygiene, and Health, Section of Veterinary General Pathology and Parasitology, Via Celoria 10, Milan 20133, Italy
stefano.comazzi@unimi.it

Elizabeth Head, Ph.D.
Sanders-Brown Center on Aging, University of Kentucky, 800 S. Limestone St, Lexington, KY 40536, USA
Elizabeth.head@uky.edu

John Kuck, D.V.M.
Aspen Animal Hospital, Nutrient Selection in the Management of Canine Atopic Dermatitis, Aspen, CO, USA
john_kuck@yahoo.com

Lester Mandelker, D.V.M., Dipl. ABVP
Fellow, American Academy of Veterinary Pharmacology and Therapeutics, Community Veterinary Hospital, Largo, FL 33770, USA
lestervet2@aol.com

Dr. Kevin J. McGraw, M.S., Ph.D.
School of Life Sciences, Arizona State University, Tempe,
AZ 85287-4501, USA
kevin.mcgraw@asu.edu

Miklós Mézes, Ph.D.
Department of Nutrition, Szent István University, H-2103 Gödöllő, Hungary
Mezes.Miklos@mkk.szie.hu

Gordon Moe, M.Sc., M.D., F.R.C.P.(C.), F.A.C.C.
Li Ka Shing Knowledge Institute, St. Michael's Hospital, University of Toronto,
Toronto ON, Canada M5B1W8
moeg@smh.toronto.on.ca

Mustafa Nazıroğlu
Department of Biophysics, Süleyman Demirel University, Morfoloji Binasi,
TR-32260 Isparta, Turkey; Neuroscience Research Center,
Suleyman Demirel University, Isparta, Turkey
mnaziroglu@med.sdu.edu.tr

Patricia A. Schenck, D.V.M., Ph.D.
Endocrine Diagnostic Section, Diagnostic Center for Population and Animal Health,
Michigan State University, 4125 Beaumont Rd, Lansing, MI, USA
schenck5@msu.edu

Peter Vajdovich, D.V.M., Ph.D. Dipl. ECVCP
Faculty of Veterinary Sciences, Department of Internal Medicine and Clinics,
Szent Istvan University, Istvan u. 2, 1143, Budapest, Hungary
Vajdovich.Peter@aotk.szie.hu

Gustavo L. Zapata, D.V.M.
Servicio Central de Laboratorio, Facultad de Ciencias Veterinarias,
Universidad Nacional de La Plata, La Plata, Buenos Aires, Argentina
zapatagu@fcv.unlp.edu.ar

Steven C. Zicker, D.V.M., Ph.D., Dipl. ACVIM, Dipl. ACVN
Pet Nutrition Center, Hill's Pet Nutrition, Inc,
P.O. Box 1658, Topeka, KS 66601, USA
steven_zicker@hillspet.com

Oxidative Stress, Free Radicals, and Cellular Damage

Lester Mandelker

Abstract Oxidative stress is a term relative to the elevated levels of reactive free radicals in an organism. Oxidative stress can occur from diminished antioxidants and/or increased production of reactive free radicals such as reactive oxygen species and/or reactive nitrogen species (ROS/RSN). The increased production of free radicals is more relevant to disease and frequently the attempted target of supplementation intervention. In many instances the body can adapt to an increase in oxidative stress by upregulation of antioxidant defense systems. If the oxidative stress can be neutralized, there is often no adverse contribution to disease pathology. If the antioxidant defense induction is inadequate or nonexistent then accompanying cellular and tissue damage often occurs. Some diseases can be caused directly by oxidative stress, however, in most diseases oxidative stress is a consequence and may often only be a secondary event. It does, however, play an important role in promoting additional tissue injury in most diseases. On the other hand, oxidative stress may have beneficial effects in activating biological pathways that alter antioxidant defenses and allow an organism to adapt. Oxidative stress is also considered necessary to promote healing and repair of tissues. Therefore, not all cases of oxidative stress are damaging. It is only when oxidative stress is excessive and inappropriate should we address it with supplementation and antioxidant therapy that reduces oxidative damage to cells, tissues, proteins, cellular membranes, and mitochondria.

Keywords Oxidative stress • Cell homeostasis • Nuclear factor kappa B (NF-kB) • Mitochondrial permeability pore transition (MPT) • Gap junctional intercellular communication (GJIC) • Glutathione

L. Mandelker (✉)
Fellow, American Academy of Veterinary Pharmacology and Therapeutics,
Community Veterinary Hospital, Largo, FL 33770, USA
e-mail: lestervet2@aol.com

L. Mandelker and P. Vajdovich (eds.), *Studies on Veterinary Medicine*,
Oxidative Stress in Applied Basic Research and Clinical Practice 5,
DOI 10.1007/978-1-61779-071-3_1, © Springer Science+Business Media, LLC 2011

Oxygen

Earth is the only planet in our solar system known to have sufficient oxygen to maintain life as we know it. As the O_2 content of the atmosphere rose, it exposed living organisms to oxygen toxicity. In its ground state (its normal configuration; O_2) molecular oxygen is relatively inert. Molecular oxygen (O_2) is the premier biological electron acceptor that serves vital roles in fundamental cellular functions. However, during normal metabolic activity, and as a consequence of various environmental disturbances such as extreme temperatures, radiation, xenobiotics, toxins, air pollutants, and various stresses and diseases, O_2 is capable of giving rise to reactive excited states such as free radicals and their derivatives [1]. With the beneficial properties of O_2 comes the inadvertent formation of reactive oxygen species (ROS) such as superoxide ($O_2^{\bullet-}$), hydrogen peroxide (H_2O_2), and hydroxyl radical ($OH^{\bullet}$). If unabated, ROS pose a serious threat, possibly causing the death of aerobic cells. Aerobic metabolism using oxygen (oxidative phosphorylation) in mitochondria is highly efficient in producing energy from organic compounds and remains the major form of energy production in virtually all animals [1]. Despite its beneficial effects there remains a fundamental problem: how to protect cells and specifically mitochondria from oxygen toxicity. To minimize the damaging effects of ROS, aerobic organisms evolved nonenzymatic and enzymatic antioxidant defenses. The latter include catalases, peroxidases, superoxide dismutases (SOD), and glutathione S-transferases. Thus, oxygen is a "double-edged sword" in that it makes life on earth possible, but in its radical form (ROS), it is highly toxic and lethal.

Oxidative Stress

Oxidative stress occurs when there is an increase in oxidant production and free radical formation that exceeds the body's ability to neutralize and eliminate these reactive radical forms. There is a great deal of evidence that oxidative stress is involved in many animal and human diseases. However, in many cases it may be that oxidative stress is a complicating factor of the pathological process and not necessarily the primary cause. The involvement of oxidative stress often depends on the nature of the disease. Every cell undergoes oxidative stress at some point in its existence and this is especially true in disease and aging tissues. All organ systems are involved with oxidative stress but when the oxidative stress is excessive it can lead to organ damage, which has a direct effect on the health of the body. The organ systems especially affected include kidneys, liver, pancreas, heart, CNS/nervous tissue, intestines, and adrenal, bone marrow, lungs, and thyroid tissues. Therefore, oxidative stress not only has an impact on cells and cellular components but on most organ systems especially the neuroendocrine system as well as the immune status and aging [2]. Common causes of oxidative stress include: infections, toxins, hypoxia/ischemia, hyperglycemia, xenobiotics (drug metabolism), hyperlipidemias, hyperproteinemias, cancer, inflammation, phagocytic and immune reactions, and elevated metabolic rates. Additionally, aging tissues often undergo oxidative stress because mitochondria fail to produce enough ATP (energy) to sustain optimum health.

Oxidative Stress and Free Radicals

In healthy animals, production of free radicals is approximately balanced by antioxidant defenses. The balance is not perfect, therefore continual damage occurs when antioxidant defenses are inadequate. Free radicals are molecules that contain one or more unpaired electrons and are capable of independent existence. Most free radicals are derivatives of oxygen (reactive oxygen species) or ROS and derivatives of nitrogen (reactive nitrogen species) or RON. They are formed wherever there is disease and cell damage and are especially produced in large quantities during infection [3]. They are even produced under normal conditions from the cell's mitochondria during energy production from the intake of oxygen. Thus, even in healthy animals up to 25% of the oxygen they breathe forms free radicals. However, in sick animals up to 75% of the oxygen may form free radicals. The presence of free radicals is not always a bad thing; in fact free radicals can act to fight infection and disease. This is the primary purpose of bacterial phagocytosis which involves an oxygen-dependent system. In bacterial phagocytosis, neutrophils show a burst of metabolic activity characterized by an increase in oxygen consumption and a subsequent increase in free radical production such as superoxide, hyroxyl radical, and hydrogen peroxide (O_2^-, OH^-, H_2O_2). This process has been termed the "respiratory burst" of phagocytosis [4]. In addition, mild elevations of free radicals drive many physiological and metabolic pathways [5] and only when there are excessive and inappropriate levels of free radicals that may accumulate during cell injury and disease do they cause oxidative damage to protein molecules, cell membranes, mitochondria, and DNA [6].

Hypoxia and Oxidative Stress

Hypoxic tissues and tissues that undergo ischemia–reperfusion cycles cause an increase in free radical production and increase in oxidative damage. The severity of the damage depends on the tissues involved and the length of ischemia. Tissues respond in several ways: the early response includes glycogen depletion causing energy levels to decrease, which depletes ATP. This causes mitochondria to dysfunction. Damaged and aging tissues also undergo oxidative stress when the mitochondria fail to produce enough energy (ATP) to sustain maximum health. This also results in mitochondrial dysfunction [7]. Mitochondrial dysfunction encompasses all the clinical diseases associated with the defective energy metabolism primarily of oxidative phosphorylation (production of ATP). These diseases also include multiple neurodegenerative conditions, cancer, and the skeletal muscle disorders and are termed mitochondrial cytopathies.

Cell Homeostasis

Cell homeostasis is defined as cells in oxidative balance. This occurs when there is a balance of antioxidants (known as reducing agents) and oxidants (known as oxidizing agents). This balance is referred to cell redox. When the balance of antioxidants to oxidants shifts to high levels of oxidants the cell undergoes oxidative stress. Oxidative stress then can also be defined as the disruption or alteration of cell redox or cell homeostasis that favors oxidants. A progressive increase in oxidative stress because of altered redox homeostasis is important in processes that regulate gene transcription in both normal and abnormal health [8]. Therefore, ROS/RSN serve as signaling molecules for the initiation and perpetuation of the inflammatory process that occurs with conditions of oxidative stress. This involves genetic regulation. Transcription factors that are directly influenced by reactive species (redox sensitive) and proinflammmatory signaling include nuclear factor kappa B (NF-kB), activated protein (AP-1), and hypoxia-inducible factor –1@ (HIF-1@) [9]. Hence, sustained oxidative stress leads to inflammation by means of upregulation of transcription factors such as NK-kB that alter genetic function and induce inflammation. Following cell injury NF-kB, which is the primary transcription factor, translocates into the nucleus and targets genes primarily concerned with cellular repair and proliferation and this often promotes inflammation [10].

Significance of NF-kB

Often referred to as the master regulator of genetic function, this signaling protein molecule is activated by cell damage and free radicals and hence is redox sensitive. It is so potent that it is contained in a protein called IkB or inhibitory kB. Before NF-kB can be activated in the cell the inhibitory protein IkB must be inactivated. The effects of stimulation of genes by NF-kB is very profound and is the driving force in cellular repair, growth, and cellular inflammation. NF-kB plays a central role in regulating genetic transcription and encoding of inflammatory cytokines, growth factors, acute phase proteins, adhesion molecules, other transcription factors, and cell death regulators. These NF-kB regulated genes are important in regulating genetic activity during critical illness, inflammatory diseases, and cancer [11]. One can clearly see the significance of the activity of NF-kB and how modulating this activity would have profound effects on inflammation, cell growth, and cancer. There are many supplements, proteins, antioxidants, minerals and vitamins, and drugs that can modulate NF-kB activity. Some work by increasing inhibitory actions of IkB, some work by inactivation of NF-kB, and some antagonize binding of NF-kB, but in the end they decrease the actions of NF-kB which has profound effects in reducing inflammation and cancer [12]. The activation of NF-kB is not always a detriment. It is the primary method that mediates cell growth and repair and protection from TNF-alpha activation-induced cell death. Many cytokines act directly or indirectly through NF-kB including transforming growth factor-B (TGF-beta), ROS, prostaglandins (PGs), leukotrienes, nitric oxide (NO), protein kinases, and certain hormones and growth factors [13].

Measure Oxidative Stress

We can measure oxidative stress in many ways: one way is to measure GSH/GSSH (reduced glutathione/oxidized glutathione). Reduced glutathione (GSH) is the primary antioxidant found in cells. Most cats and dogs that have liver and kidney disease or other chronic disease have reduced levels of glutathione. It is estimated that over 75% of cats with chronic disease have depleted levels of cellular glutathione. By supplying agents (antioxidants) that improve glutathione levels in the cells we often see a beneficial impact on the progression of many feline diseases. For example, in cats that are poisoned with acetaminophen (Tylenol®) glutathione depletion occurs in RBCs. This condition is best treated with glutathione inducers (thiol antioxidants) such as N-acetylcysteine, SAM-e, alpha lipoic acid, and/or taurine [14]. Another indicator of oxidative stress in the feline is elevation of the acute phase protein produced by the liver called serum amyloid A. This can be measured as there are assay kits available [15]. This indicator of oxidative stress and inflammation is similar to C-reactive protein that is more often used in people and dogs to measure inflammation. There are several other inflammatory chemicals that can be measured in blood or urine to confirm excessive oxidative stress. Nitric oxide and lipid perioxidation products such as malondialdehyde (MDA) are such examples.

Managing Oxidative Stress

Cells can usually tolerate mild oxidative stress, which often results in the upregulation of antioxidant synthesis in an attempt to restore cell redox (cell homeostasis) or the antioxidant–oxidant balance of the cell. Mild oxidative stress upregulates defenses so as to protect against more severe oxidative stress. This mechanism includes cellular adaptation which often involves changes in gene expression that result in elevated antioxidant defenses. However, if adaptation does not improve the antioxidant defenses, permanent cell injury occurs. In many cases, permanent injury alters the homeostasis of the cell and the cell enters a temporary prolonged altered injurious state that may or may not lead to cell death [15].

Cell Death

Cells die by necrosis or apoptosis. Necrosis is the result of cellular damage that cannot be repaired and is the result of internal or external forces or chemical or physical events that destroy cells. The progression of events as a result of necrosis is that cells swell, rupture the outer cell membranes, and eventually die and release biochemical mediators of inflammation into the systemic circulation. The cellular death that follows often induces a well-orchestrated series of cellular and biochemical events that commonly provoke a strong inflammatory response. Inflammation consists of a series of physiological reactions by the body that brings cells and molecules of the immune system to the site of cell injury [16]. The net result might appear in the form of increased blood supply, increased migration of leukocytes, and increased vascular permeability. Immediate biological mediators include PGs, leukotrienes, serotonins, histamine, platelet-activating factor, and others [17]. These are released during cell death and may also act as signaling molecules in the acute phase of inflammation. During sustained injury, other biological mediators perpetuate cellular inflammation by additionally activating chemical mediators such as TNF-alpha, substance P, acute phase proteins (C-reactive protein, Serum Amyloid A), interferons (i.e., IL-1, IL-4, etc.), adhesion molecules (VEGF), colony stimulating factors such as TGH beta, and so on [18]. This chronic response involves a more extensive commitment to inflammation. This inflammation phase can occur from continued, acute, nonspecific stimulation or from sustained immunologic stimulation. In addition, enzymes are produced by genetic stimulation (upregulation) due to chronic inflammation such as MMPs (matrix metallo-proteinases) that act to destroy the intercellular matrix between cells which promotes more inflammation and may perpetuate carcinogenesis [19].

Apoptosis

Programmed cell death (apoptosis) is different from necrosis in that it is a restrained cell death. It can be a normal physiological function of growth and development or can occur from oxidative stress to the cell and/or loss of energy production by the mitochondria. The activation of apoptosis in individual cells is based on its environment, internal metabolism, genetic information, or other various external or internal forces. One such mechanism involves the cell death protein P-53 which acts as an internal signal to induce cells to die. During this process, the cell shrinks, the nucleus condenses, and DNA fragmentation occurs [20]. However, there is no disruption of the outer membrane in contrast with necrosis and no inflammatory mediators are released into the circulation.

Apoptosis is a quiet suicide with little or no inflammation involved. Research has revealed that the mitochondria's integrity is the determining factor of whether cells live or die. Changes in both the inner and outer mitochondrial membrane lead to a disruption of the membranes, opening of the mitochondrial pores, and release of cytochrome C into the cytoplasm. This process initiates cell death by activation of the death proteins called caspases [21]. Caspases are very sensitive to the redox status of the cell and reduced levels can block their activity. Thus, alterations of intracellular redox status might trigger or block the apoptotic death proteins.

Significance of Cell Death

Knowing how cells die might allow us to find better ways of treating cell injury before cell death becomes apparent. For example, we know cell injury just prior to cell death often causes the mitochondrial pores to open (mitochondrial permeability pore transition or MPT) and this allows Ca^{2+} to enter. This starts the death pathway first beginning with mitochondrial release of cytochrome C. We have drugs that can block the pore opening such as cyclosporine. The use of cyclosporine reduces cell death by inhibiting MPT. Inhibiting MPT explains the effectiveness and why it is effective in transplant rejection disease (graft vs. host disease). Some supplements such as melatonin also inhibit MPT and are therefore beneficial in reducing many forms of mitochondrial dysfunction and cell death.

Chronic Oxidative Stress

Chronic oxidative stress can also lead to cancer. The upregulated or prolonged production of cellular oxidants has been linked to DNA mutation, cell proliferation, and cellular growth. All these are precursors to carcinogenesis [22]. Oxidative stress often damages DNA. DNA is found in both the nucleus and the mitochondria. The mitochondria is the only cell structure other than the nucleus to contain DNA (mtDNA). Nuclear DNA contains two copies of DNA, one inherited from the father and the other from the mother. On the other hand, mtDNA is different in that virtually all the mitochondria DNA mutations occur only from the mother's DNA. The free radicals and oxidative stress that damage mtDNA often reduce the ability for mitochondria to replicate, reduce the energy output by the mitochondria (ATP production), and lead to deletions, mutations, and ways of treating cell injury before cell death becomes apparent [23].

Response to Oxidative Stress

The body handles excessive oxidative stress by using both enzymatic and nonenzymatic antioxidants to neutralize damage from free radicals. A network of antioxidants is best able to handle excessive oxidative stress. So it is better to offer multiple antioxidants vs. just one or two. SOD supplements act to replenish antioxidant enzymes in the body. These are often depleted in chronic oxidative stress conditions and supplementation often helps reduce oxidative damage to tissues and cells [24]. Nonenzymatic supplementation takes the form of oral or injectable antioxidants such as reduced glutathione, Vitamin E, alpha lipoic acid, N-acetylcysteine, SAM-e, Vitamin C, flavanoids/polyphenols, and minerals including zinc and selenium. Some of these antioxidants replenish intracellular glutathione (i.e., alpha lipoic acid, N-acetylcysteine, SAM-e), which is the most important and the only available antioxidant available to the mitochondria. The body responds to oxidative stress through complex signaling pathways [25]. In this regard, these tissues are considered to be redox-sensitive as they respond to imbalances in the oxidant–antioxidant levels. Many times the body responds in a positive way to oxidative stress and in fact, this physiological mechanism drives vital cellular functions. Oxidative stress can often cause the body to adapt to the stress in a positive manner. For instance, by increasing mitochondrial numbers in tissue via genetic stimulation the body can adjust to oxidative stress in an appropriate manner. This often occurs in myocardial muscle tissue from which a cell modulates the intracellular redox state of the cell by increasing energy production.

Oxidative Stress and Cellular Communication

Cells carry on their daily activities by using signaling molecules, which are specialized proteins that act on receptors inside cells. This is referred to as intracellular communication. Intracellular communication occurs in cells via signaling molecules such as NF-kB. NF-kB is the primary intracellular signaling molecule that activates genes inside the cell and is considered the master regulator of numerous genes that involve the immune and inflammatory response. It is activated by ROS/RSN, toxins, hormones, hyperglycemia, and various inflammatory cytokines. Cells must communicate with neighboring cells to survive. This intercellular communication occurs between same cells of the same tissue. The communication between such cells occurs through very small junctions. This type of cellular communication is called gap junctional intercellular communication (GJIC). Cells share ions and electrolytes through these gap junctions made up of connexin molecules [26]. These protein molecules also contain genes. Connexin genes account for many of the cellular responses to inflammation and disease and are influenced by many of the inflammatory mediators such as PGs, leukotrienes, kinins, serotonin, platelet activating factor (PAF), TNF-alpha, adhesion molecules, interleukins, and colony growth factors (i.e., VEGF, vascular endothelial growth factor) [27]. The importance of GJIC cannot be understated. In many diseases, toxins, and physical events such as a burn GJIC are compromised. When this occurs independent cells either die or eventually become cancerous. For example, phenobarbitol is a known liver toxin. It acts in the liver to disrupt and destroy normal GJIC. This is the same mechanism of action in the brain that reduces seizures. However, in the liver the use of phenobarbitol which causes interruption of GJIC can promote carcinogenesis.

Significance of GJIC

GJIC is disrupted especially where there is extensive inflammation, oxidative stress, and cellular damage as seen with extensive injury and chronic diseases such as liver disease, kidney disease (glomerulonephritis), pancreatic disease, heart disease, bladder inflammation, and other similar conditions. Abnormal intercellular gap junctional communication has been implicated in tumor promotion, neuropathy, and angiogenesis [28]. Oxidative stress has also been implicated in similar diseases such as cancer. This appears to be a direct link to oxidative stress and GJIC as increasing oxidative stress causes more interruption of GJIC. Because there are genes located in the gap junctional areas, oxidative stress acts to stimulate many genetic functions involving inflammation and promotion of disease. There are many nutraceuticals and antioxidants that restore or improve GJIC. Carotenoids especially improve and repair GJIC [29]. DMSO is a chemical solvent that repairs cell membranes and also acts to improve GJIC. Their use in such diseases offers a novel method of altering disease progression and in this manner are considered disease-modifying agents.

Glutathione and Oxidative Stress

Glutathione is the major intracellular antioxidant by which oxidative stress is measured. It is the only antioxidant available to mitochondria. The imbalance of antioxidants to pro-oxidants can be quantified in plasma as the redox state of reduced glutathione (GSH) to oxidized glutathione (GSSG) or GSH/GSSG [30]. In health, the high efficiency of GSSG reductase often can maintain the cellular GSH pool in a predominately reduced state (intracellular GSH/GSSH >98%). However, this reduction of oxidized glutathione (GSSH) to reduced glutathione occurs only in the presence of nicotinamide adenine dinucleotide phosphate (NADPH). By supplying needed levels for this cellular reaction necessary cellular NADPH stores may be diminished [30]. NADPH is vital for scavenging toxic free radicals in the mitochondria and cytosol where it is produced. Hence, in certain situations NADPH availability might be the rate limiting factor in GSH regeneration. These complex factors reaffirm the complex interaction of vital nutrient chemicals in maintaining the balance in cell redox. GSH-Px (glutathione perioxidase) is an antioxidant enzyme that plays an essential role in stabilizing the cell redox. To achieve maximum benefits in modulating excessive oxidative stress certain nutrients, cofactors, enzymes, and vitamins should be supplied to the patient in times of physiological need.

Oxidative Stress and β-Cell Dysfunction

It has been well established that there has been a recognized link between the presence of chronic hyperglycemia and the progressive deterioration in β-cell function seen in patients and animals with diabetes. More recently, however, studies have indicated that this progressive β-cell dysfunction is a result of tissue damage induced by oxidative stress resulting from this hyperglycemia. β-cells are thought to be particularly vulnerable to oxidative stress because they contain very low levels of antioxidant enzymes [31]. In support of the hypothesis that chronic oxidative stress might play a role in the progressive β-cell dysfunction seen in type-2 diabetes are the findings that the pancreatic β-cell undergoes oxidative stress when exposed to supraphysiological concentrations of glucose and that this process can be prevented by an antioxidant. The results of a number of studies in vivo support these findings. In one study, antioxidant treatment was found to normalize plasma glucose levels and to restore insulin secretion in a diabetic rat model [32]. During the generation of oxidative stress, prolonged elevations in blood glucose levels lead to, among other things, the activation of various intracellular metabolic pathways, promoting the formation of advanced glycation end-products (AGEs), auto-oxidation, and an increase in the activity of the sorbitol pathway [33]. A number of important proteins also undergo glycation, such as the Cu,Zn-SOD, one of the most important antioxidant enzymes. Erythrocytes in patients with type-1 diabetes have been found to contain a higher percentage of glycated Cu,Zn-SOD, which is inactivated under hyperglycemic conditions compared with controls, thus leading to oxidative stress [34].

Oxidative Stress and Hypertension

Research has determined that oxidative stress plays an important role in the pathogenesis of hypertension. Several research scientists [35] showed that in response to hypertension the free radical superoxide accumulated in the extracellular space and contributed to the impairment of a normal vascular response by inhibiting the endothelial-derived relaxing factor now known as nitric oxide. In addition, it was shown by adding SOD, an antioxidant enzyme, the impaired vascular response could be restored. These findings undoubtedly prove the impact of changes in NO bioavailability and that increases in ROS reduce the bioavailability of NO resulting in increases in vascular tone. Thus, increases in ROS would be suspected to have a significant influence on blood pressure. There also appears to be a link between oxidative stress and angiotensin II. Angiotensin has been shown to generate ROS and beneficial results occur when using agents that reduce ROS such as ACEI (angiotensin-converting enzyme inhibitors) and/or angiotensin receptor blockers. These results confirm the role of angiotensin II-derived ROS in hypertension [36].

Another major source of ROS in vasculature is NADPH oxidase. Pharmacological or molecular inhibition of NADPH oxidase lowers blood pressure, reduces oxidative stress, and improves vascular response in hypertension. It has been determined that ACEI and angiotensin receptor blockers reduce expression of NADPH oxidase as does supplementation with niacinamide, a precursor to NADPH. By supplying the antioxidant niacinamide one can overcome excessive enzymatic destruction by the enzyme NADPH oxidase [37]. Antioxidant treatment in organ systems such as kidneys and CNS also have beneficial blood pressure lowering effects in hypertension.

This reaffirms that renal and central mechanisms that regulate blood pressure are also affected by oxidative stress. For example, SOD protects against increases in superoxide and endothelial dysfunction produced by angiotensin II. Glutathione peroxidase has also been shown to be of benefit in reducing hypertension and reducing vascular tone [38].

Oxidative Stress and Iron

Unbound transition metals such as iron have long been recognized as a potent source of ROS. This occurs as a result of their reaction with the superoxide ion via the Fenton reaction. The Fenton reaction promotes free iron to react with the superoxide ion, the by-product of energy production from the mitochondria to form the dangerous free radical OH− molecule [39]. In the presence of adequate SOD the reaction is reduced and less harmful H_2O_2 is formed and upon further degradation with the antioxdant enzyme catalase forms H_2O and O_2. The iron binding agent, desferrioxamine, has also been shown to decrease ROS stress from iron accumulation in mitochondria undergoing oxidative stress. These findings support the notion that therapeutic intervention with mitochondrial targeted antioxidants can reduce mitochondrial dysfunction caused by free iron and oxidative stress.

Oxidative Stress and Inflammatory Mediators

Oxidative stress aids in the breakdown of arachidonic acid (AA) in response to inflammation and cell injury. This inflammatory reaction activates phospholipases which then act on phospholipids to form AA. AA is an inflammatory molecule and the processes associated with inflammatory responses are complex and often involve ROS. There are many mediators that initiate and amplify the inflammatory response such as histamine, bradykinin, serotonin, proinflammatory cytokines [interleukin-1B (IL-1b) and tumor necrosis factor (TNF-alpha)], inflammatory cells (eosinophils, macrophages) and metabolic products of AA (thomboxane A(2), PGs, and leukotrienes) [40]. The PGs, thromboxanes, and leukotrienes constitute a rapidly growing family of compounds, all of which are oxygenated derivatives of certain polyunsaturated fatty acids, such as AA. Most of these metabolites are biologically very potent substances, displaying a wide variety of actions in many different biological systems. In addition, isoprostanes are PG-like compounds formed in vivo from the free radical-catalyzed peroxidation of essential fatty acids (primarily AA) without the direct action of cyclooxygenase (COX) enzyme. COX activity produces H_2O_2 which may nonenzymatically produce isoprostanes. A large body of evidence indicates that measurement of F2-isoprostanes, specific PG F2-like compounds derived from the nonenzymatic peroxidation of AA, is a reliable biomarker of oxidant stress in the body [41].

Oxidative Stress and Sepsis

In sepsis, there are several potential sources of ROS, including the mitochondrial respiratory electron transport chain, xanthine oxidase activation as a result of ischemia and reperfusion, the respiratory burst associated with neutrophil activation, and AA metabolism. Activated neutrophils produce superoxide as a cytotoxic agent as part of the respiratory burst via the action of membrane-bound NADPH oxidase on molecular oxygen. Neutrophils also produce the free radical nitric oxide, which can react with superoxide to produce peroxynitrite, itself a powerful oxidant, which may further proceed to form the hydroxyl radical. Under ischemic conditions followed by subsequent reperfusion, the enzyme xanthine oxidase catalyzes the formation of uric acid with the coproduction of superoxide. Superoxide release results in the recruitment and activation of neutrophils and their adherence to endothelial cells, which stimulates the formation of xanthine oxidase in the endothelium, with further superoxide production [42]. During oxidative stress, damage mediated by ROS can occur. Oxidation of DNA and proteins may take place, along with membrane damage, because of lipid peroxidation, leading to alterations in membrane permeability, modification of protein structure, and functional changes. Oxidative damage to the mitochondrial membrane can also occur, resulting in membrane depolarization and the uncoupling of oxidative phosphorylation, with altered cellular respiration [43]. This can ultimately lead to mitochondrial damage, with release of cytochrome C activation of caspases and finally apoptosis.

Oxidative Stress and Aging

Accumulating evidence supports a role of oxidative stress in aging. In as much as mitochondria are the main source of free radicals produced during energy, it was theorized some 30 plus years ago that their dysfunction could be responsible for aging due to progressive decline from molecular oxidative damage. However, the role of mitochondria in the model of senescence has been largely discarded because earlier research has failed to confirm this theory. Instead, the discovery of telomeres, the ends of chromosomes that shorten with repeated cell division, suggests that senescence is the result of the number and length of telomeres and that determines the biological clock and as such is not compatible with ROS/RSN-derived molecular damage as a cause of aging [23]. On the other hand, senescent cells display mitochondrial dysfunction characterized by lower mitochondrial membrane potential, mitochondrial DNA damage (mtDNA), and increased superoxide production [23]. This is evident by improvement of mitochondrial function by antioxidant intervention and extension of lifespan. It appears that targeting antioxidants directly to the mitochondria counteracts telomere shortening in cells undergoing mild oxidative stress [23]. Mitochondrial damage has also been shown to decrease the replicative life span of cells. From this information it appears that mitochondrial ROS/RSN do play a part in telomere-dependent replicative senescence although the level and importance have yet to be defined. Oxidative stress in aging remains one of the most popular theories of aging at the cellular levels. The imbalance of pro-oxidants to antioxidants causes excessive destructive free radical chemistry. Thiol systems (sulfur-type antioxidants) are important in the control of these processes, both by protecting against damage and serving in redox signaling mechanisms to sense danger and repair the damage. Studies by a number of research groups in collaboration with the Emory Clinical Biomarkers Laboratory show that the redox state of the central tissue antioxidant, glutathione (GSH), can be measured in plasma and provides a quantitative systemic indicator of oxidative stress [44].

Other Theories of Aging

1. Wear and tear; cell wear is a significant factor in aging tissues.
2. Waste accumulation during aging in cells disrupts normal cell functions.
3. Error catastrophe occurs from aging and damage to transcriptional and translational pathways.
4. Somatic mutations from repeated DNA damage alter genetic information.
5. Cross-linking and glycation of proteins with glucose alter their functions.
6. Dysfunction and impaired regulation of gene activation and repression mechanisms.
7. Increased metabolic rate as lifespan inversely correlates with it in mammals.
8. Neuroendocrine dysfunction controls normal physiological homeostasis.
9. Immune dysfunction; decline in immune function leads to decreased resistance to disease and infections and increases in autoimmunity [45].

Benefits of Oxidative Stress

Exposure to mild oxidative stress causes an increase in synthesis of antioxidant enzymes and other defenses. These responses help protect the cell against cell damage and more oxidative damage. This is accomplished by an increase in regulation of transcription and redox regulation of protein binding to mRNA which may increase antioxidant enzymes such as catalase activity. This is accomplished through redox-sensitive transcription factors such as NF-kB [46]. Benefits of oxidative stress include adaptation and promotion of cellular repair. Adaptation occurs when oxidative stress signals transcription factors such as NF-kB and AP-1 to upregulate antioxidant defenses and/or to increase cellular resistance to oxidative stress. This may take the form of increases in the numbers of mitochondria which would have the effect of increasing the total energy of a tissue. This commonly happens in cardiac muscle when more energy is needed to maintain muscular contractions. Hypertrophy is another adaptation that occurs in muscular tissue (skeletal, cardiac) which again allows the tissue to respond to oxidative stress by increasing the strength of contractions. The same reaction occurs when exercise is used to build muscle tissue. Skeletal muscular contraction, growth, differentiation, and adaptation all occur as a result of mild oxidative stress [47].

Cellular replacement and wound repair are other indications where transcription factors and oxidative stress are necessary. In the intestinal tract NF-kB acts to control the maintenance of tissue-immune homeostasis. In normal cells, this is very important. In pathological states where there is excessive oxidative stress, suppression of NF-kB is important in controlling excessive inflammation and disease. This dual action of NF-kB demonstrates the importance it has in regulating the immune response in both normal and disease conditions.

Summary

Despite the benefits of oxygen there remains the fundamental problem of how to protect cells and specifically mitochondria from oxygen toxicity and oxidative stress. Oxidative stress occurs during disease and the extent to which it causes significant damage depends on the antioxidant status of the animal and the disease process. When persistent and inappropriate levels occur, oxidative stress can damage cells and alter various physiological processes including an increase in aging. The body responds to oxidative stress in many ways, including induction of intracellular and intercellular signaling pathways, genetic alterations, and changes in physiological function and adaptation. These changes are necessary for cells and animals to evolve. The use of antioxidants to reduce excessive and inappropriate oxidative stress is often beneficial but not in all cases. Inappropriate and excessive antioxidant supplementation may impede the body's ability to adapt; thus during cell repair and tissue growth, excessive antioxidant supplementation may be counterproductive. Where antioxidant therapy appears beneficial is in diseases aggravated by excessive and inappropriate oxidant production.

It is much more appropriate to use multiple antioxidants than single ones. These should include both enzymatic antioxidants such as SOD and supplemental nonenzymatic antioxidants such as vitamins A, B, C, D3, and E at supraphysiological doses. Minerals such as selenium, magnesium, and zinc, along with thiol (sulfur) amino acid supplements that increase cellular glutathione (i.e., SAM-e, alpha lipoic acid, taurine, N-acetylcysteine, reduced glutathione) also benefit cells from oxidant damage. Additional benefits can be obtained with coenzyme Q10, and numerous other proteins including l-carnitine, l-carnosine, chondroitin, and glucosamine targeted for specific tissues. In addition, hormones such as melatonin, DHEA, iron binders such as lactoferrin, and omega-3 fatty acids all show the ability to reduce the damaging effects of oxidative stress on cells. Recently, it has been verified that diets high in omega-3 fatty acids reduce aging and increase longevity by altering the shortening of telomeres, a known contributor to aging. Other beneficial supplements include many polyphenols and flavonoids such as green tea extract (catechins ECGC), grape seed extract (proanthocyanidins), quercetin, reservatrol, tumeric, bilberry, and silymarin which all have various antioxidant and anti-inflammatory effects. For a more complete listing of supplements that reduce or alter oxidative stress please consult the table at the end of the book.

Adaptogens, herbal extracts are nontoxic substances that improve the body's ability to handle stress have also demonstrated the ability to reduce oxidative stress. Such adaptogens include Gingseng (eleutherococcus), licorice (Glycyrrhiza), Goldenseal (Hydrastis), Astragalus, Hawthorn, Golden Root (Rhodiola), Ashwandgha, and Schizandra and are often formulated in a combination of products but none presently exists for use in animals such as there is for people (i.e., Prime One). For a complete summary of supplements that alter oxidative stress please consult the table at the end of the book.

References

1. Miwa S, Beckman KB, Muller FL. Oxidative stress in aging. Totowa: Humana Press; 2008: pp. 4–9
2. Halliwell B, Gutterridge JMC. Free radicals in biology and medicine. Oxford: Oxford University Press; 2003: pp. 62–66
3. Southern PA, Powis G. Free radicals in medicine. 1. Clinical nature and biological reactions. Mayo Clin Proc 1988;63(4): 381–389
4. Riley PA. Free radicals in biology: Oxidative stress and the effects of ionizing radiation. Int J Radiat Biol 1994;665(1): 27–33
5. Vajdovich P. Oxidative stress: The role of mitochondria, free radicals and antioxidants. Vet Clin North Am 2008;1(1): 50–54
6. Dröge W. Free radical in the physiological control of cell function. Physiol Rev 2002;82(1): 57–89
7. Mates JM, Segura JA, Alonso J et al. Intracellular redox status and oxidative stress: Implications for cell proliferation, apoptosis and carcinogenesis. Arch Toxicol 2008;82(5): 273–299
8. Haddad JJ. Antioxidant and prooxidant mechanism in the regulation of redox(y)-sensitive transcription factors. Cell Signal 2002;14(11): 879–897
9. Chen F, Castranova V, Shi X. New Insights into the role of nuclear factor-kappa B in cell growth regulation. Am J Pathol 2001;159(2): 387–397
10. Klaunig JE, Kamendulus LM. The role of oxidative stress in carcinogenesis. Annu Rev Pharmacol Toxicol 2004;44: 239–267
11. Breithaupt TB, Vazquez A, Baez I et al. The suppression of T-cell function and NF (kappa) B expression by serine protease inhibitors is blocked by N-acetycysteine. Cell Immunol 1996;173(1): 124–130
12. Conner EM, Grisham MB. Inflammation, free radicals and antioxidants. Nutrition 1996;12(4): 274–714
13. Center SA. Metabolic, antioxidant, nutraceutical, probiotic and herbal therapies relating to the management of hepatobiliary disorders. Vet Clin North Am Small Anim Pract 2004;34(1): 67–132
14. Sasaki K, Ma Z, Khatlani TS et al. Evaluation of Serum Amyloid A (SAA) as an inflammatory marker. J Vet Med Sci 2003 Apr;65(4): 545–548
15. Milesi MA. Oxidative, diseases and antioxidants. Agro Food Ind Hi Tech 2006;25(7): 4–9
16. Halliwell B, Gutterridge JMC. Free radicals in biology and medicine. Oxford: Oxford University Press; 2003: pp. 247–251

17. Zimmerman KC, Gree DR. How cell die: Apoptosis pathways. J Allergy Clin Immunol 2001;108(4 suppl): S99–S103
18. Tsudo M. Cytokines and disease. Rinsho Byori 1994;42(8): 821–824
19. Lovschall H, Mosekilde L. Apoptosis: Cellular and clinical aspects. Nord Med 1997;112(4): 133–137
20. Susin SA, Zamzami N, Kroemer G. Mitochondria as regulators of apoptosis: Doubt no more. Biochem Biophys Acta 1998;1366(1–2): 151–165
21. Klaunig JE, Kamendulus LM. The role of oxidative stress in carcinogenesis. Annu Rev Pharmacol Toxicol 2004;44: 239–267
22. Lopez-Torres M, Barja G. Oxidative stress in aging. Totowa: Humana Press; 2008: pp. 149–158
23. Vierra HL, Kroemer G. Pathophysiology of mitochondrial cell death. Cell Mol Life Sci 1999;56(11–12): 971–976
24. Blackwell TS, Christina JW. The role of nuclear factor-kappa B in cytokine gene regulation. Am J Respir Cell Mol Biol 1997;17(1): 3–9
25. Lo CW. Genes, gene knockout, and mutations in the analysis of gap junction. Dev Genet 1999;24(1–2): 1–4
26. Yamaski H, Krutovshikh V, Mesnil M et al. Role of connexin (gap junction) genes in cell growth and carcinogenesis. CR Acad Sci III 1999;322(2–3): 151–159
27. Trosk JE, Ruch RJ. Gap junctions as targets for cancer chemoprevention and chemotherapy. Curr Drug Targets 2002;3(6): 465–482
28. Stahl W, Sies H. The role of carotenoids and retinoids in gap junctional communication. Int J Vitam Nutr Res 1998;68(6): 354–359
29. Halliwell B, Gutterridge JMC. Free radicals in biology and medicine. Oxford: Oxford University Press; 2003: pp. 140–151
30. Poitout V, Robertson RP. Minireview: Secondary beta cell failure in type 2 diabetes – a convergence of glucotoxicity and lipotoxicity. Endocrinology. 2002 Feb;143(2): 339–342
31. Martino L, Novelli M, Masini M et al. Dehydroascorbate protection against dioxin-induced toxicity in the beta-cell line INS-1E. Toxicol Lett 2009;189(1): 27–34
32. Osawa T, Kato Y. Protective role of antioxidative food factors in oxidative stress caused by hyperglycemia. Ann N Y Acad Sci 2005 Jun;1043: 440–451
33. Du X, Edelstein D, Du XL et al. Normalizing mitochondrial superoxide production blocks three pathways of hyperglycemic damage. Nature 2000;404: 787–790
34. Wei EP, Komtos HA, Christman CW et al. Superoxide generation and reversal of acetylcholine-induced cerebral arteriolar dilation after acute hypertension. Circ Res 1985;57: 781–787
35. Laursen JB, Rajagopalan S, Galis Z et al. Role of superoxide in angiotensin II induced but not catocholamine induced hypertension. Circulation 1997;95: 588–593
36. Vaziri ND, Wang XQ, Oveisi F et al. Induction of oxidative stress by glutathione depletion causes severe hypertension in normal rats. Hypertension 2000;36: 142–146
37. Halliwell B, Gutterridge JMC. Free Radicals in biology and medicine. Oxford: Oxford University Press; 2003: pp. 129–140
38. Mandelker L. The natural activities of cells, the role of reaction oxygen species and their relation to antioxidants, nutraceuticals, botanicals and other biological therapies. Vet Clin North Am Small Anim Pract 2004;34(1): 39–60
39. Heller, Koch T, Schmeck J et al. Lipid mediators in inflammatory disorders. Drugs 1998;55(4): 487–496
40. Halliwell B, Gutterridge JMC. Free Radicals in biology and medicine. Oxford: Oxford University Press; 2003: pp. 217, 307, 411
41. Macdonald J, Galley HF. Oxidative stress and gene expression in sepsis. N.R. Webster Academic Unit of Anesthesia and Intensive Care, Institute of Medical Sciences; Abstract: Medline Br J Anaest 2003;90(2): 221–232
42. Dare AJ, Phillips AR, Hickey AJ et al. A systematic review of experimental treatments for mitochondrial dysfunction in sepsis and multiple organ dysfunction syndrome. Free Radic Biol Med. 2009;47(11): 1517–1525

43. Rinaldi S, Landucci F, De Gaudio AR. Antioxidant therapy in critically septic patients. Curr Drug Targets 2009;10(9): 872–880
44. Passos J, Von Zglinicki T. Oxidative stress in aging. Totowa: Humana Press; 2008: pp. 39–50
45. Jones DR. Extracellular redox state: Refining the definition of oxidative stress in aging. Rejuvenation Res 2006 Summer;9(2): 169–181
46. Miwa S, Muller FL, Beckman KB. Oxidative stress in aging, Introduction. Totowa: Humana Press; 2008: p. 7
47. Halliwell B, Gutterridge JMC. Free radicals in biology and medicine. Oxford: Oxford University Press; 2003: pp. 619–624
48. Dröge W. Free radical in the physiological control of cell function. Physiol Rev 2002;82(1): 45–97
49. Hakim J. Pharmacologic control of intracellular signaling pathways: From research to therapy. J Cardiovasc Pharmacol 1995;25 (suppl 2): S106–S113

Use of Free Radicals and Antioxidants in Inflammatory Processes of Animals

Peter Vajdovich

Abstract The inflammatory process is complex and the role of free radicals and antioxidants can be found in each part of it, such as the hemodynamic and permeability changes, the metabolism, and the function of the cells involved in the course of events. The main purpose of the present review is to mention the most common inflammatory processes in animals, the role of free radical formation in these, and the use of antioxidants against their detrimental effects. As with severely increased oxidative stress, decreased free radical formation can cause severe diseases. Hyperbaric oxygen gives an option to overcome the disorders caused by decreased oxidative stress response which is important in the bacterial killing mechanism and wound healing. Endotoxemia, systemic inflammatory response syndrome, in its complexity causes severe deterioration of the organism due to oxidative stress. Various antioxidants can be used to prevent the severity, such as superoxide dismutase (SOD), alpha-tocopherol, glutathione and its precursors, ascorbic acid, adenosine, lactoferrin, and carotenoids, thiols, ebselen, xanthine oxidase inhibitors, inhibitors of phagocyte function, iron ion chelators, and probucol, *N*-acetylcysteine, and 21-aminosteroids. Among pulmonary and other airway inflammatory diseases recurrent airway obstruction of horses causes severe oxidative damage. Apart from the well-known antioxidants, it seems that another antioxidant taurine, via *N*-chlorotaurine formation may protect the lung from oxidant-induced injury. This session contains more data of oxidative stress and possible antioxidant therapeutic advances in selective animal inflammatory diseases, such as kidney disease, gingivitis, uveitis, cataract, dermatitis, osteoarthritis, synovitis, pancreatitis, hepatitis, colitis and giardiasis, encephalopathy, neuritis and spinal cord injury, and bovine mastitis.

Keywords Colitis • Gingivitis • Hepatitis • Hyperbaric oxygen therapy • Inflammation • Mastitis • Neuritis and spinal chord injury • Osteoarthritis • Pancreatitis • Recurrent airway obstruction • Synovitis • Systemic inflammatory response syndrome

---

P. Vajdovich (✉)
Faculty of Veterinary Sciences, Department of Internal Medicine and Clinics,
Szent Istvan University, Istvan u. 2, 1143, Budapest, Hungary
e-mail: Vajdovich.Peter@aotk.szie.hu

L. Mandelker and P. Vajdovich (eds.), *Studies on Veterinary Medicine*,
Oxidative Stress in Applied Basic Research and Clinical Practice 5,
DOI 10.1007/978-1-61779-071-3_2,

Inflammation is one of the most important and most useful of our host defense mechanisms, and without an adequate inflammatory response, none of us would be living.

"*Quae medicamenta non sanat; ferrum sanat. Quae ferrum non sanat; ignis sanat. Quae vero ignis non sanat; insanabilia reportari oportet* (Those diseases which medicines do not cure, the knife cures; those which the knife cannot cure, fire cures; and those which fire cannot cure, are to be reckoned wholly incurable)" [47]. As Hippocrates (460–370 BC) described in his aphorism, he believed that any disease could be cured by heat: burning heat and heat as both cause and consequence of inflammation. In Latin *flamma* is the light of fire. It is also one of the most common means whereby our own tissues are injured. The word inflammation itself literally means a "burning". The Roman Cornelius Celsus formulated his famous "cardinal signs" of inflammation: calor, rubor, tumor, and dolor. That means heat, redness, swelling, and pain. To this classic list of four signs must be added *functio laesa* or loss of function. This fifth sign is often attributed to Galen (AD 130–200), but it really originated with Rudolf Wirchow (1821–1902), the father of modern pathology.

The process is complex and the role of free radicals and antioxidants can be found in each part of it such as hemodynamic and permeability changes, the metabolism and function of the cells involved in the course of events. The main purpose of the present review is to mention the most common inflammatory processes in animals, the role of free radical formation in these, and the use of antioxidants against their detrimental effects.

The most important direct effect of free radicals is found in the microbicid mechanism. There are also various free radical-dependent systems involved in the formation of cytokines, inflammatory mediators, and cellular functions (i.e., platelets; see, for review [130]). Reactive oxygen species (ROS) are involved in wound healing and tissue repair mechanisms [10]. Although the synthesis of inflammatory mediators is a free radical-producing mechanism, the production of stress-induced anti-inflammatory hormones in the adrenal cortex is also a free radical-mediated process, and antioxidants can inhibit these mechanisms [6, 125].

Oxygen-Dependent Killing Mechanisms

The purpose of bacterial phagocytosis is to kill and degrade bacteria. Two different killing systems are recognized: oxygen-dependent and oxygen-independent killing systems. There are two types of oxygen-dependent killing systems known: the so-called oxygen radical system and the myeloperoxidase–halide system [12].

During phagocytosis, neutrophils show a burst of metabolic activity, characterized by a two- to threefold increase in oxygen consumption, an increase in the formation of superoxide anion (O_2^-), hydroxyl anion (OH^-), and hydrogen peroxide (H_2O_2), and an increase in cyanide-independent oxygen uptake and the amount of glucose metabolized by means of the hexose monophosphate shunt (HMPS; Stahelin 1957) [118].

The decrease of this process leads to severe microbial infections. Some hereditary defects have also been reported. The functional defects may be associated with morphological abnormalities, such as in Chédiak–Higashi syndrome, or not associated with morphological changes of neutrophils, such as in chronic granulomatous disease and granulocytopathy syndrome.

An oxygen-dependent killing mechanism is effective against microbes, but this mechanism is responsible for the harmful effects of inflammation. Experimental studies suggest that neutrophil-derived oxygen metabolites contribute to the development of tissue injury associated with a variety of disease states, including emphysema, myocardial infarction, adult respiratory distress syndrome, immune complex-mediated vasculitis, and rheumatoid arthritis [28]. Local inflammatory processes can lead to systemic inflammation and free radical and antioxidant changes can also be dispersed.

"*Altera manu fart aquam, altera ignem* (One hand carries water, the other fire)." According to this old Latin phrase, one process can lead to both beneficial and harmful processes. In addition to the harmful effects, free radicals produced by neutrophils are essential for the proper microbicide effect, but it has to be mentioned that the oxidative burst is sometimes exaggerated. Increasing oxidative burst activity is of great importance in conditions of decreased function, such as necrotizing fasciitis, or lower extremity wounds [27, 74].

Hyperbaric Oxygen in Therapy

Hyperbaric oxygen (HBO) therapy can be essential to enhance microbial killing; it is worth mentioning some details about HBO therapy in general, the clinical benefits and side effects. This session about HBO therapy is based on Slovis' [114] paper with modifications. "The art of healing comes from nature, not from the physician. Therefore the physician must start from nature, with an open mind" (Paracelsus, Philippus Theophrastus Aureolus Bombastus von Hohenheim, 1493–1541) [90]. HBO therapy is a typical practical achievement of the Paracelsus theory. One of the most essential elements in nature is oxygen.

HBO therapy involves intermittent inhalation of 100% oxygen under a pressure greater than 1 atm [29]. Increasing pressure in HBO therapy is often expressed in multiples of atmospheric pressure absolute or atmospheres absolute (ATA); 1 ATA equals 1 kg/cm^2 or 735.5 mm Hg. Most HBO treatments are performed at 2–3 ATA. In air embolism and decompression sickness, where pressure is crucial to the therapeutic effect, treatments frequently start at 6 ATA [34]. This additional pressure, when associated with inspiration of high levels of oxygen, substantially increases the level of oxygen dissolved into blood plasma. This state of serum hyperoxia is the second beneficial effect of HBO therapy [32].

Hyperoxia at sea level in room air means that approximately 97% of hemoglobin is saturated with oxygen (19.5 vol% oxygen, of which approximately 5.8 vol% is extracted by tissue). The amount of oxygen dissolved into plasma is 0.32 vol%.

An increase in pO_2 has a negligible impact on total hemoglobin oxygen content; however, it does result in an increase in the amount of oxygen dissolved directly into the plasma. With 100% inspired oxygen the amount of dissolved blood plasma oxygen increases to 2.09 vol%. At 3 ATA plasma contains 6.8 vol% oxygen, which is the level equivalent to the average tissue requirements for oxygen. Thus, HBO treatment could and has sustained life without oxygen bound to hemoglobin [13]. The cardiovascular effects of HBO include a generalized vasoconstriction and a small reduction in cardiac output [99]. This ultimately may decrease the overall blood supply to a region, but the increase in serum oxygen content results in an overall gain in delivered oxygen. In conditions such as burns, cerebral edema, and crush injuries, this vasoconstriction may be beneficial, reducing edema and tissue swelling while maintaining tissue oxygenation [85]. Hyperbaric or hyperoxic or both conditions may affect the disposition of drugs by (1) changes in the catalytic activity of drug-metabolizing enzymes, (2) hemodynamic changes, and (3) changes in membrane permeability, affecting drug distribution.

In isolated microsome preparations from rat liver, the metabolism rate of aniline, but not of amidopirin, is reduced by hyperoxia. In vivo, the clearance of salicylic acid is enhanced in the dog at 2.8 ATA and 100% O_2, but not at 6 ATA and air, for reasons that are still unknown. The disposition of theophylline, pentobarbital, or pethidine is not affected in dogs by hyperbaric or hyperoxic conditions.

In human volunteers, hyperbaric or hyperoxic or both conditions do not affect the disposition of gentamycin (2.4 bar, 100% O_2), caffeine, or lidocaine (2.5 bar, 100% O_2).

In conclusion, a single exposure to hyperbaric or hyperoxic conditions does not seem to affect single-dose pharmacokinetics of drugs metabolized and/or eliminated by the kidney (e.g., gentamycin) or by the liver with a capacity-limited clearance (e.g., pentobarbital, theophylline, and caffeine) or with a perfusion-limited clearance (e.g., pethidine, lidocaine; [101]).

HBO therapy is performed in hyperbaric chambers, where this special environment can be achieved. There are chambers available for large animals, too.

Therapeutic Effects of HBO

The typical aim in using HBO is to reverse hypoxia in order to alter ischemic effect or influence vascular reactivity [15]. Postischemic HBO affords clinical and histopathological neuroprotection after experimental cardiac arrest and resuscitation (A/R) and this neuroprotection results from improved cerebral oxygen metabolism after A/R. It was proven by an experiment when anesthetized adult female beagles underwent A/R and randomization to HBO (2.7 ATA for 60 min, 1 h after A/R) or control (PO_2 = 80–100 mm Hg; 1 ATA). Animals underwent neurological deficit scoring (NDS) 23 h after A/R.

Neuronal death (necrotic and apoptotic) in representative animals was determined stereologically in the hippocampus and cerebral neocortex. NDS improved after A/R in HBO animals. Histopathological examination revealed significantly fewer dying neurons in HBO animals; the magnitude of neuronal injury correlated well with NDS. HBO inhibited neuronal death and improved neurological outcome after A/R. It was also stated that the mechanism of HBO neuroprotection is not due to stimulation of oxidative cerebral energy metabolism [100]. Other typical use of HBO therapy is to reduce edema [84, 85] as hyperoxygenation will cause vasoconstriction, which is considered a side effect of HBO therapy. Although vasoconstriction may be present, there is more oxygen delivered to the tissues. HBO can modulate nitric oxide (NO) production [26, 86, 109]. An increase of nitric oxide leads to vasodilation whereas a decrease of nitric oxide leads to vasoconstriction. Carbon dioxide increases NO production and oxygen decreases NO production by the endothelial cells. HBO modifies growth factors and cytokine effect by regulating their levels and/or receptors [136, 144].

Enhancing wound healing is a main purpose of HBO therapy. Despite the unclear mechanisms involved in the efficacy of HBO therapy in this process it has been proven that there is a significant increase in local wound NO levels after successful HBO therapy. This suggests that this mechanism may be an important factor in promoting enhanced wound healing and wound closure [16]. Vascular endothelial growth factor (VEGF) is important for the growth and survival of endothelial cells, and is found in plasma, serum, and wound exudates. Under normobaric conditions, VEGF is stimulated by hypoxia, lactate, nitric oxide, and nicotinamide adenine dinucleotide (NAD). HBO induces production of VEGF thereby stimulating more rapid development of capillary budding and granulation formation within the wound bed. HBO induces changes in membrane proteins affecting ion exchange and gating mechanisms. HBO also promotes cellular proliferation [38, 52, 83, 109], accelerates collagen deposition, stimulates capillary budding and arborization, and therefore indicates reparative and regenerative tissue mechanisms. In spite of previous reports there are some contradictions, too. In an experiment, on day 0, two 4-cm-diameter circular sections of full-thickness skin were removed from each of two randomly selected limbs of each horse, and two 4-cm-diameter circular skin grafts were harvested from the pectoral region. The horses then received HBO therapy for 1 h daily at 1.84 ATA for 7 days. On day 21, the grafts applied on day 14 were biopsied. Histological examination of biopsy specimens revealed that grafts treated with HBO therapy developed less granulation tissue, edema, and neovascularization, but more inflammation. The superficial portion of the graft was also less viable than the superficial portion of those not treated with HBO therapy. In conclusion, the use of HBO therapy after full-thickness skin grafting of uncompromised fresh and granulating wounds of horses is not indicated [140].

HBO therapy also alleviates osteogenesis. Moreover, in a distraction osteogenesis experiment HBO appeared to accelerate ossification and vascularization of regenerated bone in the distracted area, while applying HBO to tooth movement into a distracted area [53]. HBO therapy accelerates microbial oxidative killing by providing oxygen for the phagocytosing cells and by improving select antibiotic exchange across membranes [88, 91, 92], whereas anoxia decreases the activity of several antibiotics (aminoglycosides, sulfonamides, fluoroquinolone). By raising the pO_2 of ischemic tissue to normoxic levels, it may normalize the activity of these antimicrobials. In addition, HBO may potentiate the activity of certain antimicrobials by inhibiting their biosynthetic reactions in bacteria. HBO also interferes with bacterial disease propagation by denaturing bacterial toxins and modulates the immune response system. HBO enhances de novo synthesis or activation of oxygen radical scavengers, thereby decreasing ischemia–reperfusion (I–R) injury [123, 142]. This mechanism can be described by the following.

1. HBO therapy increases the amount and activity of the free radical scavenger superoxide dismutase (SOD).
2. Decreased neutrophil adhesion and subsequent release of free radicals is an important early event leading to endothelial damage and microcirculatory failure associated with I–R Injury. HBO reversibly inhibits the ß2 integrins, therefore inhibiting the neutrophil–endothelial adhesion.

An example of the beneficial effect of HBO therapy is the treatment of necrotizing fasciitis. Compared to national reports of outcomes with "standard" regimens for necrotizing fasciitis, the author's experience with HBO therapy, adjunctive to comprehensive and aggressive management, demonstrates reduced mortality (34% vs. 11.9%), and morbidity (amputations 50% vs. 0%). The treatments were safe and no delays to surgery or interference with standard therapy could be attributed to HBO therapy [27]. Moreover, in a study of chronic diabetic patients with lower extremity wounds the authors proved the efficacy of HBO therapy. Patients managed in a state-of-the-art wound care center experienced progress toward wound healing, regardless of the treatment modality selected. Those who received HBO as part of their wound care regimen healed faster than those who received standard treatment or growth factor therapy [74].

Complications and Side Effects of HBO Therapy

Although any therapeutic application of hyperbaric oxygenation is intrinsically associated with the potential for producing mild to severe side effects, the appropriate use of hyperoxia is one of the safest therapeutics available to the practitioner [110]. CNS oxygen toxicity can occur at levels of 3 ATA for 1–2 h. Signs in humans include convulsions, nausea, dizziness, muscle twitching, anxiety, and confusion. Pulmonary oxygen toxicity is usually associated with prolonged exposure to HBO. Onset of symptoms has been noted to occur 4–6 h at 2 ATA. Symptoms include dyspnea, shortness of breath, chest tightness, and difficulties inhaling a deep breath. Possible causes for pulmonary toxicity include thickening of the alveolar membrane and pulmonary surfactant changes. Prevention of these side effects includes removal from the oxygen source when first signs occur and no 100% oxygen at pressures greater than 3 ATA. HBO retains the biochemical properties of platelets as proven by in vitro studies.

Isolated horse platelets were exposed to 100% oxygen at 2.2 atmospheres, or 100% oxygen under normobaric conditions, or air under normobaric conditions for 90 min. There were no differences in platelet SOD activity among the above-mentioned conditions, but there was a rise in SOD in all cases after 24 h. However, no catalase (CAT) activity or GSH concentration of platelets changed over time [108]. HBO is frequently used to treat conditions involving some degree of local hypoxia, but can have a direct effect on red blood cell (RBC) deformability. To investigate this, 12 normal dogs received a 10-week "clinical" course of HBO: one 90-min treatment per weekday at 2.4 ATA (243 kPa), 100% O_2. On Mondays and Fridays, a blood sample was drawn. Cells were filtered under a constant of 1.08 kPa through a precalibrated nucleopore hemafil polycarbonate membrane. Filtrate of blood cells was collected for 1 min and weighed, and the RBC "incremental volume" calculated. No significant change was seen in filtration rates, indicating that HBO itself neither improves nor impairs dog RBC deformability. Changes in other commonly measured blood parameters remained within normal values. An acute fluid shift out of RBCs and into plasma was indicated [76].

A study was performed using 1-day to 70-day-old puppies placed in a pressure chamber. The pressure of pure O_2 in the chamber was raised by 5 atmospheres (ATA, 75 psi = 6 ATA) within 10 min. The first biochemical change to take place during HBO was oxidation of mitochondrial NADH. The age of the puppy was found to affect the time to the initiation of seizures. In the puppies under the age of 24 days, the average time was 35.1 ± 5.9 min. In those puppies 24 days old and older, the average time was 5.1 ± 0.8 min. In the younger puppies, there was a later occurrence of blood vessel contractions and a longer lifespan compared to the older puppies. The comparison between the puppies of different ages during exposure to HBO showed differences in the metabolic response, hemodynamic changes, and electrical activity. These differences can partially explain the higher resistance in the younger puppies to HBO [140].

The Use of HBO in Veterinary Medicine

The use of HBO in veterinary medicine is in its infancy. The Hagyard Equine Medical Institute, McGee Medical Center, has currently treated more than 100 patients in their HBO chamber. Patients included pregnant animals as well as neonatal foals with no adverse effects noted. Patients have been pressurized from 2 to 3 ATA ranging from 60 to 90 min at treatment pressure (depth). They have used HBO as adjunctive therapy for:

- Fungal disease (fungal pneumonia)
- Thermal burns, carbon monoxide, smoke inhalation
- Closed head injuries
- Ileus
- CNS edema/perinatal asphyxia
- Peripheral neuropathies
- Sports injuries (exertional rhabdomyolysis)
- Cellulitis, compartment syndrome
- Ischemic injuries (laminitis)

Contraindications for HBO therapy are unknown for horses but may include untreated pneumothorax, high fevers (predisposition to oxygen toxicity), emphysema, and upper airway occlusions [114].

Specific Inflammatory Disorders in Animals

Inflammation in General (Endotoxemia, Systemic Inflammatory Response Syndrome)

Small natural elements such as microbes can cause severe inflammatory processes. "*Et neglecta solent incendia sumere vires*" (Neglected fires are wont to gather strength) [49]. As Horace (Horatius 65–8 BC) mentioned, neglected smaller problems can cause trouble, too. Several antioxidants are available for therapeutic use in acute inflammatory diseases, such as systemic inflammatory response syndrome (SIRS) as caused, for example, by endotoxaemia. They include natural molecules normally present in the body (SOD, alpha-tocopherol, glutathione and its precursors, ascorbic acid, adenosine, lactoferrin, and carotenoids) as well as synthetic antioxidants (such as thiols, ebselen, xanthine oxidase inhibitors, phagocyte function inhibitors, iron ion chelators, and probucol). The therapeutic efficacy of SOD, alpha-tocopherol, and ascorbic acid in the treatment of human disease has generally been unimpressive to date although dietary deficiencies of the last two molecules should certainly be avoided. Xanthine oxidase inhibitors may be of limited relevance as antioxidants for human use. Exciting preliminary results with probucol (anti-atherosclerosis), ebselen (anti-inflammatory), and iron ion chelators (against thalassemia, leukemia, malaria, stroke, traumatic brain injury, and hemorrhagic shock) need to be confirmed by controlled clinical trials. Clinical testing of NAC in HIV-1-positive subjects may also be merited [42].

It was demonstrated that the antioxidant enzyme SOD prevents inflammation and chemotaxis of neutrophils [79]. Burn is a typical damage that causes severe inflammatory and free radical-producing processes. In a study it was reported that SOD and CAT were helpful in ameliorating the damaging effects of burn.

It is known that there is marked volume loss following thermal trauma. Lipoxygenase transformation of fatty acids to fatty acid hydroperoxides is a pathway that produces superoxide anion and metabolism upon plasma volume loss. The effects of indomethacin (a cyclo-oxygenase inhibitor), orgotein (SOD-like enzyme), and CAT were studied in anesthetized dogs receiving a 15% total body surface area third-degree flame burn. The results of this study showed that indomethacin did not alter postburn plasma volume loss, orgotein reduced early plasma volume loss but did not reduce continuing loss, and CAT reduced both early and continuing plasma volume loss [46].

Not only antioxidants have anti-inflammatory properties, but several anti-inflammatory agents can also be antioxidant. Among the anti-inflammatory drugs, aspirin and salicylic acid act as antioxidants, showing a similar, but only modest, magnitude and velocity of superoxide scavenging [120].

Some other drugs, such as oxymetazoline are neither anti-inflammatory, nor antioxidant per se, but inhibit some effects of inflammation and cause decreased oxidative stress. Oxymetazoline, as a nasal decongestant, effectively reduces rhinitis symptoms. It affects arachidonic acid-derived metabolites concerning inflammatory and oxidative stress-dependent reactions. In canine alveolar macrophages, oxymetazoline suppressed pro-inflammatory reactions including 5-lipoxygenase activity, LTB_4, and respiratory burst and prevented particle-induced oxidative stress, whereas phospholipase A_2 activity and synthesis of immune-modulating PGE_2 and 15-HETE were not affected [8].

In different pathological conditions, such as SIRS, it was demonstrated that much of the cellular injury associated with SIRS is mediated by oxygen or nitrogen free radicals produced by inflammatory cells that overwhelm endogenous antioxidants. Reduced glutathione is a crucial intracellular antioxidant that becomes depleted during SIRS. Regeneration of glutathione can be achieved by acetyl-cysteine, which unlike glutathione itself penetrates cells. In animal models of sepsis and lung injury, acetylcysteine mitigates the cytopathological effects of SIRS. In humans, clinical benefit has been demonstrated in the SIRS of established fulminant hepatic failure. The animal and human studies are sufficiently encouraging to warrant formal trials to test the hypothesis that acetylcysteine therapy has a cytoprotective effect in sepsis [43].

NAC, as mentioned, is a glutathione precursor, and inhibits the in vivo effects of endotoxemia in sheep and prevents the most severe effects of endotoxic shock in dogs [7, 115]. In spite of these findings NAC failed to prevent severe effects of endotoxemia, when pigs received continuous intravenous endotoxin for 12 h. Despite the increased glutathione concentration, NAC (150 mg/kg loading dose over 1 h followed by 20 mg/kg bw/h for 11 h) did not improve systemic, pulmonary, and hepatosplanchnic hemodynamics, oxygen exchange, or metabolism. When compared with previous reports in the literature, a different timing of NAC administration and/or an ongoing or even NAC-induced aggravation of oxidative stress may account for this result [132].

Zinc also has a beneficial effect in ameliorating the effects of inflammatory disorders. It has been proposed that metalloprotein zinc mobilization mediated by hypochlorous acid (HOCl) may induce cell injury [31]. Using a dimercaptopropanol-zinc complex it was proven that, once released from thiolate bonds by HOCl, zinc can exert a significant antioxidant effect on both linolenic acid and deoxyribose oxidation induced by iron. In experimental conditions, however, the antagonism toward deoxyribose oxidation is notably less than that toward linolenic acid peroxidation, thus suggesting a more specific inhibitory effect of zinc on iron-mediated oxidant damage when polyunsaturated fatty acids (PUFA) represent the oxidizable substrate. The antioxidant effects of zinc are strictly related to the "free" form; the dimercaptopropanol-zinc complex per se is stimulatory even on biomolecular oxidant damage, apparently as a result of the pro-oxidant interaction of the thiol compound with iron. In light of these results, it may be proposed that the zinc released from thiolate bonds by HOCl could specifically limit the tissue oxidative burden in pathological conditions involving neutrophil accumulation and activation, such as inflammation and ischemia-reperfusion [67].

There are some novel anti-inflammatory and antioxidant derivates, such as the sesquiterpenoids. It was reported that the anti-inflammatory activity of avarol and avarone, sesquiterpenoid derivatives from the Mediterranean sponge *Dysidea avara*, potently inhibited carrageenan-induced paw edema as well as ear edema induced by 12-*O*-tetradecanoylphorbol acetate in mice, with effects comparable to those of indomethacin. In stimulated rat peritoneal leukocytes, avarol showed inhibition of LTB4 and thromboxane B2 release, respectively, with avarone showing a slightly lower potency. Both marine metabolites failed to show xanthine oxidase inhibitory activity or superoxide scavenging effects but were potent inhibitors of superoxide generation in rat peritoneal leukocytes. Only avarol was able to inhibit human recombinant synovial phospholipase A_2 activity. Avarol and avarone effectively control acute inflammation in experimental models after either oral or topical administration and their anti-inflammatory activity may result from inhibition of eicosanoid release and depression of superoxide generation in leukocytes [30].

Other antioxidants, such as 21-aminosteroids might also be helpful. Effects of the antioxidant, 21-aminosteroid U-74006F (tirilazad mesylate), was demonstrated on the systemic and regional hemodynamics and the oxygen extraction capabilities during endotoxic shock of dogs. Compared with the endotoxin-alone group, the U-74006F-treated dogs maintained higher mean arterial pressure, cardiac index, stroke volume index, and left ventricular stroke work index and lower pulmonary vascular resistance.

Authors also showed a higher fractional blood flow to mesenteric and renal beds [143]. Another radical scavenger, tempol, partially prevented live bacteria from causing key features of hemodynamic and metabolic derangements in porcine hyperdynamic sepsis and beneficially affected surrogate markers of sepsis-induced endothelial and coagulation dysfunction. Sepsis was induced and maintained for 24 h with continuous infusion of live *Pseudomonas aeruginosa*.

Tempol significantly attenuated reduction in mean arterial pressure. Despite comparable mesenteric macrocirculation, tempol attenuated the otherwise progressive deterioration in ileal mucosal microcirculation and prevented mucosal acidosis. Tempol also reduced nitrosative stress without significant effect on the gradual increase in plasma 8-isoprostanes, meta-stable end-products of lipid peroxidation (LPO). Tempol attenuated sepsis-induced endothelial (von Willebrand factor) and hemostatic (thrombin-antithrombin complexes, plasminogen activator inhibitor-type 1) dysfunction [77].

Nitric oxide plays an important role in systemic inflammatory responses. This was proven by the experiment in which ewes were chronically instrumented methicillin-resistant *Staphylococcus aureus* (MRSA) sepsis. This group developed impaired gas exchange and significantly increased lung lymph flow, permeability index, and bloodless wet-to-dry weight ratio (W/D ratio). The plasma nitrate/nitrite (NOx) levels, lung inducible nitric oxide synthases (iNOS) and endothelial nitric oxide synthases (eNOS), VEGF protein expressions, and poly-(ADP)-ribose (PAR) were significantly increased by the MRSA challenge [57].

Nonselective NOS inhibition in sepsis models reversed sepsis-induced derangements in hemodynamic status, but was associated with side effects such as pulmonary vasoconstriction and decreases in global oxygen delivery. Results from studies on specific inhibition of iNOS (NOS-2) and neuronal NOS (nNOS, NOS-1) in sepsis models remain inconclusive, but suggest that both isoenzymes are involved in the pathophysiological processes [66].

Different doses of L-arginine, which is a NO precursor, similarly increased mortality, and worsened shock (for reduced mean arterial pressure). These effects were associated with significant increases in plasma arginine and ornithine. In addition, serum nitrate/nitrite, liver enzymes, and blood urea nitrogen/creatinine ratios rose, whereas arterial pH and bicarbonate levels fell. NAC did not significantly decrease any of the harmful effects of L-arginine. Thus, parenteral L-arginine monotherapy was markedly harmful in animals with septic shock [58].

It has been found that lysozyme (Lzm-S), released from leukocytes, contributed to the myocardial depression that develops in a canine model of septic shock. Lzm-S binds to the endocardial endothelium, resulting in the production of nitric oxide, which, in turn, activates the myocardial soluble guanylate cyclase (sGC) pathway. In a phenylephrine-contracted canine carotid artery ring preparation, the authors found that canine Lzm-S, at concentrations similar to those found in sepsis, produced vasorelaxation. This decrease in force could not be prevented by inhibitors of NO synthase, prostaglandin synthesis, or potassium channel inhibitors and was not dependent on the presence of the vascular endothelium. However, inhibitors of the sGC pathway prevented the vasodilatory activity of Lzm-S. *Aspergillus niger* CAT, which breaks down H_2O_2, as well as other hydroxyl radical scavengers, such as hydroquinone and mannitol, prevented the effect of Lzm-S [81].

Pulmonary and Other Airway Inflammatory Diseases

Pulmonary diseases are typical life-threatening disorders. The well-known Latin saying is "*Dum spiro, spero* (while I breathe, I have hope, or in other words: I have hope while I am alive)." It is necessary to protect pulmonary tissue from the damaging effects of free radicals and inflammatory processes, as a slight problem can cause a severe decrease in oxygenation.

The mechanism of recurrent airway obstruction (RAO) in horses was investigated by measuring the membrane domain structure and oxy-redox activity in phagocytes isolated from bronchoalveolar lavage fluid (BAL) and from the blood of healthy and RAO horses by electron paramagnetic resonance (EPR). It was also found that the phagocytes were affected by oxidative stress in ROA horses proven by the altered activity of intracellular antioxidant enzymes CAT, GPx, and SOD measured in phagocytes. In comparison with polymorphonuclear leukocytes (phagocytes) from the blood of healthy horses the reduction mechanisms in BAL were faster and coincided with the merging of disordered membrane domains, whereas in horses with RAO the reduction and membrane domain structure remained unchanged. The merging of lipid domains observed in phagocytes from

BAL of healthy horses could promote cluster formation of membrane proteins or ligands, which could trigger the activation process in phagocytes of healthy horses and consequently the physiological response that probably did not happen in phagocytes of RAO horses [65].

Antioxidant supplementation is therefore warranted in horses with RAO. Antioxidant supplement tested modulated oxidant/antioxidant balance and airway inflammation of heaves-affected horses. The antioxidant treatment (vitamins E and C and selenium) significantly improved exercise tolerance and significantly reduced endoscopic inflammatory score. Plasma uric acid concentrations were significantly reduced, suggesting downregulation of the xanthine-dehydrogenase and xanthine-oxydase pathway. RBC hemolysate glutathione content showed an insignificant trend to increase, whereas plasma 8-epi-PGF_{2alpha} remained unchanged. Pulmonary markers and BAL cytology were not significantly affected by antioxidant supplementation [62].

The antioxidant ascorbic acid supplementation does not alter basic antioxidant levels apart from ascorbic acid itself. The authors reported that the concentration of lung lining fluid ascorbic acid is increased following ascorbic acid oral supplementation (20 mg/kg body weight) in an ascorbate-synthesizing species. Two weeks' supplementation with ascorbyl palmitate increased mean plasma ascorbic acid concentrations compared to control. The concentration of ascorbic acid in BAL increased in five out of six ponies following supplementation with either ascorbyl palmitate compared to control. Neither supplement altered the concentration of glutathione, uric acid, or alpha-tocopherol in plasma or BAL [24].

Apart from the well-known antioxidants, it was demonstrated that another antioxidant, taurine, via *N*-chlorotaurine formation may protect the lung from oxidant-induced injury by inhibiting production of nitrite and the release of TNF-alpha which are both known to be directly linked to tissue injury [106].

Inflammatory Kidney Diseases

Our ancestors knew that urine was a good indicator of kidney disease. *A fonte puro pura defluit aqua* (Clear water flows from a clear spring). The kidney has to work properly. If it is healthy, urine looks healthy, too. Inflammation is a typical cause of kidney failure. In a study authors examined the oxidative stress markers in azotemic dogs. They found that dogs with renal azotaemia had higher intraerythrocytic sodium content. The RBC CAT activity and glutathione content and plasma malondialdehyde (MDA) level were unaltered whereas the urinary malondialdehyde–creatinine ratio (U-MDA/Cr) increased. The U-MDA/Cr was correlated with plasma creatinine concentration, urinary protein–creatinine ratio, and fractional excretion of sodium [17].

The antioxidant vitamin E ameliorates the effects of glomerular disease. Authors reported that IgA nephropathy is one of the most common forms of glomerular disease. Nearly 25% of affected patients progress to end-stage renal disease over a 20–25-year follow-up period. IgA-containing immune complexes stimulate oxygen-free radical production by mesangial cells in vitro. The excessive oxidant stress may mediate glomerular injury in this disorder. IgA nephropathy with mild kidney inflammation was induced in Lewis rats by oral immunization with 0.1% bovine gamma-globulin (BGG)-containing drinking water. Vitamin E-treated rats gained more live weight and had a lower incidence of hematuria and proteinuria, and reduced renal plasma flow was restored to normal value, compared to untreated rats with IgA nephropathy. Glomerular hypertrophy occurred in animals with IgA nephropathy, but less so in those receiving vitamin E supplementation. Renal cortical MDA content was reduced in rats fed the vitamin E-enriched diet. Finally, renal transforming growth factor-beta 1 gene expression was reduced by 34% in rats with IgA nephropathy receiving vitamin E treatment. It can be concluded that experimental IgA nephropathy is associated with increased renal oxidant injury. Dietary treatment with the antioxidant vitamin E attenuated renal functional and structural changes in experimental glomerulopathy [124].

A potential treatment option is providing a very low protein diet supplemented with amino and keto acids and vitamins A, C, and E (VLPD) to patients with chronic renal failure (CRF). It was demonstrated that VLPD has a protective role against LPO of erythrocytes in patients with CRF. Peuchant [94] have evaluated parameters indicative of LPO of RBC at baseline in patients with CRF compared to controls, and the effects of a very low protein diet supplemented with amino and keto acids and vitamins A, C, and E over a 6-month period. The presence of peroxidative damage in CRF patients before the administration of VLPD was demonstrated by elevated levels of free MDA and decreased levels of PUFA, as compared to controls. Similarly, RBC vitamin E content was decreased whereas enzymatic activities were unaltered. Peuchant suggested that VLPD-reduced erythrocyte LPO was proven by (a) decreased levels of free and total RBC MDA, (b) increased levels of PUFA, and (c) increased levels of vitamins A and E as compared to prediet results. However, antioxidant enzyme activities were not modified [94].

Moreover, 21-aminosteroid (21-AS) and mannitol have synergistic effects in the case of kidney damage. Salahudeen [102] reported that, employing Raman spectroscopy, they confirmed in vitro the ability of 21-AS to inhibit iron-induced fatty acid peroxidation. The 21-AS was then administered to rats developing renal failure from glycerol-induced rhabdomyolysis. Although 21-AS inhibited rhabdomyolysis-induced plasma and renal LPO, renal protection was incomplete. Applying fluid-alkaline-mannitol (FAM) diuresis to inhibit cast formation afforded better renal protection. However, when these therapies were combined to inhibit both LPO and cast formation, there was synergistic renal functional protection. This was accompanied by a maximum inhibition of renal and plasma LPO, as well as renal tubular necrosis and cast formation. Compared to combination therapy, FAM therapy alone, despite identical volume, was accompanied by a higher tubular necrosis and cast formation [102].

Kidney transplantation is a great challenge in humans. The canine autotransplantation method is a good model to evaluate the technical problems and ischemia–reperfusion (I/R) injury in this intervention. Antioxidant supplementation seems to be effective in alleviating the reperfusion injury caused by oxidative stress. In a study authors found that antioxidant enzyme activity in blood plasma was significantly lower in the control group than in the ascorbic acid supple-mented (cold normal saline and ascorbic acid in a dose of 100 mg/kg was perfused followed by autotransplantation) treatment group. The histopathology findings revealed the treatment group to have less damage than the control group. The results of this study suggest that ascorbic acid alone might play a role in attenuating I/R injury and assist in the recovery of the renal function in a renal transplantation model [69].

Gingivitis

To keep teeth healthy could be important in the past, as they were considered as weapons. As Lucretius (99–55 BC) said: "Arma antiqua manus ungues dentesque fuerunt (Ancient weapons were hands, nails, and teeth)" [73]. Inflammatory processes of the oral region such as gingiva are common in humans, and many animal models are used to get adequate information about the development of the disorders and possible treatment options. This disease group can cause severe harm in canine patients, too. Antioxidants provide a hopeful alternative way of treatment.

Cu/Zn-SOD has moderate anti-inflammatory activity which could be enlarged by encapsulation into liposomes. The greater efficiency of liposomal SOD compared to the free enzyme is explained by an increased fixation to cell walls as well as with improved tissue penetration. It was shown that liposome-encapsulated SOD penetrates into the cell probably by endocytosis [126]. The phagocytes are present in abundance in inflamed periodontal tissues [9]. The hypothesis that highly reactive radicals may be responsible for the initial degradation of extracellular matrix components seen in periodontal disease is in agreement with Misaki et al. [82] who showed pronounced better healing of gingivae wounds in rats after i.v. application of SOD [82]. SOD also prevents amplification and interrupts cascade processes of free radical formation, that is, single superoxide molecule interaction with lipids, H_2O_2, flavins to other radicals [80]. Moreover, production of free radicals from freshly cancelous bone specimens was also detected [71] and bone resorption was prevented by application of SOD [98]. Free radical processes are one of the mechanisms by which bone is destroyed under inflammatory circumstances. Drugs that inhibit free radical production or scavenge those radicals may be useful therapeutic agents for local bone destruction [35]. In spite of beneficial effects of this kind of supportive periodontal therapy, we must also be aware that inappropriate concentration of SOD might lead to tissue damage by excessive production of H_2O_2 [75]. However, high production of H_2O_2 might be also beneficial due to antibacterial effect on periodontal pathogens as indicated by Hillman and Socransky [45]. Replacement therapy may also find a practical application in the prevention and cure of certain periodontal diseases. Hydrogen peroxide-producing streptococci are invariably found in plaque taken from healthy gingiva; they are rarely found in samples from active disease sites of patients with juvenile or refractory periodontitis. Naturally occurring oral bacteria are promising effector strains for the replacement therapy of dental infectious diseases [45].

It was proven that scaling and root planing with subgingival application of liposome-encapsulated SOD suppresses periodontal inflammation and stimulates periodontal healing and alveolar bone apposition on experimentally induced periodontitis in beagle dogs. Further studies are in progress to evaluate the optimal concentration of SOD, CAT, or a combination of both enzymes for suppression of periodontal inflammation [93].

Nitric oxide is a free radical produced in host tissues by constitutive and inducible forms of the enzyme nitric oxide synthase. Nitric oxide plays physiological roles, but it is also involved in the pathophysiology of several inflammatory conditions, such as gingivitis. Local increases in iNOS and reactive nitrogen products have also been demonstrated in humans and animals with periodontal disease. A placebo-controlled preclinical investigation examined the effect of two mercapto-alkylguanidines, mercapto-ethylguanidine (MEG) and guanidine-ethyldisulfide (GED), which are iNOS inhibitors and reactive nitrogen radical scavenging compounds, on the development of experimental gingivitis in beagle dogs. Experimental gingivitis was then induced, with cessation of plaque control and institution of a soft diet over 8 weeks. Beagles randomly received 0.3% MEG, 0.3% GED, or placebo (vehicle) gels, topically applied twice daily to premolar teeth. At weeks 2, 3, 4, and 8, gingival index scores were significantly lower for MEG and GED groups compared to the placebo group. In addition, MEG and GED gels significantly reduced gingival bleeding responses by 8 weeks. The data from this preclinical study indicate that mercapto-alkylguanidines, topically administered, may significantly reduce experimental gingivitis in the beagles [89].

Uveitis and Canine Cataract

We know that "*In oculis animus habitat* (The soul dwells in the eyes)," therefore much attention must be paid to the inflammatory processes of the eye. It can be important to use antioxidants in ocular disorders, too. The experimental autoimmune disease elicited by a large dose of retinal S antigen in guinea pigs is characterized by massive necrotizing uveitis and retinitis. It was proven that SOD prevents S-antigen-induced experimental autoimmune uveoretinitis (EAU) in guinea pigs, and bovine serum albumin-induced passive Arthus type uveitis in rabbits. These results suggest that superoxide may play a role in causing tissue damage in animal models of ocular inflammation and possibly in Behcet disease [139].

Similarly to the previous experiment, S-antigen-induced EAU in guinea pigs treated with the antioxidants SOD, CAT, and sodium benzoate resulted in marked reduction of uveal inflammation. The attenuated inflammation was characterized by a relatively well-preserved retina and retinal pigment epithelium along with a reduction of subretinal exudate and vitreous inflammation [97].

Not only antioxidant enzymes, but also other antioxidant substances, such as deferoxamine can be used in treating uveitis. Because photoreceptors contain a high proportion of PUFA, deferoxamine, in turn, will act to ameliorate the experimental autoimmune uveitis-mediated retinal degeneration. Treatment of experimental uveitis in Lewis rats with an iron chelator, deferoxamine mesylate, resulted in a marked reduction in choroidal inflammation and suppression of retinal damage [96]. Moreover, in vitro addition of 10 mM deferoxamine, the free radical generation of inflamed retina was suppressed by nearly 40% [138]. Other antioxidant substances were also helpful in uveitis. The lipoxygenase inhibitors phenidone and nordihydroguaiaretic acid, and dimethyl sulfoxide decreased fibrin production at 0.5 and 1 h after induction of uveitis. Phenidone and nordihydroguaiaretic acid also inhibited the initial increase in intraocular pressure early in the course of inflammation [25].

Babizhayev [145], Williams and Munday [137] have proven that *N*-acetyl carnosine can treat immature lens opacities and nuclear sclerosis in dogs. The dogs were treated three times daily with topical 2% *N*-acetyl carnosine in a buffered vehicle containing glutathione, cysteine, ascorbate, L-taurine, and riboflavin (Ocluvet®, Practivet, Phoenix, AZ, USA). The examination was performed prior to treatment and at 2, 4, and 8 weeks during treatment, by direct and indirect ophthalmoscopy and slit-lamp biomicroscopy after pharmacological pupil dilation. This study demonstrated some marginal reduction in lens opacification of canine cataract. Lens opacification was improved with treatment in eyes with immature cataract or nuclear sclerosis whereas in eyes with mature cataract or cataract with associated intraocular inflammatory pathology less reduction was seen [137].

Dermatitis

"*Floreat curator cutis* (May the manager of the skin prosper)," said old Latin. To treat skin problems is essential. There are many antioxidants used in various human and animal inflammatory skin disease, such as silibinin, silymarine, and other flavonoids, furfuryl palmitate, pentoxifylline, melatonin, vitamin E, retinol, selenium salts, NAC, Ginkgo biloba extract, topically applied SOD, beta-carotene, bucillamine, and so on. Still there are open questions and quite a few basic contradictions even in connection with the application of the most commonly used lipophilic antioxidant vitamin E in skin disorders. It has been used for more than 50 years in clinical and experimental dermatology. Although a large number of case reports were published, there is still a lack of controlled clinical studies providing a rationale for clinical indications and dosage. In contrast, advances in basic research on the physiology, mechanism of action, penetration, bioconversion, and photoprotection of vitamin E in human skin have led to the development of numerous new formulations for use in cosmetics and skin care products. Although its current use is largely limited to cosmetics, controlled clinical studies for indications such as atopic dermatitis or prevention of photocarcinogenesis are needed to evaluate the clinical benefit of vitamin E [122].

In a study of feline perforating dermatitis authors applied methyl-prednisolone acetate (20 mg/cat im) and vitamin C (100 mg/kg twice a day for 2 months). Vitamin C administration failed to resolve the disease. In two cases, methyl-prednisolone acetate was used to manage relapsing episodes and vitamin C helped to reduce glucocorticoid requirements [2].

Osteoarthritis, Synovitis

Painless bones are essential for living a comfortable life. Ovid (43 BCE–18 CE) needed to keep his bones comfortable in his grave, too. "*Nasonis molliter ossa cubent* (May the bones of Naso lie gently)" [87]. The development of osteoarthritis is highly dependent on the antioxidant system. In addition to matrix metalloproteinases, ROS are the main causative agents of cartilage degradation. To prevent ROS toxicity, chondrocytes possess a well-coordinated enzymatic antioxidant system formed principally by SODs, CAT, and glutathione peroxidase (GPX). Mathy-Hartert [78] proved the dysregulation of the antioxidant enzymatic system due to inflammatory mediators. Bovine chondrocytes were cultured in monolayer in the absence or presence of IL-1beta or IL-6. To study the signal transduction pathway, inhibitors of mitogen-activated protein kinases (MAPK; PD98059, SB203580, and SP600125) and nuclear factor (NF)-kappaB inhibitors (BAY11-7082 and MG132) were used. Mn-SOD and GPX activities were dose- and time-dependently increased by IL-1beta. In parallel, IL-1beta markedly enhanced Mn-SOD and GPX gene expressions, but decreased Cu/Zn SOD, EC-SOD (which is a copper-containing enzyme also known as SOD3; it has the major role in vessel walls; Itoh 2009 [55]) and CAT gene expressions. Induction of SOD enzymatic activity and Mn-SOD mRNA expression were inhibited by NF-kappaB inhibitors but not by MAPK inhibitors. IL-1beta, and to a lesser extent IL-6, dysregulates enzymatic antioxidant defenses in chondrocytes. These changes could lead to a transient accumulation of H_2O_2 in mitochondria, and consequently may cause mitochondrial damage. These changes contribute to an explanation of the mitochondrial dysfunction observed in osteoarthritis chondrocytes [78]. Therefore the use of antioxidants to treat osteoarthritis might be the crucial point in treatment options.

There are hundreds of plants yielding drugs with antioxidant properties used to treat arthritis or osteoarthritis. Until now, little information has been available on the antioxidative status of chondrocytes. Nevertheless, most of the literature suggests that this antioxidant is good and that an antioxidant is even more beneficial in treatment of this disease group. For instance, in vitro experiments in guinea pigs in which a Glynn–Dumonde synovitis was induced with BGG suggest that desferrioxamine inhibits iron-catalyzed LPO when it is poorly saturated with iron, but loses this effect when it is iron saturated [11]. Moreover, it was also stated that hydroxyl radical scavengers can be beneficially used in arthritis. Santos and Tipping [104] reported that the contribution of ROS, in particular hydroxyl radical ($OH^{-\cdot}$), to joint inflammation was examined in rats developing adjuvant arthritis

(AA) by treatment with ROS scavengers dimethylthiourea (DMTU) and DMSO. Both DMTU and DMSO significantly reduced the clinical evidence of arthritis; synovial fluid cell accumulation was also significantly reduced compared with disease control [104].

It was also reported that many trials are known to treat arthritis by various antioxidant plant substances. Numerous agents derived from plants can suppress these cell signaling intermediates, including curcumin (from turmeric), resveratrol (red grapes, cranberries, and peanuts), tea polyphenols, genistein (soy), quercetin (onions), silymarin (artichoke), guggulsterone (guggul), boswellic acid (salai guggul), and withanolides (ashwagandha). Indeed, several preclinical and clinical studies suggest that these agents have potential for arthritis treatment. Although gold compounds are no longer employed for the treatment of arthritis, the large numbers of inexpensive natural products that can modulate inflammatory responses, but lack side effects, constitute "goldmines" for arthritis treatment [60]. As one of the most important and useful antioxidants, SOD seems to be effective in treating arthritis in various clinical trials, but the manner of application is problematic. The efficacy of SOD decreases a lot until is penetrating to the site where it is really needed. Therefore liposome-encapsulated SOD is applied and seems to be beneficial, but its activity is still not enough. A new way is the application of acylated superoxide dismutase (Ac-SOD) enzymosome in the liposomal enzymatic system. In this case Ac-SOD is inserted in the lipid bilayer of liposomes, unlike when SOD is located in the aqueous compartment of liposomes. Cruz [146], Gaspar (2006) have found that Ac-SOD enzymosomes are nanocarriers combining the advantages of expressing enzymatic activity in intact form and thus being able to exert a therapeutic effect even before liposome disruption, as well as acting as a sustained release of the enzyme [37].

Despite the many good examples there are numerous contradictory findings known in connection with the efficacy of antioxidant treatment of arthritis. The modification of the antioxidant defense system in this disease group remains unknown. Some antioxidant supplements or drugs with antioxidant properties have been developed to reinforce the cellular antioxidant status. However, until now, there has been no consistent evidence that additional antioxidant supply is efficient to relieve arthritis symptoms or to prevent structural changes in arthritis cartilage [44]. Similarly to the aforementioned skin problems, it was found that clinical trials testing the efficacy of vitamin E in the treatment of osteoarthritis and inflammatory arthritis have been methodologically weak and have produced contradictory findings. There is presently no convincing evidence that melatonin, Amazonian vine, *Withania somnifera* root powder, bucillamine, SOD, glucosamine, chondroitin sulfate, avocado–soybean unsaponifiables, selenium, vitamin A, vitamin C, the combination of selenium compounds, and the like are effective in the treatment of any type of arthritis [18, 33]. Most of these review articles highlight the need for additional randomized, placebo-controlled trials to further define the utility of nutraceuticals and any drug with antioxidant property in arthritis treatment.

The development of osteoarthritic structural changes in the anterior cruciate ligament was examined in a dog model. It was induced by anterior cruciate ligament transection of the right knee in dogs. The dogs were treated with avocado–soybean unsaponifiables (10 mg/kg per day). The treatment reduced the loss of subchondral bone volume and calcified cartilage thickness as compared to placebo-treated dogs. Immunohistochemical analysis of cartilage revealed that avocado–soybean unsaponifiables significantly reduced the level of inducible nitric oxide synthase and matrix metalloproteinase- (MPP-)13 in cartilage. The beneficial effect of the treatment was suggested as a consequence of the inhibition of inducible nitric oxide synthase and MMP-13, which are key mediators of the structural changes that take place in osteoarthritis [14].

Pancreatitis

Pancreatitis is a life-threatening disease. This section gives some points to overcome this disorder, according to the theory of Cicero: "*Conqueri fortunam adversam, non lamentari decet* (You should fight against misfortune instead of mourning about it;" Marcus Tullius Cicero 106–43 BC; [131]). Antioxidants can also be used in the treatment of acute pancreatitis. Similarly to other diseases, a wide scale of antioxidants was examined in this disease. Here we focus on some important and interesting studies. The efficacy of antioxidant enzyme treatment was evaluated. Schoenberg et al. [105] reported that treatment with SOD (100,000 U/kg/h) and CAT (400,000 U/kg/h) prevented LPO and reduced zymogen degranulation and tissue necrosis, but tissue edema and inflammatory response were not affected in both models (cerulein, sodium taurocholate) of acute pancreatitis [105]. Other drugs with atypical antioxidant properties were beneficial in preventing inflammatory and free radical producing processes. BN 52021-platelet activating factor (PAF) receptor antagonist had a beneficial effect against cerulein-induced acute pancreatitis in rats depending on the prevention of inflammatory cell activation and subsequent generation of oxygen radicals within pancreatic tissue.

There was significant reduction of pancreas edema, diminution of hyperamylasemia, lack of SOD activity depletion, and inhibition of LPO in pancreatic tissue. These changes were accompanied by significant reduction of acinar cell vacuolization and remarkable inhibition of infiltration with inflammatory cells in the interacinar space [23]. The xanthine oxidase inhibitor allopurinol was also examined. It was found that addition of allopurinol to the treatment protocol in acute pancreatitis might improve the pathological score, bacterial translocation, and oxidative stress parameters due to decreasing ROS formation caused by the inflammatory process of the pancreas as there can be a continuous hypoxia reperfusion injury in this disease [54]. Efficacy of other antioxidants, such as resveratrol, melatonin, and pentoxifylline was also examined in acute pancreatitis. Resveratrol seemed to ameliorate the severe effects of this disease. Szabolcs [119] suggested that the beneficial effects of resveratrol on acute pancreatitis seem to be mediated by the antioxidant effect of resveratrol or by an NF-kappa-B-independent anti-inflammatory mechanism. In connection with melatonin, Szabolcs [119] suggested that it is an antioxidant able to counteract some of the L-arginine-induced changes during acute pancreatitis, and may therefore be helpful in the supportive therapy of patients with acute necrotizing pancreatitis. Pentoxifylline was reported as a methylxanthine derivative with rheological and marked anti-inflammatory properties. It inhibits the production of proinflammatory cytokines. Gomez-Cambronero [39] found that using pentoxifylline attenuated inflammatory responses within the pancreas in acute pancreatitis.

In spite of the wide-scale trial for commonly used antioxidants such as selenium, NAC, and vitamin C in the treatment of acute pancreatitis (and there are many studies that report beneficial effects of these compounds), Siriwardena [113] in a study provide no evidence to justify continued use of these drugs in severe acute pancreatitis. These authors also point out the harmful side effects of these derivatives [113].

Chronic pancreatitis was beneficially treated by a mixture of antioxidants, including selenium, beta-carotene, L-methionine, and vitamins C and E. Treatment with these compounds was associated with significant improvements in quality of life in terms of pain, physical and social functioning, and general health perception. Kirk [61] concluded that treatment with antioxidants may improve quality of life and reduce pain in patients suffering from chronic pancreatitis [61].

Hepatitis

There are various drugs used in liver diseases: more advice, more success. Nevertheless, we have to be aware that "*Facile omnes, cum valemus, recta consilia aegrotis damus* (We all find it easy to give the right advice to the sick when we are well;" Marcus Terentius Varro (116–27 BC) [121]. It was stated that in copper toxicosis (CT) of Bedlington terriers, non-copper-associated chronic extrahepatic cholestasis (EC), or chronic hepatitis (CH), the dogs have reduced protection against oxidative stress, opening a rationale to use antioxidants as possible therapy [117]. The treatment of idiopathic chronic hepatitis consists of controlling inflammation (prednisone, azathioprine), reversing fibrosis (colchicine), and protecting against oxidant damage (vitamin E, ursodeoxycholic acid, S-adenosylmethionine) (Honeckman 2003) [148]. The following concerning the use of antioxidants in the case of hepatopathies is based upon Honeckman's (2003) article.

Ursodeoxycholic acid is a hydrophilic bile acid with anti-inflammatory, immunomodulating, and choleretic properties. It displaces hydrophobic bile acids that can cause oxidative injury to hepatocytes. Ursodeoxycholic acid may have direct antioxidant effects by increasing glutathione and metallothionein [128]. Ursodeoxycholic acid may also decrease collagen formation [48]. It is useful in humans with primary biliary cirrhosis [111].

Zinc is not only useful for copper-associated hepatopathy but may also be helpful in other forms of chronic hepatitis in dogs because of its antioxidant and antifibrotic effects [70].

Vitamin E is an antioxidant that is recommended at a dose of 50–600 IU per day [128]. Oxidant injury to hepatic mitochondria has been reported in Bedlington terriers with copper storage hepatopathy [116]. There is in vitro evidence suggesting that vitamin E may be useful in preventing oxidative damage to hepatocytes exposed to copper or hydrophobic bile acids [129]. In human studies, vitamin E was helpful in treating patients with chronic hepatitis B [4]. Although there are no controlled studies of its use in canine chronic hepatitis, the use of vitamin E seems reasonable based on experimental studies. Because absorption of fat soluble vitamins may be reduced in cholestatic liver disease, a water-soluble form may be preferred [127].

S-adenosylmethionine is a nutraceutical that has recently been marketed for use in dogs with liver problems. It has anti-inflammatory and antioxidant effects, and also plays a role in cellular replication and protein synthesis [128]. It has been useful for the treatment of acetaminophen toxicity [135], but there are no controlled studies of its use in canine chronic hepatitis.

Silymarin, an antioxidant extracted from milk thistle, has also been recommended for chronic hepatitis [127]. Although silibinin (the main isomer in silmaryin) has been proven to be useful for treating amanita mushroom toxicity in dogs and humans [134], there are currently no controlled studies on its use in canine chronic hepatitis. Most studies of milk thistle in humans have shown little or no improvement in biochemical markers, histology of liver biopsies, or survival [56].

Colitis and Giardiasis

Improper eating often comprises the origin of intestinal disease. Therefore, the motto of this section is "*Edas ut vivas, ut edas, noli vivere* (Eat in order to live, and do not live for eating;" Caecilius Balbus 1 BC). There have been numerous experiments done to prove the efficacy of antioxidants in intestinal diseases, too. In a study the effects of gastrointestinal mucosal injury were examined, which was induced by exposing gastric mucosa with bile and a luminal pH of 1 in cats; moreover live *Escherichia coli* was infused i.v. Although hemodynamic responses (mean arterial pressure and cardiac output) were significantly improved in methylprednisolone-pretreated cats, misoprostol/SOD/CAT administered with a nasogastric tube reduced late hypotension. Pulmonary arterial pressure rose to approximately 200% of basal in both methylprednisolone and misoprostol/SOD/CAT administration [5].

It was proven that 5-aminosalicylic acid-containing drugs but, more effectively, specific scavengers, have been found to reduce the intestinal inflammatory process. It is beneficial to use new anti-inflammatory drugs for inflammatory bowel disease specifically designed to scavenge toxic oxygen metabolites [133].

Keshavarzian [59] performed a wide-scale study in connection with the efficacy of different drugs of the antioxidant effect in experimental colitis on rats by acetic acid administration at 2.5%, which produced moderate inflammation. They found that specific superoxide anion scavenger methoxypolyethylene glycol–SOD, and reactive oxygen metabolite scavenger, sulfasalazine, significantly decreased the severity of inflammation. The xanthine oxidase inhibitors, tungsten and pterin-aldehyde, failed to improve inflammation but another xanthine oxidase inhibitor, allopurinol, a compound with known superoxide anion scavenging effect, did limit the inflammation. Inhibition of hydroxyl radical production by deferoxamine or lowering hydroxyl radical values by a scavenger, dimethyl-sulfoxide, did not affect the severity of inflammation. According to the authors, these data suggest that (1) reactive oxygen metabolites play an important role in experimental colitis, (2) the xanthine oxidase pathway is not a major source of reactive oxygen metabolites in colitis, and (3) tissue injury in experimental colitis is not caused by generation of hydroxyl radicals [59].

Other experiments also prove the benefit of using antioxidants in the early stage of induced colitis in rats. Antioxidant enzymes (such as SOD, GPX) and an antioxidant, alpha-tocopherol, were significantly decreased with the severity of colonic damage. Mn-SOD at a dose of 50,000 U/kg attenuated this colitis when preadministered subcutaneously 1 h before the induction of colitis [141].

Gross [40] used the aminosalicylates in the treatment of inflammatory bowel disease. They found that this compound ameliorated the effects of enteritis. They stated that as aminosalicylates are potent antioxidants, this fact underscores the importance of reactive oxygen metabolites in this disease [40].

Sener [107] performed a really contemporary examination in connection with erdosteine, which is a sulfhydryl-containing antioxidant. The treatment by this compound reversed increases in the colonic luminol and lucigenin chemiluminescence values, macroscopic and microscopic damage scores, MDA and collagen levels, myeloperoxidase activity and DNA fragmentation, decrease in tissue glutathione level, elevated serum cytokines and lactate dehydrogenase activity, and histopathological alterations induced by trinitrobenzene sulphonic acid (TNBS) in Sprague–Dawley rats. The authors suggest that erdosteine protects the colonic tissue via its radical scavenging and antioxidant activities [107].

The antioxidant silymarin can be used beneficially together with traditional treatment. Silymarin, in supplement with antiprotozoal drugs (metronidazole), can beneficially influence the therapy of canine giardiasis. Ten days post-treatment the efficacy of metronidazole plus silymarin (91%) was significantly different in comparison with that of metronidazole (75%). Signs of side effects and decreased body weight were not observed in silymarin plus metronidazole-treated dogs. Poor appetite and intermittent vomiting signs and decrease in body weight were observed in two dogs of the metronidazole-treated group. Two weeks after metronidazole treatment, serum concentration of GOT, GPT, and NH3 were significantly increased in comparison with those treated with silymarin [21].

Encephalopathy, Neuritis, and Spinal Cord Injury

The central nervous sytem can also suffer from inflammatory processes. "*Feriunt summos fulmina montes* (Lightning strikes the highest moutain tops;" Horace 65–8 BC) [50]. The iron chelator deferoxamine can be used in treating neuritis. It has been found that conjugated deferoxamine reduces disruption of the blood–brain barrier and the OH^- radical generated from perivascular H_2O_2 may possibly play a role in alterations of vascular permeability in experimental optic neuritis [41].

Some authors who examined neonates with hypoxic ischemic encephalopathy used a nonantioxidant compound phenobarbital in the therapy, but they found a decrease in LPO products (Gathwala [147], Singh [112]). Singh [112] suggested that phenobarbital in the dose of 20 mg/kg i.v. given within 6 h of life in term and near-term neonates with hypoxic ischemic encephalopathy, was associated with a decrease in lipid peroxides, antioxidant enzymes, and antioxidant vitamins. They demonstrated that there was a trend toward lower levels of cerebrospinal fluid (CSF), MDA content, and activities of SOD and GPx and blood vitamin content of A and E in babies with normal outcome as compared to babies with adverse outcome (death or neurologically abnormal at discharge). Moreover, CSF levels of MDA and activities of SOD and GPx were significantly lower in the group receiving phenobarbital [112].

A new way to treat neuritis is the antioxidant gene therapy which was tested in mice with experimental autoimmune encephalomyelitis (EAE), a strategy designed to treat patients at risk for axonal degeneration and persistent visual loss from optic neuritis and multiple sclerosis. Qi [95] cloned the human extracellular superoxide dismutase (EC-SOD) or CAT gene into recombinant adeno-associated virus (AAV). Animals were sensitized for EAE, followed by serial contrast-enhanced MRI for 6 months, and then were euthanized. The effects of ECSOD and CAT modulation on the EAE optic nerve were gauged by computerized analysis of optic nerve volume, myelin fiber area, axonal cell loss, and retinal ganglion cell (RGC) loss. One month after intraocular injections, transgene expression increased 4-fold for AAV-ECSOD and 3.3-fold for AAV-CAT. Six months after intraocular injections and EAE sensitization, combination therapy with ECSOD and CAT decreased RGC loss by 29%, optic nerve demyelination by 36%, axonal loss by 44%, and cellular infiltration by 34% compared with the contralateral control eyes inoculated with AAV-GFP. It was concluded that viral-mediated delivery of antioxidant genes provides long-lasting suppression against neuronal and axonal loss associated with permanent visual disability in patients with optic neuritis and multiple sclerosis [95].

In spinal cord traumas antioxidants can be beneficially used, but still there are some contradictory findings in connection with their routine application. Anderson [3] reported that antioxidants can cause some visible biochemical and histopathological effects in decreasing severity of spinal cord traumas, but they do not diminish the clinical signs, that is, paralysis. Compression trauma of the cat spinal cord induces a very rapid alteration in the lipid metabolism of cellular membranes, including lipid hydrolysis with release of fatty acids including arachidonate, production of biologically active eicosanoids, and loss of cholesterol. This disturbance of cellular membranes can directly damage cells and can lead to the secondary development of tissue ionic imbalance, ischemia, edema, and inflammation with neuronophagia. Pretreatment with either the synthetic glucocorticoid methylprednisolone sodium succinate (MPSS) or the antioxidants vitamin E and selenium (Se) completely prevented the loss of cholesterol and partially inhibited lipolysis and prostanoid production. Treatment with only MPSS significantly reduced the postinjury tissue necrosis and paralysis [3].

Moreover, according to Coates et al. [22], the antioxidant 21-aminosteroid compound was also significantly effective in this disorder. Although significant differences in some portions of the neurological and histopathological examinations were observed, clinical efficacy for the 21-aminosteroid compound (U74389G) could not be established in acute-compressive spinal cord trauma at the second lumbar spinal cord segment (100 g, 300 s) of dogs [22].

Lucas [72] examined the level of different LPO products, such as hydroxyalkenals (4-HAs) and MDA after spinal cord injury. They used a specific treatment against them, such as the GSH precursor γ-glutamylcysteine. These authors mention that 4-HAs are much more reactive than MDA and are considered among the most toxic LPO products. After spinal cord injury, spinal cord irrigation with γ-glutamylcysteine preserved GSH and reduced 4-HAs below naive levels but had no effect on MDA. By 24 h after spinal cord injury, MDA returned to naive levels but 4-HAs were still elevated. Once again, γ-glutamylcysteine treatment reduced 4-HAs elevation [72].

Bovine Mastitis

We can say that "*Aliena capella gerit distentius uber* (The udder of another's gout is more distended;" Horace 65–8 BC), which means that another's gout is better than ours, but not in the case when this distension is caused by mastitis [51]. An important role played by acute phase response and oxidative status in inflammation of the mammary gland in cows as mastitis is associated with significantly higher concentrations of inflammatory and oxidative mediators in the cells and blood.

Aitken [1] proved that gene expression of the pro-oxidant, 15-lipoxygenase 1, which is known to increase during times of oxidative stress, also increased dramatically in mammary tissue from early lactation (EL, 15–28 day in milk) cows. Expression of the pro-inflammatory cytokines, IL-1beta, IL-6, and IL-8 did not change significantly during the periparturient period. Close positive correlations were found between several antioxidant enzymes (cytosolic GPX, thioredoxin reductase 1, and heme oxygenase-1) and vascular adhesion molecules (intercellular vascular adhesion molecule-1, vascular cell adhesion molecule-1) suggesting a protective response of these antioxidants to an enhanced proinflammatory state. Ability to control oxidative stress through the antioxidant enzymes in the future may modify the proinflammatory state of periparturient cows and reduce incidence and severity of some diseases such as mastitis [1]. Local as well as systemic inflammation might play important roles in increased mammary oxidative stress [63]. The protective effects of antioxidants in a context of neutrophil-induced damage to mammary epithelial cells were evaluated in vitro using a coculture model of activated bovine neutrophils and a bovine mammary epithelial cell line (MAC-T cells). When incubated with neutrophils activated by lipopolysaccharides and phorbol 12-myristate 13-acetate, MAC-T cells released large amounts of lactate dehydrogenase indicating significant cell damage. The addition of DMTU or bathocuproine disulfonic acid did not reduce the damage whereas catechin, deferoxamine, or glutathione ethyl ester significantly reduced neutrophil-induced cytotoxicity in a dose-dependent manner. The effect of deferoxamine, an iron chelator, on the growth of *E. coli* and the ability of bovine neutrophils to phagocytose these bacteria were then assessed in vitro. Data showed that deferoxamine did not interfere with the phagocytic activity of neutrophils but inhibited growth of the bacteria. The results suggest that antioxidants may be effective tools for protecting mammary tissue against neutrophil-induced oxidative stress during bovine mastitis [68].

Chaiyotwittayakun [20] reported a study in which they used vitamin C in the treatment of mastitis. Mastitis was induced in eight nonpregnant Holstein cows by intramammary infusion of endotoxin. Two doses of 25 g of ascorbic acid were administered intravenously at 3- and 5-h postendotoxin challenge. The authors reported an increased milk production recovery (9% higher, $P < 0.02$) and tended to reduce the extent of rumen stasis [20].

The plasma selenium levels indicate the amount of circulating selenoproteins and selenoenzymes. These are important for the maintenance of the redox system, modulating the immune system and also for thyroid hormone metabolism. The thyroid gland is among the human tissues with the highest Se content per mass unit similar to other endocrine organs and the brain. Not only are all three deiodinases selenoenzymes, but within the thyroid gland there are several other selenoenzymes, which are important for the maintenance of normal thyroid function [36]. Selenoproteins involved in cellular antioxidative defense systems and redox control, such as the GPx and the thioredoxin reductase (TxnRd) family, are involved in protection of the thyroid gland from excess hydrogen peroxide and ROS produced by the follicles for biosynthesis of thyroid hormones. In addition, the three key enzymes involved in activation and inactivation of thyroid hormones, the iodothyronine deiodinases (DIO1,2,3), are selenoproteins with development, cell-, and pathology-related expression patterns [64]. The trace element selenium is also in the prosthetic group of selenium-dependent GPX. Therefore it affects the innate and the adaptive immune responses of the mammary gland through cellular and humoral activities.

Substantial research has been carried out on the effect of selenium on the immune function of the mammary gland and subsequent improvement in bovine udder health and mastitis control. Levels higher than current recommendations and Se-yeast (0.3 mg/kg diet dry matter) can potentially be used to enhance our capacity to modulate the physiological mechanisms of the bovine mammary gland to respond to infection [103]. In spite of this replacement of sodium selenite with an organic source of Se (Se yeast) in diets not suboptimal in basal Se concentrations did not improve Se status, uterine health, fertilization, or embryo quality in early lactation dairy cows [19].

References

1. Aitken SL, Karcher EL, Rezamand P, Gandy JC, VandeHaar MJ, Capuco AV, Sordillo LM. Evaluation of antioxidant and proinflammatory gene expression in bovine mammary tissue during the periparturient period. J Dairy Sci. 2009;92(2):589–98.
2. Albanese F, Tieghi C, De Rosa L, Colombo S, Abramo F. Feline perforating dermatitis resembling human reactive perforating collagenosis: clinicopathological findings and outcome in four cases. Vet Dermatol. 2009;20(4):273–80.

3. Anderson DK, Saunders RD, Demediuk P, Dugan LL, Braughler JM, Hall ED, Means ED, Horrocks LA. Lipid hydrolysis and peroxidation in injured spinal cord: partial protection with methylprednisolone or vitamin E and selenium. Cent Nerv Syst Trauma. 1985;2(Winter 4):257–67.
4. Andreone P, Fiorino S, Cursaro C, et al. Vitamin E as treatment for chronic hepatitis B – results of a randomized controlled pilot trial. Antivir Res 2001;49:75–81.
5. Arvidsson S, Fält K, Haglund U. Feline E. coli bacteremia – effects of misoprostol/scavengers or methylprednisolone on hemodynamic reactions and gastrointestinal mucosal injury. Acta Chir Scand. 1990;156(3):215–21.
6. Asaba K, Iwasaki Y, Asai M, Yoshida M, Nigawara T, Kambayashi M, Hashimoto K. High glucose activates pituitary proopiomelanocortin gene expression: possible role of free radical-sensitive transcription factors. Diabetes Metab Res Rev. 2007;23(4):317–23.
7. Bakker J, Zhang H, Depierreux M, van Asbeck S, Vincent JL. Effects of N-acetylcysteine in endotoxic shock. J Crit Care. 1994;9(4):236–43.
8. Beck-Speier I, Dayal N, Karg E, Maier KL, Schumann G, Semmler M, Koelsch SM. Oxymetazoline inhibits proinflammatory reactions: effect on arachidonic acid-derived metabolites. J Pharmacol Exp Ther. 2006;316(2):843–51. Epub 2005 Oct 12.
9. Berglundh T, Lindhe J. Gingivitis in young and old dogs. J Clin Periodontol 1993;20:179–85.
10. Bishop A. Role of oxygen in wound healing. J Wound Care. 2008;17(9):399–402, Review.
11. Blake DR, Hall ND, Bacon PA, Dieppe PA, Halliwell B, Gutteridge JM. Effect of a specific iron chelating agent on animal models of inflammation. Ann Rheum Dis. 1983;42(1):89–93.
12. Bochsler PN, Slauson DO. Chapter 4: Inflammation and repair of tissue. In: Slauson DO, Cooper BJ, eds, Mechanisms of Disease: A Textbook of Comparative General Pathology, Third edition. MO, St. Louis: Elsevier (2002); pp 141–245.
13. Boerema I, Meyne NG, Brummelkamp WK, et al. Life without blood: a study of the influence of high atmospheric pressure and hypothermia on dilution of blood. J Cardiovasc Surg. 1960;1:133–46.
14. Boileau C, Martel-Pelletier J, Caron J, Msika P, Guillou GB, Baudouin C, Pelletier JP. Protective effects of total fraction of avocado/soybean unsaponifiables on the structural changes in experimental dog osteoarthritis: inhibition of nitric oxide synthase and matrix metalloproteinase-13. Arthritis Res Ther. 2009;11(2):R41. Epub 2009 Mar 16.
15. Bouachour G, Cronier P, Gouello JP et al. Hyperbaric oxygen therapy in the management of crush injuries: a randomized double blind placebo-controlled clinical trial. J Trauma. 1996;41:333–9.
16. Boykin JV Jr, Baylis C. Hyperbaric oxygen therapy mediates increased nitric oxide production associated with wound healing: a preliminary study. Adv Skin Wound Care. 2007;20(7):382–8.
17. Buranakarl C, Trisiriroj M, Pondeenana S, Tungjitpeanpong T, Jarutakanon P, Penchome R. Relationships between oxidative stress markers and red blood cell characteristics in renal azotemic dogs. Res Vet Sci. 2009;86(2):309–13. Epub 2008 Jul 29.
18. Canter PH, Wider B, Ernst E. The antioxidant vitamins A, C, E and selenium in the treatment of arthritis: a systematic review of randomized clinical trials. Rheumatology (Oxford) 2007;46(8):1223–33.
19. Cerri RL, Rutigliano HM, Lima FS, Araújo DB, Santos JE. Effect of source of supplemental selenium on uterine health and embryo quality in high-producing dairy cows. Theriogenology. 2009;71(7):1127–37. Epub 2009 Feb 1.
20. Chaiyotwittayakun A, Erskine RJ, Bartlett PC, Herd TH, Sears PM, Harmont RJ. The effect of ascorbic acid and L-histidine therapy on acute mammary inflammation in dairy cattle. J Dairy Sci. 2002;85(1):60–7.
21. Chon SK, Kim NS. Evaluation of silymarin in the treatment on asymptomatic Giardia infections in dogs. Parasitol Res. 2005;97(6):445–51.
22. Coates JR, Sorjonen DC, Simpson ST, Cox NR, Wright JC, Hudson JA, Finn-Bodner ST, Brown SA. Clinicopathologic effects of a 21-aminosteroid compound (U74389G) and high-dose methylprednisolone on spinal cord function after simulated spinal cord trauma. Vet Surg. 1995;24(2):128–39.

23. Dabrowski A, Gabryelewicz A, Chyczewski L. The effect of platelet activating factor antagonist (BN 52021) on cerulein-induced acute pancreatitis with reference to oxygen radicals. Int J Pancreatol. 1991;8(1):1–11.
24. Deaton CM, Marlin DJ, Smith NC, Roberts CA, Harris PA, Kelly FJ, Schroter RC. Pulmonary bioavailability of ascorbic acid in an ascorbate-synthesising species, the horse. Free Radic Res. 2003;37(4):461–7.
25. Dziezyc J, Millichamp NJ, Rohde BH, Baker JS, Chiou GC. Effects of lipoxygenase inhibitors in a model of lens-induced uveitis in dogs. Am J Vet Res. 1989;50(11):1877–82.
26. Elayan IM, Milton J. Effect of hyperbaric oxygen treatment on nitric oxide and oxygen free radicals in rat brain. J Neurophysiol. 2000;83:2022–9.
27. Escobar SJ, Slade JB Jr, Hunt TK, Cianci P. Adjuvant hyperbaric oxygen therapy (HBO_2) for treatment of necrotizing fasciitis reduces mortality and amputation rate. Undersea Hyperb Med. 2005;32(6):437–43.
28. Fantone JC, Ward PA. Polymorphonuclear leukocyte-mediated cell and tissue injury: oxygen metabolites and their relations to human disease. Hum Pathol. 1985;16(10): 973–8. Review.
29. Feldmeier JJ. Hyperbaric Oxygen Therapy 2003: Indications and Results. The Hyperbaric Oxygen Therapy Committee Report. Kensington, MD: Undersea and Hyperbaric Medical Society; 2003.
30. Ferrándiz ML, Sanz MJ, Bustos G, Payá M, Alcaraz MJ, De Rosa S. Avarol and avarone, two new anti-inflammatory agents of marine origin. Eur J Pharmacol. 1994;253(1–2):75–82.
31. Fliss H, Ménard M. Hypochlorous acid-induced mobilization of zinc from metalloproteins. Arch Biochem Biophys. 1991;287(1):175–9.
32. Francis TJ, Dutka AJ, Hallenbeck JM. Pathophysiology of decompression sickness. In: Bove AA, Davis JC, eds, Diving Medicine. Philadelphia, PA: WB Saunders (1990); pp 170–87.
33. Frech TM, Clegg DO. The utility of nutraceuticals in the treatment of osteoarthritis. Curr Rheumatol Rep. 2007;9(1):25–30. Review.
34. Gabb G, Robin ED. Hyperbaric oxygen: a therapy in search of diseases. Chest. 1987;92:1074–82.
35. Garrett IR, Boyee BE, Orefto ROC, Bonewald L, Poser J, Mundy GR. Oxygen-derived free radicals stimulate osteociastic bone resorption in rodent bone in vitro and in vivo. J Clin Invest 1990;85:632–9.
36. Gärtner R. Selenium and thyroid hormone axis in critical ill states: an overview of conflicting view points. J Trace Elem Med Biol. 2009;23(2):71–4. Epub 2009 Feb 25.
37. Gaspar MM, Boerman OC, Laverman P, Corvo ML, Storm G, Cruz ME. Enzymosomes with surface-exposed superoxide dismutase: in vivo behaviour and therapeutic activity in a model of adjuvant arthritis. J Control Release. 2007;117(2):186–95. Epub 2006 Nov 10.
38. Gimbel ML, Hunt TK. Wound healing and hyperbaric oxygenation. In: EP Kindwall (ed.), Hyperbaric Medicine Practice. Flagstaff, AZ: Best Publishing (2002); pp 169–204.
39. Gomez-Cambronero L, Camps B, de La Asuncion JG, Cerda M, Pellin A, Pallardo FV, Calvete J, Sweiry JH, Mann GE, Vina J, Sastre J. Pentoxifylline ameliorates cerulein-induced pancreatitis in rats: role of glutathione and nitric oxide. J Pharmacol Exp Ther. 2000;293(2):670–6. L 2000.
40. Gross V, Arndt H, Andus T, Palitzsch KD, Scholmerich J. Free radicals in inflammatory bowel diseases pathophysiology and therapeutic implications. Hepatogastroenterology. 1994;41(4):320–7. Review.
41. Guy J, McGorray S, Qi X, Fitzsimmons J, Mancuso A, Rao N. Conjugated deferoxamine reduces blood-brain barrier disruption in experimental optic neuritis. Ophthalmic Res. 1994;26(5):310–23.
42. Halliwell B. Drug antioxidant effects: A basis for drug selection? Drugs. 1991;42(4):569–605.
43. Henderson A, Hayes P. Acetylcysteine as a cytoprotective antioxidant in patients with severe sepsis: potential new use for an old drug. Ann Pharmacother. 1994;28(9):1086–8.
44. Henrotin Y, Kurz B. Antioxidant to treat osteoarthritis: dream or reality? Curr Drug Targets. 2007;8(2):347–57. Review.

45. Hillman JD, Socransky SS. Replacement therapy of the prevention of dental disease. Adv Dent Res. 1987;1(1):119–25.
46. Hilton JG. Effects of alterations of polyunsaturated fatty acid metabolism upon plasma volume loss induced by thermal trauma. J Trauma. 1980;20(8):663–9.
47. Hippocrates. Aphorisms. Section VII. Translated by Francis Adams, The Internet Classics Archive by Daniel C. Stevenson, Web Atomics. World Wide Web presentation is copyright (C) 1994–2000, Daniel C. Stevenson, Web Atomics. Point: 87, 400 BC.
48. Hitt ME. Chronic active hepatitis in dogs – experiences with aggressive treatment efforts. In: 2001 Proceedings, 19th Annual American College of Veterinary Internal Medicine Forum. Denver, CO, American College of Veterinary Internal Medicine, 2001, pp 775–7.
49. Horatius Quintus Flaccus. (Horace). Epistles I.18, 80. 20–14 BC. H. Rushton Fairclough, Horace. Satires, Epistles, and Ars Poetica. The Harvard University Press, Cambridge, MA 1978
50. Horatius Quintus Flaccus. (Horace). Odes I, 11.8. 23–13 BC H. Rushton Fairclough, Horace. Satires, Epistles, and Ars Poetica The Harvard University Press, Cambridge, MA 1978.
51. Horatius Quintus Flaccus. (Horace). Satires, 35–20 BC H. Rushton Fairclough, Horace. Satires, Epistles, and Ars Poetica The Harvard University Press, Cambridge, MA 1978.
52. Hunt TK, Niinikoski J, Zederfeldt BH. Oxygen in wound healing enhancement: cellular effects of Oxygen, In: Davis JC, Hunt TK, eds, Hyper Baric Therapy. Bethesda, MD: The Undersea Medical Society, 1977;111–22.
53. Inokuchi T, Kawamoto T, Aoki K, Aoki A, Nagahama K, Baba Y, Suzuki S, Shibayama M, Mano Y, Ohya K, Moriyama K. The effects of hyperbaric oxygen on tooth movement into the regenerated area after distraction osteogenesis. Cleft Palate Craniofac J. 2009;14:1.
54. Isik AT, Mas MR, Yamanel L, Aydin S, Comert B, Akay C, Erdem G, Mas N. The role of allopurinol in experimental acute necrotizing pancreatitis. Indian J Med Res. 2006;124(6):709–14.
55. Itoh S, Ozumi K, Kim HW, Nakagawa O, McKinney RD, Folz RJ, Zelko IN, Ushio-Fukai M, Fukai T. Novel mechanism for regulation of extracellular SOD transcription and activity by copper: role of antioxidant-1. Free Radic Biol Med. 2009;46(1):95–104.
56. Jacobs BP, Dennehy C, Ramirez G, et al: Milk thistle for the treatment of liver disease – a systematic review and meta-analysis. Am J Med 2002;113:506–15.
57. Jonkam CC, Bansal K, Traber DL, Hamahata A, Maybauer MO, Maybauer DM, Cox RA, Lange M, Connelly RL, Traber LD, Djukom CD, Salsbury JR, Herndon DN, Enkhbaatar P. Pulmonary vascular permeability changes in an ovine model of methicillin-resistant Staphylococcus aureus sepsis. Crit Care. 2009;13(1):R19.
58. Kalil AC, Sevransky JE, Myers DE, Esposito C, Vandivier RW, Eichacker P, Susla GM, Solomon SB, Csako G, Costello R, Sittler KJ, Banks S, Natanson C, Danner RL. Preclinical trial of L-arginine monotherapy alone or with N-acetylcysteine in septic shock. Crit Care Med. 2006;34(11):2719–28.
59. Keshavarzian A, Morgan G, Sedghi S, Gordon JH, Doria M. Role of reactive oxygen metabolites in experimental colitis. Gut. 1990;31(7):786–90.
60. Khanna D, Sethi G, Ahn KS, Pandey MK, Kunnumakkara AB, Sung B, Aggarwal A, Aggarwal BB. Natural products as a gold mine for arthritis treatment. Curr Opin Pharmacol. 2007;7(3):344–51. Epub 2007 May 1. Review.
61. Kirk GR, White JS, McKie L, Stevenson M, Young I, Clements WD, Rowlands BJ. Combined antioxidant therapy reduces pain and improves quality of life in chronic pancreatitis. J Gastrointest Surg. 2006;10(4):499–503.
62. Kirschvink N, Smith N, Fievez L, Bougnet V, Art T, Degand G, Marlin D, Roberts C, Genicot B, Lindsey P, Lekeux P. Effect of chronic airway inflammation and exercise on pulmonary and systemic antioxidant status of healthy and heaves-affected horses. Equine Vet J. 2002;34(6):563–71.
63. Kleczkowski M, Kluciińskil W, Jakubowski T, Sikora J. Dependence between acute phase response, oxidative status and mastitis of cows. Pol J Vet Sci. 2006;9(2):151–8.
64. Köhrle J, Gärtner R. Selenium and thyroid. Best Pract Res Clin Endocrinol Metab. 2009;23(6):815–27.

65. Kramaric P, Pavlica Z, Koklic T, Nemec A, Erzen NK, Sentjurc M. Membrane switch hypothesis. 2. Domain structure of phagocytes in horses with recurrent airway obstruction. J Chem Inf Model. 2005;45(6):1708–15.
66. Lange M, Enkhbaatar P, Nakano Y, Traber DL. Role of nitric oxide in shock: the large animal perspective. Front Biosci. 2009;14:1979–89.
67. Lapenna D, De Gioia S, Ciofani G, Mezzetti A, Consoli A, Di Ilio C, Cuccurullo F. Hypochlorous acid-induced zinc release from thiolate bonds: a potential protective mechanism towards biomolecules oxidant damage during inflammation. Free Radic Res. 1994;20(3):165–70.
68. Lauzon K, Zhao X, Bouetard A, Delbecchi L, Paquette B, Lacasse P. Antioxidants to prevent bovine neutrophil-induced mammary epithelial cell damage. J Dairy Sci. 2005;88(12):4295–303.
69. Lee JI, Son HY, Kim MC. Attenuation of ischemia-reperfusion injury by ascorbic acid in the canine renal transplantation. J Vet Sci. 2006;7(4):375–9.
70. Leveille CR, Arias IM. Pathophysiology and pharmacologic modulation of hepatic fibrosis. J Vet Intern Med 1993;7:73–84.
71. Lewis DD, Church DF, Hosgood G and van Ee RT. Investigation of oxygen-derived free radical generation in cancellous bone specimens obtained from dogs. AJVR 1994;55:1608–12
72. Lucas JH, Wheeler DG, Guan Z, Suntres Z, Stokes BT. Effect of glutathione augmentation on lipid peroxidation after spinal cord injury. J Neurotrauma. 2002;19(6):763–75.
73. Lucretius Titus Carus. De rerum natura. Concordances of Lucretius, 5.1284 50 BC.
74. Lyon KC. The case for evidence in wound care: investigating advanced treatment modalities in healing chronic diabetic lower extremity wounds. J Wound Ostomy Continence Nurs. 2008;35(6):585–90.
75. Mao GD, Thomas ED, Lopaschuk GD, Poznansky MJ. Superoxide dismutase (SOD)-catalase conjugates. J Biol Chem 1993;268:416–20.
76. Martindale VE, McKay K. Hyperbaric oxygen treatment of dogs has no effect on red cell deformability but causes an acute fluid shift. Physiol Chem Phys Med NMR. 1995;27(1):45–53.
77. Matejovic M, Krouzecky A, Martinkova V, Rokyta R Jr, Radej J, Kralova H, Treska V, Radermacher P, Novak I. Effects of tempol, a free radical scavenger, on long-term hyperdynamic porcine bacteremia. Crit Care Med. 2005;33(5):1057–63.
78. Mathy-Hartert M, Hogge L, Sanchez C, Deby-Dupont G, Crielaard JM, Henrotin Y. Interleukin-1beta and interleukin-6 disturb the antioxidant enzyme system in bovine chondrocytes: a possible explanation for oxidative stress generation. Osteoarthr Cartil. 2008;16(7):756–63.
79. McCord JM, Wong K, Stokes SH, Petrone WF, English D. Superoxide and inflammation: a mechanism for the anti-inflammatory activity of superoxide dismutase. Acta Physiol Scand Suppl. 1980;492:25–30.
80. Michelson AM, Puget K. Anti inflammatory activity of superoxide dismutases: comparison of enzymes from different sources in different modeis in rats: mechanism of action. Free Radic Res Commun. 1986;2:43–56.
81. Mink SN, Kasian K, Santos Martinez LE, Jacobs H, Bose R, Cheng ZQ, Light RB. Lysozyme, a mediator of sepsis that produces vasodilation by hydrogen peroxide signaling in an arterial preparation. Am J Physiol Heart Circ Physiol. 2008 Apr;294(4):H1724–35. Epub 2008 Feb 8.
82. Misaki H, Suzuki M, Yoshie H, Se Hara K. The effect of superoxide dismutase on the inflammation induced by periodontal pathogenic bacteria and wound healing of gingival incision. J Jpn Assoc Peniodontol 1990;32, 93–100.
83. Niinikoski J. Effect of oxygen supply on wound healing and formation of experimental granulation tissue. Acta Physiol Scand Suppl 1969;334:1.
84. Nylander G, Nordstrom H, Eriksson E. Effects of hyperbaric oxygen on edema formation after a scald burn. Burns, 1984:10(3):193–6.
85. Nylander G, Lewis DH, Nordstrom HR. Reduction of post ischemic edema with hyperbaric oxygen. Plast Reconstr Surg. 1985;76:596–601.

86. Oury TD, Ho Y, Piantadosi CA. Extracellular superoxide dismutase, nitric oxide, and central nervous system O_2 toxicity. Proc Natl Acad Sci 1992;89:9715–19.
87. Ovidius Publius Naso. Grave carving. Tristia. 3.3.73-76. 10 CE.
88. Pakman LM. Inhibition of Pseudomonas aeruginosa by hyperbaric oxygen. I. Sulfonamide activity enhancement and reversal. Infect Immun. 1971;4:479–87.
89. Paquette DW, Rosenberg A, Lohinai Z, Southan GJ, Williams RC, Offenbacher S, Szabo C. Inhibition of experimental gingivitis in beagle dogs with topical mercaptoalkylguanidines. J Periodontol. 2006;77(3):385–91.
90. Paracelsus – Collected Writings Vol. I (1926) edited by Bernhard Aschner, p 110.
91. Park MK. Effects of hyperbaric oxygen in infectious diseases: basic mechanisms. In: Kindwall EP, ed, Hyperbaric Medicine Practice. Flagstaff, AZ: Best Publishing (2002); pp 205–44.
92. Park MK, Muhvich KH, Myers RAM. Hyperoxia prolongs the aminoglycoside-induced post antibiotic effect in pseudomonas aeruginosa. Antimicrob Agents Chemother. 1991;35:691–5.
93. Petelin M, Pavlica Z, Ivanusa T, Sentjurc M, Skaleric U. Local delivery of liposome-encapsulated superoxide dismutase and catalase suppress periodontal inflammation in beagles. Clin Periodontol. 2000;27(12):918–25.
94. Peuchant E, Beauvieux MCD, Dubourg L, Thomas MJ, Perromat A, Aparicio M, Clerc M, Combe C. Antioxidant effects of a supplemented very low protein diet in chronic renal failure. Free Radic Biol Med. 1997;22(1–2): 313–20
95. Qi X, Sun L, Lewin AS, Hauswirth WW, Guy J. Long-term suppression of neurodegeneration in chronic experimental optic neuritis: antioxidant gene therapy. Invest Ophthalmol Vis Sci. 2007;48(12):5360–70.
96. Rao NA, Romero JL, Fernandez MA, Sevanian A, Marak GE Jr. Effect of iron chelation on severity of ocular inflammation in an animal model. Arch Ophthalmol. 1986;104(9):1369–71.
97. Rao NA, Sevanian A, Fernandez MA, Romero JL, Faure JP, de Kozak Y, Till GO, Marak GE Jr. Role of oxygen radicals in experimental allergic uveitis. Invest Ophthalmol Vis Sci. 1987;28(5):886–92.
98. Ries WL, Key LL, Rodriguez RM. Nitroblue tetrazolium reduction and bone resorption by osteoclasts in vitro inhibited by a manganese-based superoxide dismutase mimic. J Bone Miner Res 1992;7:931–9.
99. Risber J, Tyssebotn I. Hyperbaric exposure to a 5 ATA He-[N.sub.2]-[O.sub.2] atmosphere affects the cardiac function and organ blood flow distribution in awake trained rats. Undersea Biomed Res. 1986;13:77–90.
100. Rosenthal RE, Silbergleit R, Hof PR, Haywood Y, Fiskum G. Hyperbaric oxygen reduces neuronal death and improves neurological outcome after canine cardiac arrest. Stroke. 2003 May;34(5):1311–16.
101. Rump AF, Siekmann U, Kalff G. Effects of hyperbaric and hyperoxic conditions on the disposition of drugs: theoretical considerations and a review of the literature. Gen Pharmacol. 1999;32(1):127–33.
102. Salahudeen AK, Wang C, Bigler SA, Dai Z, Tachikawa H. Synergistic renal protection by combining alkaline-diuresis with lipid peroxidation inhibitors in rhabdomyolysis: possible interaction between oxidant and non-oxidant mechanisms. Nephrol Dial Transplant. 1996;11(4):635–42.
103. Salman S, Khol-Parisini A, Schafft H, Lahrssen-Wiederholt M, Hulan HW, Dinse D, Zentek J. The role of dietary selenium in bovine mammary gland health and immune function. Anim Health Res Rev. 2009;10(1):21–34. Epub 2009 Feb 6. Review.
104. Santos L, Tipping PG. Attenuation of adjuvant arthritis in rats by treatment with oxygen radical scavengers. Immunol Cell Biol. 1994;72(5):406–14.
105. Schoenberg MH, Büchler M, Baczako K, Bültmann B, Younes M, Gasper M, Kirchmayr R, Beger HG. The involvement of oxygen radicals in acute pancreatitis. Klin Wochenschr. 1991;69(21–23):1025–31.
106. Schuller-Levis GB, Levis WR, Ammazzalorso M, Nosrati A, Park E. Mycobacterial lipoarabinomannan induces nitric oxide and tumor necrosis factor alpha production in a macrophage cell line: down regulation by taurine chloramine. Infect Immun. 1994;62(10):4671–4.

107. Sener G, Aksoy H, Sehirli O, Yüksel M, Aral C, Gedik N, Cetinel S, Yeğen BC. Erdosteine prevents colonic inflammation through its antioxidant and free radical scavenging activities. Dig Dis Sci. 2007;52(9):2122–32.
108. Shaw FL, Handy RD, Bryson P, Sneyd JR, Moody AJ. A single exposure to hyperbaric oxygen does not cause oxidative stress in isolated platelets: no effect on superoxide dismutase, catalase, or cellular ATP. Clin Biochem. 2005;38(8):722–6.
109. Sheffield PJ, Smith AP. Physiological and pharmacological basis of hyperbaric oxygen therapy. In: DJ Bakker, ed., Hyperbaric Surgery: Perioperative Care. Flagstaff, AZ: Best Publishing (2002); pp 63–110.
110. Sheffield JC, Sheffield PJ, Ziemba AL. Complications rates of HBO treatments: a 21-year analysis. Proceedings of XIV International Congress on Hyperbaric Medicine. Flagstaff, AZ: Best Publishing, 2002.
111. Sifton DW, ed. Physicians' Desk Reference (ed 56). Montvale, NJ: Medical Economics Co. Inc.; 2002.
112. Singh D, Kumar P, Majumdar S, Narang A. Effect of phenobarbital on free radicals in neonates with hypoxic ischemic encephalopathy – a randomized controlled trial. J Perinat Med. 2004;32(3):278–81.
113. Siriwardena AK, Mason JM, Balachandra S, Bagul A, Galloway S, Formela L, Hardman JG, Jamdar S. Randomized, double-blind, placebo-controlled trial of intravenous antioxidant (n-acetylcysteine, selenium, vitamin c) therapy in severe acute pancreatitis. Gut. 2007; 56(10):1439–1444.
114. Slovis N. Review of Hyperbaric Medicine. Veterinary Hyperbaric Medicine Society. University of Tennessee, College of Veterinary Medicine. http://www.vet.utk.edu/vhms/review.html. 2006.
115. Snapper JR, Bernard GR, Brigham KL. In vivo oxidants and pulmonary inflammation. Bull Eur Physiopathol Respir. 1986;22(1):257s–60s.
116. Sokol RJ, Twedt D, McKim JM, et al. Oxidant injury to hepatic mitochondria in patients with Wilson's disease and Bedlington terriers with copper toxicosis. Gastroenterology 1994;107:1788–98.
117. Spee B, Arends B, van den Ingh TS, Penning LC, Rothuizen J. Copper metabolism and oxidative stress in chronic inflammatory and cholestatic liver diseases in dogs. J Vet Intern Med. 2006;20(5):1085–92.
118. Stahelin H, Karnovsky ML, Farnham AE, Suter E. Studies on the interaction between phagocytes and tubercle bacilli: III. Some metabolic effects in guinea pigs associated with infection with tubercle bacilli. J Exp Med. 1957;105(3):265–77.
119. Szabolcs A, Varga IS, Varga C, Berko A, Kaszaki J, Letoha T, Tiszlavicz L, Sari R, Lonovics J, Takacs T. Beneficial effect of resveratrol on cholecystokinin-induced experimental pancreatitis. Eur J Pharmacol. 2006;532(1–2):187–93.
120. Taubert D, Berkels R, Grosser N, Schroder H, Grundemann D, Schomig E. Aspirin induces nitric oxide release from vascular endothelium: a novel mechanism of action. Br J Pharmacol. 2004;143(1):159–65. Epub 2004 Aug 2.
121. Terentius Varro, Marcus. http://audiolatinproverbs.blogspot.com/2007/04/facile-omnes-cum-valemus-recta-consilia.html. Latin Via Proverbs, Dictum sapienti sat est. April 08, 2007.
122. Thiele JJ, Hsieh SN, Ekanayake-Mudiyanselage S. Vitamin E: critical review of its current use in cosmetic and clinical dermatology. Dermatol Surg. 2005;31(7 Pt 2):805–13; discussion 813. Review.
123. Thom SR, Mendiguren I, Hardy K, Bolotin T. Inhibition of human neutrophil beta2-integrin-dependent adherence by hyperbaric O_2. Am J Physiol. 1997;26:82–6.
124. Trachtman H, Chan JC, Chan W, Valderrama E, Brandt R, Wakely P, Futterweit S, Maesaka J, Ma C. Vitamin E ameliorates renal injury in an experimental model of immunoglobulin A nephropathy. Pediatr Res. 1996;40(4):620–6.
125. Tsiakitzis K, Kourounakis AP, Tani E, Rekka EA, Kourounakis PN. Stress and active oxygen species – effect of alpha-tocopherol on stress response. Archiv der Pharmazie 2005;338(7):315–21.

126. Turrens JF, Craps JD, Freeman BA. Protection against oxygen toxicity by intravenous injection of liposomal-en trapped catalase and superoxide dismu tase. J Clin Invest 1984;73:87–95.
127. Twedt DC. Nutraceuticals for liver dysfunction. In: 2003 Proceedings, North American Veterinary Conference. Orlando, FL, Eastern States Veterinary Association, 2003, pp 434–6.
128. Twedt DC. Antioxidants and liver disease. In: 2001 Proceedings, 19th Annual American College of Veterinary Internal Medicine Forum. Denver, CO, American College of Veterinary Internal Medicine, 2001a, pp 13–15.
129. Twedt DC. A review of traditional and not so traditional therapies for liver disease. In: 2001 Proceedings, 19th Annual American College of Veterinary Internal Medicine Forum. Denver, CO, American College of Veterinary Internal Medicine, 2001b, pp 610–12
130. Vajdovich P. Free radicals and antioxidants in inflammatory processes and ischemia-reperfusion injury. Vet Clin North Am Small Anim Pract. 2008 Jan;38(1):31–123, v. Review.
131. Váradi Á. Preliminary data of life and health insurance in the Roman law (Collegium funeraticium and collegium tenuiorum) Hungarian Med J 2008;2(4):533–8.
132. Vassilev D, Hauser B, Bracht H, Iványi Z, Schoaff M, Asfar P, Vogt J, Wachter U, Schelzig H, Georgieff M, Brückner UB, Radermacher P, Fröba G. Systemic, pulmonary, and hepatosplanchnic effects of N-acetylcysteine during long-term porcine endotoxemia. Crit Care Med. 2004;32(2):525–32.
133. Verspaget HW, Mulder TP, Veer A, Peña AS, Lamers CB. Reactive oxygen metabolites and colitis: a disturbed balance between damage and protection, A selective review. Scand J Gastroenterol Suppl. 1991;188:44–51.
134. Vogel G, Tuchweber B, Trost W, et al. Protection by silibinin against Amanita phalloides intoxication in beagles. Toxicol Appl Pharmacol 1984;73:355–62
135. Wallace KP, Center SA, Hickford FH, et al. S-adenosyl-l-methionine (SAMe) for the treatment of acetaminophen toxicity in a dog. J Am Anim Hosp Assoc 2002;38:246–54.
136. Wang X, Ding I, Xie H. Hyperbaric oxygen and basic fibroblast growth factor promote growth of irradiated bone. Growth Factors 1995;12(1): 29–35.
137. Williams DL, Munday P. The effect of a topical antioxidant formulation including N-acetyl carnosine on canine cataract: a preliminary study. Vet Ophthalmol. 2006;9(5):311–16.
138. Wu GS, Walker J, Rao NA. Effect of deferoxamine on retinal lipid peroxidation in experimental uveitis. Invest Ophthalmol Vis Sci. 1993;34(11):3084–9.
139. Yamada M, Shichi H, Yuasa T, Tanouchi Y, Mimura Y. Superoxide in ocular inflammation: human and experimental uveitis. J Free Radic Biol Med. 1986;2(2):111–17.
140. Yoles E, Zurovsky Y, Zarchin N, Mayevsky A. The effect of hyperbaric hyperoxia on brain function in the newborn dog in vivo. Neurol Res. 2000;22(4):404–8.
141. Yoshikawa T, Takahashi S, Kondo M. Possible role of free radicals in the chronic inflammation of the gut. EXS. 1992;62:353–8.
142. Zamboni WA, Wong HP. Effect of hyperbaric oxygen on neutrophil concentration and pulmonary sequestration in reperfusion injury. Arch Surg 1996;131:756–60.
143. Zhang H, Spapen H, Manikis P, Rogiers P, Metz G, Buurman WA, Vincent JL. Tirilazad mesylate (U-74006F) inhibits effects of endotoxin in dogs. Am J Physiol. 1995;268(5 Pt 2):H1847–55.
144. Zhao LL, Davidson JD, Wee SC. Effect of hyperbaric oxygen and growth factors on rabbit ear ischemic ulcers. Int J Radiat Oncol Biol Phys 1998;40(1):189–96.
145. Babizhayev MA. Biological activities of the natural imidazole-containing peptidomimetics n-acetylcarnosine, carcinine and L-carnosine in ophthalmic and skin care products. Life Sci. 2006;78(20):2343–57.
146. Cruz ME, Manuela Gaspar M, Bárbara M, Martins F, Luísa Corvo M. Liposomal superoxide dismutases and their use in the treatment of experimental arthritis. Methods Enzymol. 2005;391:395–413.
147. Gathwala G, Marwah A, Gahlaut V, Marwah P. Effect of High-dose Phenobarbital on Oxidative Stress in Perinatal Asphyxia: An Open Label Randomized Controlled Trial. Indian Pediatr. 2010 Nov 30. pii: S097475590900675-1.
148. Honeckman A. Current concepts in the treatment of canine chronic hepatitis. Clin Tech Small Amin Pract. 2003 Nov;18(4):239–44.

Oxidative Stress and Calcium Metabolism

Patricia A. Schenck

Abstract Reactive oxygen species (ROS) are a causal factor in aging, atherosclerosis, carcinogenesis, and infarction (Sheweita and Khoshhal, Curr Drug Metab 8:519–525, 2007). In bone metabolism, ROS play a dual role, with different actions under physiological and pathological conditions (Sontakke and Tare, Clin Chim Acta 318:145–148, 2002). Reactive oxygen species include hydroxyl radicals, superoxide anion radicals, hydrogen peroxide, and nitric oxide; ROS lead to oxidation of enzymes and protein oxidation and degradation. The effects of ROS are eliminated by antioxidant enzymes such as superoxide dismutase (SOD), glutathione peroxidase, and catalase. Oxidative stress occurs when the effects of ROS are greater than the antioxidant mechanisms in biological systems, and cellular damage occurs. The full effect of oxidative stress on bone remodeling is unknown, but there is evidence that both pro-oxidants and antioxidants play a role in skeletal integrity in health and disease.

Keywords Parathyroid hormone • Calcitonin • Calcitriol • Nitric oxide (NO) • Nuclear factor-kappa B (NF-kB) • NADPH oxidase • Klotho gene • RANKL pathway • Calcium metabolism

Calcium is required for many intracellular and extracellular functions, as well as for skeletal support. Ionized calcium is required for enzymatic reactions, membrane transport and stability, blood coagulation, nerve conduction, neuromuscular transmission, muscle contraction, vascular smooth muscle tone, hormone secretion, bone formation and resorption, control of hepatic glycogen metabolism, and cell growth and division.

Calcium in serum or plasma exists in three fractions: ionized, complexed, and protein-bound [1]. In healthy dogs, protein-bound, complexed, and ionized calcium

P.A. Schenck (✉)
Endocrine Diagnostic Section, Diagnostic Center for Population
and Animal Health, Michigan State University, 4125 Beaumont Rd, Lansing, MI, USA
e-mail: schenck5@msu.edu

L. Mandelker and P. Vajdovich (eds.), *Studies on Veterinary Medicine*,
Oxidative Stress in Applied Basic Research and Clinical Practice 5,
DOI 10.1007/978-1-61779-071-3_3,

accounts for 34%, 10%, and 56% of total serum calcium concentration [2]. Ionized calcium is the biologically active fraction of calcium.

Calcium ions bind directly to cell membrane receptors that are specific for ionized calcium [3]. Normal homeostatic control mechanisms usually maintain serum ionized calcium concentration within a narrow range by integrated interactions among ionized calcium, phosphorus, parathyroid hormone (PTH), vitamin D metabolites, and calcitonin. Intestine, kidney, and bone are the major target organs affected by calcium regulatory hormones and the skeleton provides a major supply of calcium and phosphorus when intestinal absorption and renal reabsorption are inadequate to maintain normal serum calcium concentration.

PTH is synthesized and secreted by the chief cells in the parathyroid gland in response to a decrease in ionized calcium concentration. The major biological effects of PTH are to increase the serum calcium concentration, to increase tubular reabsorption of calcium with a decreased loss of calcium in the urine, to increase bone resorption and numbers of osteoclasts on bone surfaces, and to accelerate the formation of 1,25-dihydroxyvitamin D (calcitriol) by the kidney. In bone, the increase in calcium concentration results from an interaction of PTH with receptors on osteoblasts that stimulate increased calcium release and direct an increase in osteoclastic bone resorption [4]. The response of bone to PTH is biphasic. The rapid effects of PTH result in an increased flow of calcium from deep bone to the surface through the action of an osteocyte–osteoblast pump. The longer-term effects of PTH are not dependent on the continuous presence of hormone. Osteoclasts are responsible for the long-term actions of PTH on bone resorption and remodeling [5].

Calcitonin is synthesized by C cells in the thyroid gland, and is secreted in response to hypercalcemia. The effects of calcitonin in calcium homeostasis are minor. Calcitonin primarily acts on bone where it inhibits the resorption of calcium and phosphorus [1].

The most important vitamin D metabolites are derived from cholecalciferol (vitamin D_3 of animal origin). Of these, the metabolites 25-hydroxyvitamin D (calcidiol) and 1,25-dihydroxyvitamin D (calcitriol) are the most important. These metabolites may also be derived from ergocalciferol (vitamin D_2 of plant origin), and are equally bioactive. Dogs and cats are dependent on dietary vitamin D as they inefficiently photosynthesize vitamin D in their skin. In the liver, vitamin D is hydroxylated to 25-hydroxyvitamin D. The 25-hydroxyvitamin D is further hydroxylated to form calcitriol in the proximal tubule of the kidney.

Calcitriol is the active metabolite of vitamin D, and acts to increase serum calcium and phosphorus concentrations. Calcitriol stimulates osteoclastic calcium mobilization and resorption of bone, increases calcium, phosphorus, and magnesium absorption in the intestine, and increases renal tubular resorption of calcium and phosphorus by the kidney [1]. Calcitriol is necessary for normal bone development because it regulates the production of alkaline phosphatase, collagen type I, osteocalcin, and osteopontin produced by osteoblasts [6, 7]. Calcitriol is necessary for normal bone resorption because it promotes differentiation of monocytic hematopoietic precursors in the bone marrow into osteoclasts [8].

Bone Modeling and Remodeling

Bone consists of a structured mineral matrix that contains heterogeneous populations of chondrocytes, osteoblasts, osteocytes, osteoclasts, endothelial cells, monocytes, macrophages, lymphocytes, and hemopoietic cells. The process of bone remodeling is the major activity of bone cells in the adult [9]. Bone multicellular units (BMUs; bone remodeling units) occur on the surface of trabecular bone as Howship lacunae, or as haversian systems in cortical bone. Osteoclasts are activated by the interaction of osteoblast lineage cells with hematopoietic precursors, but they can also be activated by inflammatory cells. After the formation of osteoclasts, there is a bone resorption phase of short duration, followed by a brief reversal phase. At this point, bone formation has not started, but the bone surface is covered by mononuclear cells. The formation phase is initiated either by osteoclast factors or factors released from the bone matrix. The formation phase is long, and involves production of matrix by osteoblasts. These cells become flat lining cells, become embedded in the bone as osteocytes, or undergo apoptosis [10].

In bone remodeling, bone resorption requires contact between osteoblasts and osteoclasts. Macrophage colony stimulating factor (M-CSF) stimulates differentiation of hematopoietic progenitors which express RANK (receptor activator of nuclear factor-κB). Osteoclasts are stimulated by RANK interaction. Bone resorption stimulates COX2 activity which amplifies the response to RANK ligand (RANKL). In disease states, inflammatory and malignant cells increase the production of osteoclasts by producing M-CSF, RANKL, parathyroid hormone related protein (PTHrP), cytokines, and prostaglandins [11].

Because the resorption phases are short, but the formation phase requires a longer period of time, an increase in the rate of bone remodeling results in a loss of bone mass. In addition, the increase in unfilled Howship lacunae and Haversian canals will weaken the bone. Excess resorption also decreases the trabecular structure so bone formation cannot occur. The inadequate formation of bone during remodeling is of major importance in the pathogenesis of osteoporosis.

Nitric Oxide

Nuclear factor-κB (NF-κB) is an oxidative stress-responsive transcription factor and is important in the production of osteoclasts. Thus free radicals can increase bone resorption through activation of NF-κB. A number of risk factors for the development of osteoporosis (smoking, hypertension, and diabetes mellitus) have been shown to increase the circulating concentrations of free radicals, which are involved in bone resorption [12].

Nitric oxide (NO) is a free radical that is involved in vascular relaxation, platelet aggregation, neurotransmission, and immune regulation [13]. NO is generated from arginine by nitric oxide synthase (NOS), and also has important effects on bone cell function. The constitutive form of NOS (cNOS) is widely expressed in bone marrow stromal cells, osteoblasts, osteocytes, and osteoclasts, whereas the inducible form (iNOS) is produced in response to inflammation. NO influences bone remodeling through influence on osteoblast and osteoclast activity.

NO can exhibit a dual role in bone metabolism [14]. When osteoblasts and osteocytes are placed under mechanical stress, they will produce iNOS-mediated NO that inhibits prostaglandin-induced bone resorption, which contributes to an increase in bone. Osteocytes secrete more NO than do osteoblasts under stress, and cytokines present in inflammation further increase NO production. Subsequently, osteoblast growth and differentiation are inhibited, leading to apoptosis. Physiologically, osteoclasts also express cNOS which is essential for normal function. cNOS-induced NO inhibits osteoclast activity and mobility. The effects of NO on osteoclastic bone resorption are dose-dependent. Low doses of NO enhance interleukin-1 (IL-1)-induced bone resorption, via the iNOS pathway. High doses of NO inhibit osteoclast activity and formation, with apoptosis of osteoclast progenitor cells.

Oxidative stress can have an impact on the effects of NO on bone cells. Superoxide anion can transform NO into peroxynitrile anion, which is highly reactive. This decreases the concentration of NO that affects the osteoblast–osteoclast balance, causing disturbances in bone metabolism.

NADPH Oxidase and Tartrate-Resistant Acid Phosphatase

NADPH oxidase (Nox) catalyzes the generation of superoxide anion from oxygen, and can also generate superoxide in osteoclasts. Nox4 is present in kidney tissue and osteoclasts. In osteoclasts, Nox4 affects bone resorption, and if Nox4 activity is decreased, osteoclastic superoxide production and bone resorption are inhibited [13].

An increase in superoxide anion production is associated with an increase in bone resorption, and a decrease in superoxide anion is associated with inhibition of bone resorption. Inasmuch as the superoxide anion is located on the outer membrane of osteoclasts, it is thought that the superoxide anion itself contributes to bone resorption.

Osteocalcin is a bone matrix protein that can break into smaller pieces when exposed to superoxide anion. The smaller pieces of protein are more susceptible to degradation and enzyme digestion, thus superoxide anion contributes to the degradation of bone matrix. Superoxide anion also stimulates the formation of osteoclasts which leads to an increase in bone resorption.

Tartrate-resistant acid phosphatase (TRACP) is an enzyme primarily expressed in macrophages and osteoclasts. In osteoclasts it is located in vesicles that transport bone matrix degradation products from bone lacuna to the basolateral membrane. TRACP can generate ROS (reactive oxygen species), which can destroy bone matrix; thus, TRACP may be important in bone resorption via ROS.

The Klotho Gene

The Klotho gene was identified in 1997, and acts as an aging-suppressor gene. Klotho binds to fibroblast growth factor (FGF) receptors and acts as a coreceptor for FGF23 which suppresses vitamin D synthesis in the kidney. The Klotho protein regulates insulin, IGF-1, and Wnt, and protects cells from oxidative stress by stimulating the expression of antioxidant proteins [15]. Klotho-deficient mice show ectopic calcification in the gastric mucosa, trachea, aorta, and arteries in the kidney. These mice have significantly elevated serum calcium, phosphorus, and vitamin D levels which play a role in ectopic calcification.

Klotho-deficient mice also display a decrease in bone density, primarily in the femur, tibia, and vertebrae. These mice have a decrease in both bone formation and resorption; the decrease in formation is greater than the decrease in resorption, leading to a net bone loss. The decreased bone mineral density with reduced bone turnover is characteristic of osteoporosis.

In mammals, Fork-head box O (FOXO) transcription factor is negatively regulated by insulin/IGF-1 signaling [16]. Antioxidant enzymes such as catalase and mitochondrial manganese–superoxide dismutase (SOD2) are upregulated by FOXOs [17]. Increased catalase and SOD2 production reduce oxidative stress by removal of ROS. Klotho-overexpressing mice have increased expression of SOD2 and decreased markers of oxidative stress, demonstrating increased resistance to oxidative stress. Klotho-deficient mice demonstrate the impact of oxidative stress on certain aspects of bone remodeling and calcium metabolism.

Oxidative Stress and Parathyroid Hormone

One of the major effects of PTH is to stimulate calcitriol synthesis in the kidney. PTH also increases the mRNA for cytochrome P450 of mitochondrial 1α-hydroxylase (CYP1α) [18], which increases the synthesis of CYP1α protein. This protein combines with ferredoxin and ferredoxin reductase on the inner mitochondrial membrane to produce 1α-hydroxylase activity that is required in the production of calcitriol. In humans and rats, the ability of PTH to increase calcitriol production decreases with age [19, 20]. This CYP1α protein may be inactivated by oxidative damage, because free radical production also occurs on the inner mitochondrial membrane. Studies with hydrogen peroxide have demonstrated a minor decrease in CYP1α protein, yet a significant decrease in calcitriol production. It appears that the protein is being produced, but the activity is reduced with oxidative damage [18].

Oxidative Stress and Osteoporosis

Osteoporosis is an important pathological problem in humans, and is characterized by low bone mass with altered architecture of bone, leading to increased bone fragility with an increased risk of fractures. In osteoporosis, there are multiple pathogenetic mechanisms leading to a loss of bone mass and deterioration of skeletal structure, with complex interactions among regulators of bone cell function [9]. Interactions of bone cells with estrogen, growth factors, PTH, vitamin D, cytokines, prostaglandins, leukotrienes, collagen, NO, and leptin are important for a proper balance of bone formation and resorption.

A normal balance between oxidants and antioxidants is needed to maintain equilibrium between osteoblasts and osteoclasts [12]. With age-related bone loss, oxidative stress leads to an increase in osteoblast and osteocyte apoptosis, with a decrease in osteoblast numbers and a decrease in bone formation via Wnt/β-catenin signaling [21]. Oxidative stress also inhibits osteoblast differentiation; osteoblasts normally produce antioxidants to protect against ROS. It is possible that in osteoporosis, osteoclasts produce excessive ROS which overwhelms the natural antioxidant systems. This increase in ROS leads to bone loss and osteoporosis. Superoxide anion generation increases from the increased osteoclast activity, and there is inhibition of SOD and glutathione peroxidase activity as well, leading to a net bone loss.

In a study of humans with and without osteoporosis, oxidative stress was an independent risk factor for osteoporosis [22]. Increases in oxidative stress were responsible for osteoporosis occurring in relatively young men [23], and links have been shown between increased oxidative stress and reduced bone mineral density in older men and women [24]. Older women with osteoporosis have been shown to have markedly decreased circulating antioxidants, suggesting the relationship between oxidative stress and osteoporosis [25].

Lipoprotein oxidation may also contribute to the development of osteoporosis [26]. When lipoproteins are circulating at increased levels, they are more prone to oxidation. Oxidized LDL induces RANKL-dependent osteoclastic differentiation, leading to bone resorption. There may also be direct interactions of oxidized lipids on bone cells via receptor-mediated responses or through generation of cytokines that may produce observed effects of oxidized lipids. The generation of inflammatory cytokines in response to oxidized lipids may cause the observed resorptive effects of oxidized lipid on bone.

The Role of Antioxidants in Skeletal Health

Some of the most convincing evidence for the role of oxidative stress in bone resorption comes from research evaluating the role of antioxidants in pathological conditions. Polyphenols are antioxidants that are similar to tocopherols (vitamin E). Polyphenols can reduce hydrogen peroxide-induced oxidative stress, and can protect osteoblasts from hydrogen peroxide-induced cell damage. High levels of polyphenols are present in green tea, and recent studies have shown that liver glutathione peroxidase activity increases after green tea polyphenol supplementation. Green tea polyphenol supplementation reduces the loss of bone mineral density, and decreased the formation of oxidative stress-induced calcium stone formation in rats because of antioxidative effects. In osteoporosis, green tea polyphenols act via inhibition of cyclooxygenase-2 (COX-2), lipoxygenase, and inducible NO synthase. The protective effects of green tea polyphenols are due to the decrease in lipid peroxidation, oxidative stress, and production of NO radicals by inhibiting NO synthase [21].

Ascorbic acid can also decrease bone resorption. Ascorbic acid stimulates osteoblast differentiation, and inhibits RANKL-induced differentiation of osteoclast precursors which reduces the rate of bone resorption [12]. Subclinical vitamin deficiencies are common in elderly patients that are affected with osteoporosis. Deficiencies of vitamins K, C, or B are risk factors for the development of osteoporosis and bone fracture. Increased vitamin D intake is needed for maintenance of bone health. The lowest incidence of osteoporosis occurs in Mediterranean countries; thus the "Mediterranean" diet that is rich in fruits and vegetables containing naturally occurring antioxidants can contribute to a "bone-sparing" effect [13].

Flavonols are phytoestrogens found in onions, beans, fruits, red wine, and tea. Flavonols increase nuclear ERbeta protein and decrease ERalpha protein in osteoclast precursors. The anti-resorption properties of flavonols are mediated by the ER proteins via inhibition of RANK protein.

Dietary fatty acids can have various effects on calcium metabolism and bone health in animals and humans [27]. Animals with essential fatty acid (EFA) deficiency exhibit a loss of bone calcium, with marked bone demineralization. These animals also develop calcium deposition in the kidney and aorta; subsequent studies showed that fish oil administration could reduce this calcium deposition. Eskimos who consume a diet rich in fish oils have a very low incidence of atherosclerosis and other diseases related to calcium metabolism (such as nephrocalcinosis) [28].

Fish oils are rich in eicosapentaenoic acid (EPA) and docosahexaenoic acid (DHA) which are long-chain omega-3 fatty acids. Dietary fish oils may have a beneficial effect on osteoporosis [29]. Supplementation with fish oil decreased bone turnover and increased bone mineral density in elderly patients. In healthy men, the highest circulating concentrations of DHA were associated with peak bone mineral density and bone accrual [30]. In a study of gonad-intact rats, rats receiving a diet highest in fish oils had the highest bone mineral content and bone mineral density compared to rats fed diets high in omega-6 fatty acids [31]. Rats fed the fish oil diet had the lowest bone NO production, whereas rats fed omega-6 fatty acids had higher bone prostaglandin E_2 production. Thus fish oil supplementation conferred a protective effect on age-induced bone loss. In another rat study, dietary fish oil supplementation decreased osteoclastic activity and alveolar bone resorption [32]. When female rats were fed a diet low in calcium, bone weight was significantly reduced; this decrease was inhibited when EPA was supplemented in the diet [33]. In rats fed omega-3 deficient diets, supplementation with DHA was most effective at maintaining bone architecture and modeling [34].

The dietary ratio of omega-6 to omega-3 fatty acids is also important in skeletal health. A ratio of 5:1 omega-6:DHA significantly elevated the DHA concentration in bone, and led to lower bone loss in ovariectomized rats [35]. Diets with this ratio produced the lowest concentrations of bone resorption markers and favored bone conservation in ovariectomized rats. In piglets, higher circulating concentrations of omega-3 fatty acids were associated with a lower level of bone resorption.

A proper ratio of omega-6:omega-3 fatty acids can also reduce bone resorption and demineralization caused by fatty acid deficiency [27].

Omega-6 fatty acids increase the production of inflammatory cytokines and ROS such as NO [36]. Omega-3 fatty acids appear to reduce the release of leukotriene B_4 and interleukin-1. Increased osteoclastic activity and bone resorption seen in inflammation-induced osteoporosis is linked to activation of the iNOS pathway with an increase in NO. Omega-3 fatty acids elevate the production of constitutive NO and suppress induced NO by inhibiting the inflammatory cytokines. This helps to maintain bone mass under normal and inflammatory conditions.

Reactive oxygen species and oxidative stress have an impact on calcium metabolism and bone formation and resorption. Antioxidant therapy in bone-resorptive disorders has the potential to improve skeletal health. Further investigation is needed to evaluate dietary antioxidants as an adjunctive therapy to prevent or minimize bone loss.

References

1. Schenck PA, Chew DJ, Nagode LA, et al. Disorders of calcium: hypercalcemia and hypocalcemia. In: DiBartola SP, ed. *Fluid Therapy in Small Animal Practice*. 3rd ed. St. Louis: Elsevier, 2005;122–194.
2. Schenck PA, Chew DJ, Brooks CL. Fractionation of canine serum calcium, using a micropartition system. *Am J Vet Res* 1996;57:268–271.
3. Brown EM, Hebert SC. Calcium-receptor-regulated parathyroid and renal function. *Bone* 1997;20:303–309.
4. Morris SA, Bilezikian JP. Signal transduction in bone physiology: messenger systems for parathyroid hormone. In: Bilezikian JP, Raisz LG, Rodan GA, eds. *Principles of Bone Biology*. New York: Academic, 1996;1203–1215.
5. Canalis E, Hock JM, Raisz LG. Anabolic and catabolic effects of parathyroid hormone on bone and interactions with growth factors. In: Bilezikian JP, Marcus R, Levine MA, eds. *The Parathyroids*. New York: Raven Press, 1994;65–82.
6. Aubin JE, Heersche JN. Vitamin D and osteoblasts. In: Feldman D, ed. *Vitamin D*. New York: Academic, 1997;313–328.
7. St. Arnaud R, Glorieux FH. Vitamin D and bone development. In: Feldman D, ed. *Vitamin D*. San Diego: Academic, 1997;293–303.
8. Suda T, Takahashi N. Vitamin D and osteoclastogenesis. In: Feldman D, ed. *Vitamin D*. San Diego: Academic, 1997;329–340.
9. Raisz LG. Pathogenesis of osteoporosis: concepts, conflicts, and prospects. *J Clin Invest* 2005;115:3318–3325.
10. Seeman E. Bone modeling and remodeling. *Crit Rev Eukaryot Gene Expr* 2009;19:219–233.
11. Deschaseaux F, Sensebe L, Heymann D. Mechanisms of bone repair and regeneration. *Trends Mol Med* 2009;15:417–429.
12. Sheweita SA, Khoshhal KI. Calcium metabolism and oxidative stress in bone fractures: role of antioxidants. *Curr Drug Metab* 2007;8:519–525.
13. Banfi G, Iorio EL, Corsi MM. Oxidative stress, free radicals and bone remodeling. *Clin Chem Lab Med* 2008;46:1550–1555.
14. Sontakke AN, Tare RS. A duality in the roles of reactive oxygen species with respect to bone metabolism. *Clin Chim Acta* 2002;318:145–148.
15. Kuro-o M. Klotho as a regulator of oxidative stress and senescence. *Biol Chem* 2008; 389:233–241.

16. Brunet A, Bonni A, Zigmond MJ, et al. Akt promotes cell survival by phosphorylating and inhibiting a Forkhead transcription factor. *Cell* 1999;96:857–868.
17. Kops GJ, Dansen TB, Polderman PE, et al. Forkhead transcription factor FOXO3a protects quiescent cells from oxidative stress. *Nature* 2002;419:316–321.
18. Armbrecht HJ, Boltz MA, Ritter CS, et al. Parathyroid hormone stimulation of the renal 25-hydroxyvitamin D-1alpha-hydroxylase – effect of age and free radicals. *J Steroid Biochem Mol Biol* 2007;103:330–333.
19. Kinyamu HK, Gallagher JC, Petranick KM, et al. Effect of parathyroid hormone (hPTH-[1-34]) infusion on serum 1,25-dihydroxyvitamin D and parathyroid hormone in normal women. *J Bone Miner Res* 1996;11:1400–1405.
20. Friedlander J, Janulis M, Tembe V, et al. Loss of parathyroid hormone-stimulated 1,25-dihydroxyvitamin D3 production in aging does not involve protein kinase A or C pathways. *J Bone Miner Res* 1994;9:339–345.
21. Shen CL, Yeh JK, Cao JJ, et al. Green tea and bone metabolism. *Nutr Res* 2009;29:437–456.
22. Sanchez-Rodriguez MA, Ruiz-Ramos M, Correa-Munoz E, et al. Oxidative stress as a risk factor for osteoporosis in elderly Mexicans as characterized by antioxidant enzymes. *BMC Musculoskelet Disord* 2007;8:124.
23. Polidori MC, Stahl W, Eichler O, et al. Profiles of antioxidants in human plasma. *Free Radic Biol Med* 2001;30:456–462.
24. Basu S, Michaelsson K, Olofsson H, et al. Association between oxidative stress and bone mineral density. *Biochem Biophys Res Commun* 2001;288:275–279.
25. Maggio D, Barabani M, Pierandrei M, et al. Marked decrease in plasma antioxidants in aged osteoporotic women: results of a cross-sectional study. *J Clin Endocrinol Metab* 2003;88:1523–1527.
26. Parhami F. Possible role of oxidized lipids in osteoporosis: could hyperlipidemia be a risk factor? *Prostaglandins Leukot Essent Fatty Acids* 2003;68:373–378.
27. Baggio B. Fatty acids, calcium and bone metabolism. *J Nephrol* 2002;15:601–604.
28. Buck AC, Davies RL, Harrison T. The protective role of eicosapentaenoic acid [EPA] in the pathogenesis of nephrolithiasis. *J Urol* 1991;146:188–194.
29. Watkins BA, Li Y, Lippman HE, et al. Omega-3 polyunsaturated fatty acids and skeletal health. *Exp Biol Med (Maywood)* 2001;226:485–497.
30. Hogstrom M, Nordstrom P, Nordstrom A. n-3 Fatty acids are positively associated with peak bone mineral density and bone accrual in healthy men: the NO2 Study. *Am J Clin Nutr* 2007;85:803–807.
31. Shen CL, Yeh JK, Rasty J, et al. Protective effect of dietary long-chain n-3 polyunsaturated fatty acids on bone loss in gonad-intact middle-aged male rats. *Br J Nutr* 2006;95:462–468.
32. Iwami-Morimoto Y, Yamaguchi K, Tanne K. Influence of dietary n-3 polyunsaturated fatty acid on experimental tooth movement in rats. *Angle Orthod* 1999;69:365–371.
33. Sakaguchi K, Morita I, Murota S. Eicosapentaenoic acid inhibits bone loss due to ovariectomy in rats. *Prostaglandins Leukot Essent Fatty Acids* 1994;50:81–84.
34. Li Y, Greiner RS, Salem N, Jr., et al. Impact of dietary n-3 FA deficiency on rat bone tissue FA composition. *Lipids* 2003;38:683–686.
35. Watkins BA, Li Y, Seifert MF. Dietary ratio of n-6/n-3 PUFAs and docosahexaenoic acid: actions on bone mineral and serum biomarkers in ovariectomized rats. *J Nutr Biochem* 2006;17:282–289.
36. Watkins BA, Li Y, Lippman HE, et al. Modulatory effect of omega-3 polyunsaturated fatty acids on osteoblast function and bone metabolism. *Prostaglandins Leukot Essent Fatty Acids* 2003;68:387–398.

TRPM2 Cation Channels and Oxidative Stress-Induced Neuronal Cell Death

Mustafa Nazıroğlu

Abstract Of all the organs in the body, the central nervous system is especially sensitive to free radical damage. Its high consumption of oxygen, rich content of easily oxidizable fatty acids, relatively low content of antioxidant enzymes and antioxidants, and the presence of high levels of iron make it a prime substrate for damage by ROS. The Na^+ and Ca^{2+}-permeable melastatin-related transient receptor potential 2 (TRPM2) cation channels can be gated either by ADP-ribose (ADPR) in concert with Ca^{2+} or by hydrogen peroxide (H_2O_2), an experimental model for oxidative stress, binding to the channel's enzymatic Nudix domain. Because the mechanisms that lead to TRPM2 inhibiting in response to ADPR and H_2O_2 are not understood in neuronal cells, we reviewed the effects of ADPR and oxidative stress in neurological cells such as microglia, hippocampus, and brain as well as neurological diseases such as bipolar diseases. It was observed that TRPM2 cation channels in microglia and hippocampal cells were gated both by ADPR and H_2O_2. In addition, H_2O_2 seems to be responsible for activation of TRPM2 in neurological diseases. Genetic defects may have an important role in the etiology of bipolar diseases. Experimental studies with respect to patch-clamp and Ca^{2+} imaging, inhibitor roles of antioxidants are also summarized in the review.

Keywords TRPM2 • Ca^{2+} • Neurological cells • Oxidative stress • ADP-ribose • Glial cells

---

M. Nazıroğlu (✉)
Department of Biophysics, Süleyman Demirel University, Morfoloji Binasi, TR-32260 Isparta, Turkey
and
Neuroscience Research Center, Suleyman Demirel University, Isparta, Turkey
e-mail: mnaziroglu@med.sdu.edu.tr

L. Mandelker and P. Vajdovich (eds.), *Studies on Veterinary Medicine*, Oxidative Stress in Applied Basic Research and Clinical Practice 5,
DOI 10.1007/978-1-61779-071-3_4,

Abbreviations

ADPR	Adenosine diphosphatase ribose
CHO	Chinese hamster ovary
CNS	Central nervous system
DRG	Dorsal rood ganglion
GSH	Glutathione
GSH-Px	Glutathione peroxidase
HEK	Human embryonic kidney
NO	Nitric oxide
NOS	Nitric oxide synthase
PARG	Poly(ADP-ribose) glycohydrolase
PARP-1	Poly(ADP-ribose) polymerase
ROS	Reactive oxygen species
TRP	Transient receptor potential

Introduction

Of all the organs in the body, the brain may be the most vulnerable to oxidative stress and become exposed to reactive oxygen species (ROS) continuously generated via the autooxidation of polyunsaturated fatty acids (PUFAs). The central nervous system (CNS) is also especially sensitive to free radical damage. Its high consumption of oxygen, rich content of easily oxidizible fatty acids, relatively low content of antioxidant enzymes and antioxidants, and the presence of high levels of iron make it a prime substrate for damage by ROS.

Compared with other organs, the brain consumes a large fraction (20%) of the oxygen that the body takes in, suggesting a metabolic rate. This high oxygen consumption and possible metabolic rate may increase the amount of free radical produced in the brain [3]. The brain contains membranes composed of proteins and an abundant amount of phospholipids. These phospholipids contain oxidizable PUFAs, such as arachidonic acid and docohexaenoic acid [12]. These PUFAs are vulnerable to attack by free radicals because they contain hydrogen ions held together by weak double bonds that serve as a target for ROS damage [42]. It has been suggested that oxidative damage resulting in alterations of brain phospholipids may play a role in neurological diseases [31, 32].

Brain and neurological cells are protected by antioxidants against peroxidative damage [12, 31]. Glutathione peroxidase (GSH-Px) catalyzes the reduction of hydrogen peroxide to water. GSH-Px can also remove organic hydroperoxides. Glutathione (GSH) is a hydroxyl radical and singlet oxygen scavenger and participates in a wide range of cellular functions [47]. Vitamin E (α-tocopherol) is the most important antioxidant in the lipid phase of cells. Vitamin E acts to protect cells against the effects of free radicals, which are potentially damaging by-products of the body's metabolism [12, 31]. Vitamin C (ascorbic acid), as well as being a free radical scavenger, also transforms vitamin E to its active form [34]. The brain ascorbic acid concentration is extremely low compared to body tissues such as the liver and kidney [34]. Vitamin A (retinol) serves as a prohormone for retinoids and is involved with signal transduction at cytoplasmic and membrane sites [51].

TRPM2, TRPM6, and TRPM7 share a feature unique among known channels, by having a functional enzyme moiety in the C-terminal domain. In TRPM2, this is a type of Nudix hydrolase (NUDT9-H) that can bind to and hydrolyze adenosine diphosphate (ADP) ribose (ADPR), although not as effectively as other known Nudix ADPR-hydrolases ([23]; Fig. 1). Binding of ADPR to NUDT9-H activates the channel, allowing the passage of cations down their electrochemical gradient. Because TRPM2 is a plasma membrane channel, Ca^{2+} and Na^{+} will flow into the cell when TRPM2 opens. ADPR is the most potent physiological activator of TRPM2, but other less potent activators have been proposed [13, 15]. These include nicotinamide adenine dinucleotide (NAD^{+}), oxidants such as H_2O_2, and cyclic ADPR (cADPR). NAD^{+} and H_2O_2 increase intracellular levels of ADPR and could thus gate TRPM2 either indirectly [46] or directly [13, 35] (see Table 1).

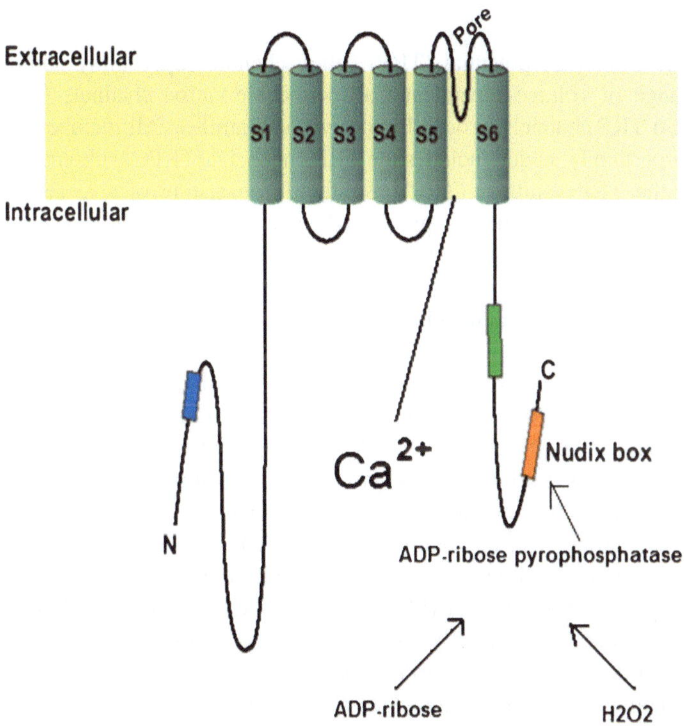

Fig. 1 Cellular responses following the opening of transient receptor potential melastatin 2 (TRPM2) cation channels and the associated increase in intracellular calcium concentration. The channels are containing C (Nudix box) and N sections. The Nudix box domain has ADP-ribose pyrophosphatase activity. The TRPM2 channels can be gated by H_2O_2 and ADP-ribose via activation ADP-ribose pyrophosphatase

Table 1 Role of inhibitor and gate in oxidative stress and ADPR on TRPM channels in different neuronal cells

Activator	Cells	Effects	Reference
ADPR	Rat microglia	Activator unknown	[21, 40]
Oxidative stress	Striatal neurons	Activator	[17, 48]
	C13 microglia cells	Activator	[9]
Both ADPR and oxidative stress	Cortical neurons	Activator	[19]
		No effects	[21]
ADPR	Pyramidal neurons and CA1 interneurons of hippocampus	Insufficient	[41]
Oxidative stress	Pyramidal neurons and CA1 interneuron of hippocampus	Activator	[41]
Oxidative stress	Rat primary striatal cultures	Activator	[8]
Oxidative stress	Astroglia	Activator	[2]

ADPR adenosine diphosphoribose

TRP Superfamily

The transient receptor potential (TRP) cation channel superfamily compromises a diverse range of voltage-dependent Ca^{2+}-permeable cation channels. There are 28 mammalian TRP channels, grouped into seven subfamilies. All members of the TRP channel superfamily, which includes the TRP cononcial (TRPC) subfamily consisting of seven, the TRP vanilloid (TRPV) subfamily consisting of six, TRP melastatin (TRPM) subfamily consisting of eight, TRP polycystein (TRPP) subfamily consisting of three, TRP mucolipin (ML) subfamily consisting of three, and the TRP ankyrin (TRPA) subfamily consisting of one are poorly characterized, but are attracting increasing interest because of their involvement in several human diseases. TRP channels have a basic structure similar to voltage-gated potassium channels, with homo- or heterotetrameric arrangements around a central ion conducting pore between the fifth and sixth segments of the channel pore [5]; the fourth transmembrane segment is not positively charged. The N-termini of TRPV and TRPC, but not those of TRPM channels, contain multiple ankyrin binding repeats. The C-terminal part of the sixth segment in TRPC and TRPM channels includes the "TRP domain," a conserved stretch of 25 amino acids starting with the nearly invariant "TRP box" that is missing in TRPV channels. In addition, all TRP channels have multiple regulatory protein interaction sites. Multiple protein kinase A (PKA) and C (PKC) putative phosphorylation sites have been identified and partially tested for function. Phosphatidylinositide 3-kinase SH2-recognition domains have also been identified in several TRP channels [53].

TRP channels contribute to changes in cytosolic free Ca^{2+} $[Ca^{2+}]i$ by acting as Ca^{2+} entry channels in the plasma membrane directly or by changing membrane potentials, modulating the driving forces for the Ca^{2+} entry channels. All functionally characterized TRP channels are permeable to Ca^{2+} with the exceptions of TRPM4 and TRPM5, which are only permeable to monovalent cations but not to Ca^{2+} or Mg^{2+} (see Fig. 2). Two mammalian TRPs, TRPV5 and TRPV6, are highly Ca^{2+}

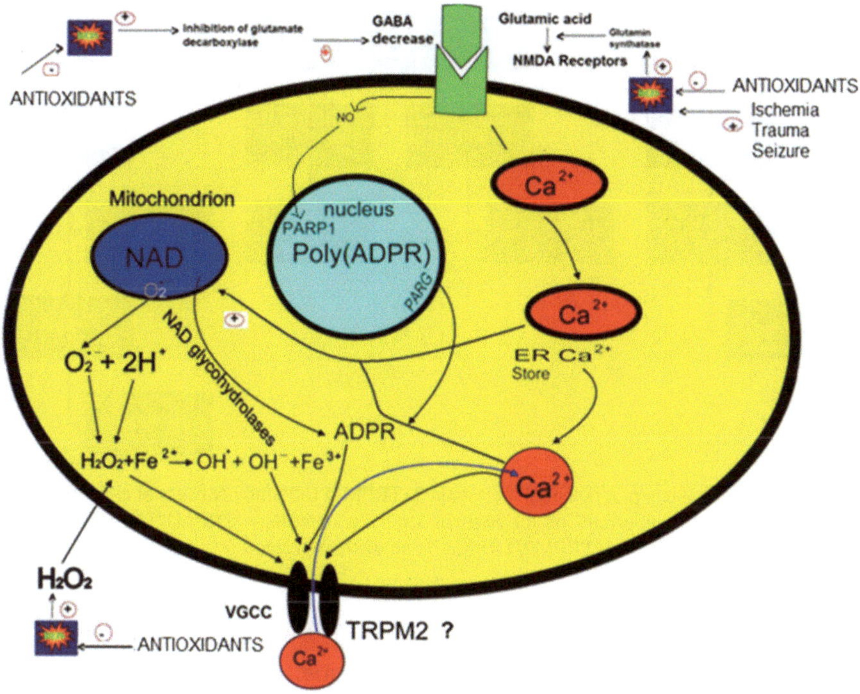

Fig. 2 Cells regulate intracellular Ca^{2+} levels lightly and excessive Ca^{2+} loads can lead to inappropriate activation of processes that normally operate at low levels, causing metabolic derangements and eventual cell death. For example, excess elevations in intracellular Ca^{2+} may activate enzymatic degradation, induce formation of reactive oxygen species, or disrupt normal mitochondrial function leading to oxidative stress and bioenergetic failure. Excessive Ca^{2+} load, in particular via *N*-methyl-D-aspartate (NMDA) receptors (NMDAR), is toxic to neurons in neurodegenerative diseases. NMDAR-mediated Ca^{2+} entry triggers a neurotoxic signal cascade involving the activation of neuronal nitric oxide (NO) synthase (nNOS), formation of toxic reactive oxygen species (ROS) and NO, and activation of the proapoptotic protein poly(ADP-ribose) polymerase (PARP-1). Antioxidants regulate Ca^{2+} influx into cytosol by inhibition of ROS. Sustained depolarization of mitochondrial membranes and enhanced ROS production activate transient receptor potential melastatin 2 (TRPM2), voltage-gated Ca^{2+} channels (VGCC), and Ca^{2+} influx increases by activation of TRPM2 via ROS. The molecular pathway may be a cause of neurological symptoms and thus represents a fruitful subject for further study

permeable. TRPM6 and TRPM7 are highly permeable to Mg^{2+}. At least three TRP channels, TRPV1, TRPML1, and TRPP3, are highly permeable to H^+ ions [38].

TRPM

The TRPM2 subfamily consists of eight members, subdivided into three subgroups on the basis of sequence homology: TRPM1/TRPM3, TRPM4/TRPM5, and TRPM6/7, with TRPM2/TRPM8 being distinct proteins ([31]; Fig. 3).

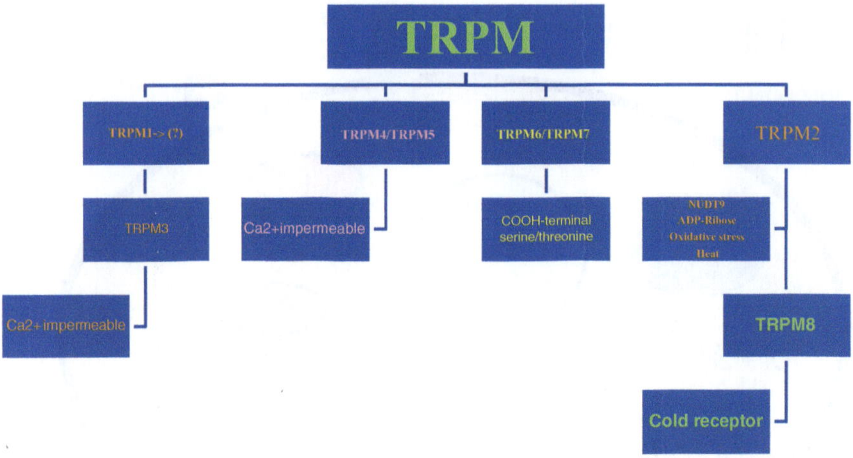

Fig. 3 The transient receptor potential melastatin 2 (TRPM2) subfamily consists of eight members, subdivided into three subgroups on the basis of sequence homology: TRPM1/TRPM3, TRPM4/TRPM5, and TRPM6/7, with TRPM2/TRPM8 being distinct proteins

Three members of this subfamily carry an entire functional enzyme in their COOH-termini: TRPM2 contains a functional NUDT9 homology domain, exhibiting ADPR pyrophophatase activity (Fig. 1), whereas both TRPM6 and TRPM7 contain a functional COOH-terminal serine/threonine kinase. Except for TRPM1 which has not been functionally characterized until now, all TRPM channels are cation channels, although the Ca^{2+} permeability is diverse, ranging from highly permeable (TRPM6/TRPM7 and splice variants of TRPM3) to Ca^{2+} impermeable (TRPM4 and TRPM5). TRPM4 and TRPM5 are heat-sensitive, Ca^{2+} activated channels [39]. TRPM2 is activated ADPR, H_2O_2, and heat. TRPM3 channels are, much like TRPM6 and TRPM7, regulated intracellular Mg^{2+} levels. They show considerable constitutive activity although TRPM3 activation was also reported after cell swelling and by sphingosine. TRPM8 is the infamous cold receptor [6, 13, 37, 39, 56].

TRPM2 Channels

One of the TRP channels is TRPM, so named because the founding member, TRPM1, was first described in connection with metastatic melanomas [6]. TRPM2 was formerly known as TRPC2 and LTRPC2. TRPM2 channels were first identified in 1998 and subsequently recognized as members of the TRPM family [5]. All eight TRPM family members have been linked to disease, either by functional studies in mouse models or by genetic evidence [39].

Molecular Mechanisms in the Activation of TRPM2

Three extracellular signals are known to activate TRPM2: oxidative stress, ADPR/ NAD^+ metabolism, and tumor necrosis factor alpha [13]. The channel may also be temperature sensitive with body temperature acting as "an endogenous co-activator" for TRPM2 [52].

Oxidative Stress

ROS including superoxide anion, hydrogen peroxide (H_2O_2), and singlet oxygen act as subcellular messengers in such complex processes as mitogenic signal transduction, gene expression, and regulation of cell proliferation when they are generated excessively or when enzymatic and nonenzymatic defense systems are impaired [31, 32]. Although many intra- and extracellular molecules may participate in neuronal injury and cell apoptosis, accumulation of oxidative stress due to excessive generation of ROS appears to be a potential factor for cell damage and death [12].

Oxidative stress for which application of H_2O_2 is an experimental paradigm, induces TRPM2 currents and an increase in free intracellular Ca^{2+} [Ca^{2+}] in various cell types transfected with TRPM2 [36], as well as in pancreatic β-cells [18], neutrophil granulocytes [14], and U937 monocytes [44, 56]. However, the exact mediator molecules for H_2O_2-induced TRPM2 channel activation remain to be identified. The ROS sensitivity of TRPM2 can be mediated by NAD^+ [13] or ADPR [54] released from mitochondria or through protein oxidation [11]. H_2O_2 and cADPR have been proposed to potentiate the effect of ADPR at lower concentrations and to gate the TRPM2 channel directly at higher concentrations [27].

Role of Oxidative Stress and ADP-Ribose/NAD Products in the Activation of TRPM2

Ca^{2+} influx via TRPM2 is thought to occur via production of ADP ribose (ADPR). ADPR may arise from a mitochondrial source or alternatively via activation of poly(ADPR). ADPR may also arise from a mitochondrial source [11].

NAD^+ has been reported to stimulate TRPM2 [13]. TRPM2 contains a characteristic structural feature known as a Nudix domain in its C-terminal cytosolic tail [44]. A Nudix domain is a consensus region that is known to be present in a class of pyrophosphatases that degrade nucleoside diphosphates ([31]; Fig. 1). Indeed, the Nudix domain of TRPM2 cleaves ADPR, a breakdown product of NAD^+ and cyclic ADPR, representing an intracellular second messenger that stimulates calcium release mediated by ryanodine receptors [33]. ADPR is hydrolyzed by TRPM2, however, it also activates TRPM2 and induces TRPM2 currents during infusion of ADPR by the patch pipette [13, 44]. NAD^+ has been reported to stimulate TRPM2

[13, 23, 25]. In our recent study, we observed the activation of TRPM2 by NAD^+, which supports the idea that this substance activates TRPM2 directly [35, 36].

Previously published data indicated that TRPM2 can be gated by H_2O_2 and it has been suggested to gate the channel independently of ADPR [13, 54]. Because Perraud et al. [44] demonstrated that reducing the ADPR concentration within mitochondria largely suppresses H_2O_2-mediated activation of TRPM2, the gating mechanism of H_2O_2 is primarily based on its ability to release ADPR from mitochondria, which then proceeds to activate TRPM2. However, Kolisek et al. [20] recently demonstrated that H_2O_2 may be a self-sufficient stimulus for TRPM2 activation, as it can initiate the release of ADPR from mitochondria and at the same time function as a potentiating cofactor of ADPR. In our recent study, we observed that the H_2O_2-mediated activation of TRPM2 appears to result from a direct gating mechanism, because TRPM2 activation by H_2O_2 was relatively rapid in whole cell recordings, and the compound triggered single-channel activity in excised membrane patches [35]. After adding H_2O_2 to the bath, cell inside-out records were washed by intracellular buffer and then H_2O_2 was given again. The channel was activated again by a second administration of H_2O_2 although no intracellular component was present. From, this we concluded that H_2O_2-induced, single-channel activity observed in excised membrane patches is likely caused by a direct gating mechanism.

Methodological problems have led to different reports on the activation of TRPM2 by NAD^+, ADPR, and H_2O_2. For example, in a previous study [54] using HEK293 cells as an expression system, we demonstrated characteristic currents through the splice variant of TRPM-ΔC induced by H_2O_2. In later experiments we could also stimulate them with NAD^+, although stimulation with H_2O_2 was not possible in the CHO cell expression system using wild-type TRPM2 with a sufficient consistency [20].

A second problem with different results arises from the use of different cell types to study TRPM channel activation by H_2O_2, ADPR, and its metabolites. On the other hand, the channel activation by H_2O_2 appears to represent a cell-specific process in cells with endogenous expression of TRPM2. For example, the TRPM2 channel is activated by H_2O_2 in HEK293 cells [13, 54], CHO cells [37], CRI-GI rat insulinoma cell lines [18], and rat primary striatal cultures [8], but not in human neutrophil granulocytes [14]. Kolisek et al. [20] used cADPR and the compound stimulated TRPM2 in HEK293 cells; that channel was opened by cADPR. Later, the report was supported by the results of Gasser et al. [10] and TRPM2 channels in Jurkat cells were opened by cADPR. However, a more recent study did not support the results for neutrophil granulocytes [15] and TRPM2 channels did not open by stimulation with cADPR.

Role of TRPM2 Channels in Neurological Cells

Northern blotting and quantitative PCR techniques indicated that TRPM2 is broadly expressed in the CNS. However, as this evidence was derived from homogenized tissue samples, it does not allow expression in neurons to be distinguished from that in glia. The importance of making such a distinction is highlighted by a recent study that failed to identify TRPM2 transcripts or functional channels in cerebellar granule cells and astrocytes [21]. Rather, TRPM2 was detected in microglial cells leading to the suggestion that the CNS distribution of TRPM2 corresponds to its expression in nonneuronal cells [22, 41, 45].

Role of TRPM2 Channels and Oxidative Stress in Microglia and Astroglia Cells

The NADPH-oxidase in phagocytic cells such as microglia is an electron transport system that catalyzes the reduction of oxygen to superoxide radical. Under physiological conditions, the system contributes to the elimination of pathogens but under chronic inflammatory conditions such as scleroderma or liver fibrosis it is thought to induce neurodegeneration by the massive accumulation of superoxide radicals. Monocyte superoxide, in high glucose media, is released by the NADPH-oxidase but not by the mitochondrial respiratory chain, and antioxidants such as α-tocopherol inhibits superoxide release via inhibition of protein kinase C (PKC)-a. The assembly of a functional NADPH-oxidase complex at the plasma membrane depends on the phosphorylation and subsequent translocation of several cytosolic subunits ($p40^{phox}$, $p47^{phox}$, $p67^{phox}$, and Rac1/2; [4]). In microglia cells, antioxidants inactivate PKC via the phosphotase-mediated pathway (PP1 or PP2A) and, as a consequence, blocks the phosphorylation-dependent translocation of $p67^{phox}$ to the plasma membrane. As a result, the production of superoxide radical by the microglial NADPH-oxidase system is substantially inhibited, offering a partial explanation for the beneficial effect of antioxidants such as α-tocopherol on a variety of neurodegenerative diseases [31].

Like other activated macrophages, microglia remove bacteria and cellular debris and produce a diverse range of mediators of the inflammatory cascade including arachidonic acid derivates and H_2O_2. Thus microglia cells are a key factor in the immune defense and tissue repair in the CNS. Kraft et al. [21] described that novel calcium influx pathway in microglia cells coupled to hydrogen peroxide and ADPR and provided evidence that this pathway involved TRPM2 although they failed to detect TRPM2 in cultured cerebellar granule neurons. Recently, Ohana ct al. [40] investigated expression of putative Ca^{2+}-permeable TRPM2 channels in rat-cultured microglia cells by quantitative real-time RT-PCR. They detected transcripts in the rat-cultured microglia cells for TRPM2 genes.

Although the role of astroglia in the progression of neurodegenerative disease is still relatively unknown, their importance in regulating the normal and abnormal neuronal environment is attracting increasing attention. Through Ca^{2+} signaling cascades, astroglia control gene expression, neuronal differentiation, and programmed cell death, which are all integral to developmental and degenerative processes. Under conditions of oxidative stress, glial cells provide energy support for neurons, exert a protective function by scavenging and detoxifying ROS, and direct neuronal resistance or vulnerability to degeneration through Ca^{2+}-dependent secretion of trophic or inflammatory factors [30]. There are few reports on TRPM2 in astroglia cells. Bond and Greenfield [2] reported that the additive effects of L-VGCC blockade and TRPM2 inhibition during oxidative stress significantly enhanced recovery from protein synthesis suppression and repressed subsequent compensatory protein overexpression. These results indicated that Ca^{2+} signaling is integral to astroglial transcriptional and translational responses to oxidative stress.

TRPM2 Channel in Hippocampal Neurons

TRPM2 is expressed in diverse cell types and despite convincing evidence for high expression in the mammalian brain, much of this signal is attributed to strong expression in nonneuronal cells. Thus, the existence of functional TRPM2 channels in neurons is controversial at best. Recently, Olah et al. [41] reported that functional TRPM2 channels are highly expressed in pyramidal hippocampus neurons, including those of CA1 interneurons in hippocampal slices. They also reported that ADPR alone is insufficient to gate TRPM2 in hippocampal neurons. They concluded that concomitant influx of Ca^{2+} through voltage-dependent Ca^{2+} channels and/or NMDAR is necessary to fully activate TRPM2 channels.

TRPM2 Channels in Dorsal Rood Ganglion Cells

There are several types of sensory neurons in the dorsal rood ganglion (DRG), with responsiveness to different external and internal stimuli. These stimuli, including nociceptive, thermal, or mechanical, activate different receptors and ion channels that are present in the nerve terminals at the sensory receptive fields and their expression in selective subsets of DRG neurons determines the response profile of nonselective cation channels that play important functions in sensory neurons. TRPM8 is the only TRPM channel with a clearly assigned function in DRG neurons. It is activated by innocuous cool stimuli and responds to menthol and icilin with intracellular Ca^{2+} elevations [43]. Recently, TRPM2 channels were firstly detected in DRG mouse cells [50]. They also observed that levels of TRPM2 channels in DRG cells were significantly higher in lumbar tissue than in thoracic tissue in the adult mouse although they did not observe the channels in the DRG cells of embryonic day 12–12 weeks of age.

Role of TRPM2 Channels in Oxidative Stress-Induced Neurological Diseases

TRPM2 is highly expressed in the brain, in both microglia and neuronal cells, but its biological role in these cells still needs to be understood. There are few studies on the role of TRPM2 cation channels in neurological diseases and these studies suggested that Ca^{2+} influx via TRPM2 is necessary for microglia and other phagocytes to mount effective inflammatory and clearance responses [21].

Western Pacific Amyotrophic Lateral Sclerosis and Parkinson Dementia

Among neurodegenerative diseases, Amyotrophic Lateral Sclerosis (ALS) and Parkinson Dementia (PD) are ideal for studying the relative contributions of genes and environment in disease etiology because they occur in geographically separate foci among three genetically different, homogeneous groups of people [16]. Intensive research conducted over the years led to the identification of two candidate environmental factors: (1) severely low levels of Ca^{2+} and Mg^{2+} in the soil and drinking water [58]; and (2) the putative neurotoxin-methylamino-L-alanine derived from the cycad plant, a traditional food source in Guam [49]. These findings led to the hypothesis that prolonged exposure to such an environment could be involved in the pathogenesis of Western Pacific ALS and PD [49]. Supporting this hypothesis are observations that afflicted individuals exhibit altered Ca^{2+} metabolism, and reports of neuronal damage in experimental models of animals fed diets that mimic the mineral composition in the disease foci environment [57].

TRPM2 has been implicated in cell death induced by oxidants [13, 35]. The presence of low Mg^{2+} and high transition metals in the Western Pacific ALS and PD foci create conditions of increased oxidative stress [16]. As a channel with a high expression profile in microglia and neuronal cells that could sense and respond to oxidative stress, TRPM2 is an attractive candidate susceptibility gene for these disorders. Hermosura et al. [16] reported the presence of heterozygous TRPM2P1018L, a variant of TRPM2 in the pathogenesis of Western Pacific ALS and PD. They also observed that P1018L forms functional channels that activate in response to H_2O_2 and ADPR. However, in the presence of physiological concentrations of extracellular Ca^{2+}, P1018L channels inactivate quickly and are thus unable to allow sustained ion influx. In intact cells, this is manifested as an attenuated Ca^{2+} rise in response to H_2O_2. Defective Ca^{2+} handling is implicated in many diseases, including neurodegeneration. Of particular interest is a recent report that described another Pro-to-Leu substitution, this time in CalHM1, a putative Ca^{2+}-permeable channel [7]. CalHM1P86L increases risk for Alzheimer's disease.

Table 2 Some genetic properties of TRPM2 in different neuronal cells

Activator	Diseases	Reference
Oxidative stress sensor in microglia cells	Bipolar-I and II	[24, 28, 55]
Oxidative stress and ADPR in rat primary striatal cultures	Amyloid beta-peptide-induced Alzheimer's disease	[8]
ADPR	Nonsyndromic hereditary Alzheimer's disease	[8, 29]

ADPR adenosine diphosphoribose

In functional studies, cells expressing CalHM1P86L proteins exhibited attenuated Ca^{2+} permeability, reduced cytosolic Ca^{2+} levels, and increased amyloid deposition. The similarity in the effects of CalHM1P86L and RPM2P1018L suggests that attenuation of intracellular Ca^{2+} rises and its effects on downstream signaling pathways may contribute to the pathophysiological mechanism in neurodegenerative diseases ([16]; Table 2).

Bipolar Disorders

Although the C-terminal cytosolic tail mediates the interaction with ADPR, the role of the N-terminal tail that is also located within the cytosol has not been defined on a molecular level. A number of N-terminal truncated isoforms of TRPM2 have been identified and in some cases they regulate the function of the full-length channel [45]. In general, the N terminus is indispensable. Already the deletion of a stretch of 20 amino acid residues (Δ537–556), as has been found in the TRPM2-ΔN splice variant in neutrophils, abolishes any channel function [54]. This ΔN stretch comprises several structural elements, which may explain why TRPM2-ΔN is dysfunctional. First, it contains an IQ-like sequence motif that represents a CaM binding domain [1]. Interestingly, a further IQ-like motif is found immediately downstream of the ΔN stretch. Second, the ΔN stretch contains two PxxP motifs that are characteristic for sites enabling interaction with other proteins [26]. For TRP channels the significance of endogenous PxxP motifs until now was only investigated in the TRPC channel subfamily [59]. Although so far no proteins are known that form functional protein complexes with TRPM2, TRPM2 is unusually rich in PxxP motifs compared to other channels of the TRP superfamily. As a further remarkable property of the ΔN stretch, the exchange of a single amino acid residue (D543E) was correlated with the presence of bipolar disorder in members of a family with a high incidence of this disease [28, 55]. Recently, it has been reported that no functional role can be attributed to any of the structural motifs within the ΔN stretch that may be a spacer segment for other functional sites in the N terminus [24].

Conclusions

The evidence for association of TRPM2 variants with neurological diseases such as bipolar disorders found that genetic variation of TRPM2 influences the susceptibility to the diseases. In addition, the identification of TRPM2 as a key component of the neurological Ca^{2+} entry pathway in response to ROS sheds new light on the physiology and pathophysiology of the neurological cells and brain [60]. Because there is substantial evidence for the deteriorating role of oxidative stress in neurological and brain dysfunction, manipulating TRPM2 function in the neurological cells may be highly useful in the future for experimental therapies of brain and neurological dysfunctions.

References

1. Bähler M, Rhoads A. 2002. Calmodulin signaling via the IQ motif. FEBS Lett 513:107–113.
2. Bond CE, Greenfield SA. 2007. Multiple cascade effects of oxidative stress on astroglia. Glia 55:1348–1361.
3. Butterfield DA. 2003. Amyloid beta-peptide [1–42]-associated free radical-induced oxidative stress and neurodegeneration in Alzheimer's disease brain: mechanisms and consequences. Curr Med Chem 10:2651–2659.
4. Chanock SJ, Roesler J, Zhan S, Hopkins P, Lee P, Barrett DT, Christensen BL, Curnutte JT, Görlach A. 2000. Genomic structure of the human p47-phox (NCF1) gene. Blood Cells Mol Dis 26:37–46.
5. Clapham DE. 2003. TRP channels as cellular sensors. Nature 426:517–524.
6. Clapham DE. 2007. Snapshot: mammalian TRP channels. Cell 129:220.
7. Dreses-Werringloer U, Lambert JC, Vingtdeux V, Zhao H, Vais H, Siebert A, Jain A, Koppel J, Rovelet-Lecrux A, Hannequin D, Pasquier F, Galimberti D, Scarpini E, Mann D, Lendon C, Campion D, Amouyel P, Davies P, Foskett JK, Campagne F, Marambaud P. 2008. A polymorphism in CALHM1 influences Ca^{2+} homeostasis, Abeta levels, and Alzheimer's disease risk. Cell 133:1149–1161.
8. Fonfria E, Marshall IC, Boyfield I, Skaper SD, Hughes JP, Owen DE, Zhang W, Miller BA, Benham CD, McNulty S. 2005. Amyloid beta-peptide(1–42) and hydrogen peroxide-induced toxicity are mediated by TRPM2 in rat primary striatal cultures. J Neurochem 95:715–723.
9. Fonfria E, Mattei C, Hill K, Brown JT, Randall A, Benham CD, Skaper SD, Campbell CA, Crook B, Murdock PR, Wilson JM, Maurio FP, Owen DE, Tilling PL, McNulty S. 2006. TRPM2 is elevated in the tMCAO stroke model, transcriptionally regulated and functionally expressed in C13 microglia. J Recept Signal Transduct Res 26:179–198.
10. Gasser A, Glassmeier G, Fliegert R, Langhorst MF, Meinke S, Hein D, Kruger S, Weber K, Heiner I, Oppenheimer N, Schwarz JR, Guse AH. 2006. Activation of T cell calcium influx by the second messenger ADP-ribose. J Biol Chem 281:2489–2496.
11. Guse AH. 2005. Second messenger function and the structure-activity relationship of cyclic adenosine diphosphoribose (cADPR). FEBS J 272:4590–4597.
12. Halliwell B. 2006. Oxidative stress and neurodegeneration: where are we now? J Neurochem 97:1634–1658.
13. Hara Y, Wakamori M, Ishii M, Maeno E, Nishida M, Yoshida T, Yamada H, Shimizu S, Mori E, Kudoh J, Shimizu N, Kurose H, Okada Y, Imoto K, Mori Y. 2002. LTRPC2 Ca^{2+}-permeable channel activated by changes in redox status confers susceptibility to cell death. Mol Cell 9:163–173.

14. Heiner I, Eisfeld J, Luckhoff A. 2003. Role and regulation of TRP channels in neutrophil granulocytes. Cell Calcium 33:533–540.
15. Heiner I, Eisfeld J, Warnstedt M, Radukina N, Jungling E, Luckhoff A. 2006. Endogenous ADP-ribose enables calcium-regulated cation currents through TRPM2 channels in neutrophil granulocytes. Biochem J 398:225–232.
16. Hermosura MC, Cui AM, Go RC, Davenport B, Shetler CM, Heizer JW, Schmitz C, Mocz G, Garruto RM, Perraud AL. 2008. Altered functional properties of a TRPM2 variant in Guamanian ALS and PD. Proc Natl Acad Sci U S A 105:18029–18034.
17. Hill K, Tigue NJ, Kelsell RE, Benham CD, McNulty S, Schaefer M, Randall AD. 2006. Characterization of recombinant rat TRPM2 and a TRPM2-like conductance in cultured rat striatal neurons. Neuropharmacology 50:89–97.
18. Inamura K, Sano Y, Mochizuki S, Yokoi H, Miyake A, Nozawa K, Kitada C, Matsushime H, Furuichi K. 2003. Response to ADP-ribose by activation of TRPM2 in the CRI-G1 insulinoma cell line. J Membr Biol 191:201–207.
19. Kaneko S, Kawakami S, Hara Y, Wakamori M, Itoh E, Minami T, Takada Y, Kume T, Katsuki H, Mori Y, Akaike A. 2006. A critical role of TRPM2 in neuronal cell death by hydrogen peroxide. J Pharmacol Sci 101:66–76.
20. Kolisek M, Beck A, Fleig A, Penner R. 2005. Cyclic ADP-ribose and hydrogen peroxide synergize with ADP-ribose in the activation of TRPM2 channels. Mol Cell 18:61–69.
21. Kraft R, Grimm C, Grosse K, Hoffmann A, Sauerbruch S, Kettenmann H, Schultz G, Harteneck C. 2004. Hydrogen peroxide and ADP-ribose induce TRPM2-mediated calcium influx and cation currents in microglia. Am J Physiol Cell Physiol 286:C129–C137.
22. Kraft R, Harteneck C. 2005. The mammalian melastatin-related transient receptor potential cation channels: an overview. Pflugers Arch 451:204–211.
23. Kühn FJ, Heiner I, Luckhoff A. 2005. TRPM2: a calcium influx pathway regulated by oxidative stress and the novel second messenger ADP-ribose. Pflugers Arch 451:212–219.
24. Kühn FJP, Kühn C, Nazıroğlu M, Lückhoff A. 2009. Role of an N terminal splice segment in the activation of the cation channel TRPM2 by ADP-Ribose and hydrogen peroxide. Neurochem Res 34:227–233.
25. Kühn FJ, Lückhoff A. 2004. Sites of the NUDT9-H domain critical for ADP-ribose activation of the cation channel TRPM2. J Biol Chem 279:46431–46437.
26. Li SS. 2005. Specificity and versatility of SH3 and other proline-recognition domains: structural basis and implications for cellular signal transduction. Biochem J 390:641–653.
27. Massullo P, Sumoza-Toledo A, Bhagat H, Partida-Sánchez S. 2006. TRPM channels, calcium and redox sensors during innate immune responses. Semin Cell Dev Biol 17:654–666.
28. McQuillin A, Bass NJ, Kalsi G, Lawrence J, Puri V, Choudhury K, Detera-Wadleigh SD, Curtis D, Gurling HM. 2006. Fine mapping of a susceptibility locus for bipolar and genetically related unipolar affective disorders, to a region containing the C21ORF29 and TRPM2 genes on chromosome 21q22.3. Mol Psychiatry 11:134–142.
29. Morale MC, Serra PA, L'episcopo F, Tirolo C, Caniglia S, Testa N, Gennuso F, Giaquinta G, Rocchitta G, Desole MS, Miele E, Marchetti B.1998. Molecular cloning of a novel putative Ca^{2+} channel protein (TRPC7) highly expressed in brain. Genomics 54:124–131.
30. Morale MC, Serra PA, L'episcopo F, Tirolo C, Caniglia S, Testa N, Gennuso F, Giaquinta G, Rocchitta G, Desole MS, Miele E, Marchetti B. 2006. Estrogen, neuroinflammation and neuroprotection in Parkinson's disease: glia dictates resistance versus vulnerability to neurodegeneration. Neuroscience 138:869–878.
31. Nazıroğlu M. 2007c. New molecular mechanisms on the activation of TRPM2 channels by oxidative stress and ADP-ribose. Neurochem Res 32:1990–2001.
32. Nazıroğlu M. 2007b. Molecular Mechanisms of vitamin E on intracellular signaling pathways in brain. In Reactive Oxygen Species and Diseases. Ed. Laszlo Goth, Research Signpost Press: Kerala, India. pp. 239–256.
33. Nazıroğlu M. 2009. Role of selenium on calcium signaling and oxidative stress-induced molecular pathways in epilepsy. Neurochem Res 34:2181–2191.

34. Nazıroğlu, M. 2010. Effects of vitamin C on oxidative stress-induced molecular pathways in epilepsy. In Vitamin C: Nutrition, Side Effects and Supplements. Ed. Christina M. Jackson, Nova Science Publishers, Inc. ISBN: 978-1-61728-754-1.
35. Nazıroğlu M, Lückhoff A. 2008a. A calcium influx pathway regulated separately by oxidative stress and ADP-ribose in TRPM2 channels: single channel events. Neurochem Res 33: 1256–1262.
36. Nazıroğlu M, Lückhoff A. 2008b. Effects of antioxidants on calcium influx through TRPM2 channels in transfected cells activated by hydrogen peroxide. J Neurol Sci 270:152–158.
37. Nazıroğlu M, Lückhoff A, Jungling E. 2007a. Antagonist effect of flufenamic acid on TRPM2 cation channels activated by hydrogen peroxide. Cell Biochem Funct 25:383–387.
38. Nilius B. 2007. TRP channels in disease. Biochem Biophys Acta 1772:805–812.
39. Nilius B, Owsianik G, Voets T, Peters JA. 2007. Transient Receptor Potential cation channels in disease. Physiol Rev 87:165–217.
40. Ohana L, Newell EW, Stanley EF, Schlichter LC. 2009. The Ca^{2+} release-activated Ca^{2+} current (I_{CRAC}) mediates store-operated Ca^{2+} entry in rat microglia. Channels 3:129–139.
41. Olah ME, Jackson MF, Li H, Perez Y, Sun HS, Kiyonaka S, Mori Y, Tymianski M, MacDonald JF. 2009. Ca^{2+}-dependent induction of TRPM2 currents in hippocampal neurons. J Physiol 587(Pt 5):965–979.
42. Özmen I, Nazıroğlu M, Alici HA, Sahin F, Cengiz M, Eren İ. 2007. Spinal morphine administration reduces the fatty acid contents in spinal cord and brain in rabbits due to oxidative stress. Neurochem Res 32:19–25.
43. Peier AM, Moqrich A, Hergarden AC, Reeve AJ, Andersson DA, Story GM, Earley TJ, Dragoni I, McIntyre P, Bevan S, Patapoutian A. 2002. A TRP channel that senses cold stimuli and menthol. Cell 108:705–715.
44. Perraud AL, Fleig A, Dunn CA, Bagley LA, Launay P, Schmitz C, Stokes AJ, Zhu Q, Bessman MJ, Penner R, Kinet JP, Scharenberg AM. 2001. ADP-ribose gating of the calcium-permeable LTRPC2 channel revealed by Nudix motif homology. Nature 411:595–599.
45. Perraud AL, Schmitz C, Scharenberg AM. 2003. TRPM2 Ca^{2+} permeable cation channels: from gene to biological function. Cell Calcium 33:519–531.
46. Perraud AL, Takanishi CL, Shen B, Kang S, Smith MK, Schmitz C, Knowles HM, Ferraris D, Li W, Zhang J, Stoddard BL, Scharenberg AM. 2005. Accumulation of free ADP-ribose from mitochondria mediates oxidative stress-induced gating of TRPM2 cation channels. J Biol Chem 280:6138–6148.
47. Rayman MP. 2009. Selenoproteins and human health: insights from epidemiological data. Biochim Biophys Acta 1790:1533–1540.
48. Smith MA, Herson PS, Lee K, Pinnock RD, Ashford ML. 2003. Hydrogen-peroxide-induced toxicity of rat striatal neurons involves activation of a nonselective cation channel. J Physiol 547:417–425.
49. Spencer PS, Nunn PB, Hugon J, Ludolph AC, Ross SM, Roy DN, Robertson RC. 1987. Guam amyotrophic lateral sclerosis-parkinsonism-dementia linked to a plant excitant neurotoxin. Science 237:517–522.
50. Staaf S, Franck MC, Marmigère F, Mattsson JP, Ernfors P. 2010. Dynamic expression of the TRPM subgroup of ion channels in developing mouse sensory neurons. Gene Expr Patterns 10:65–74.
51. Tafti M, Ghyselinck NB. 2007. Functional implication of the vitamin A signaling pathway in the brain. Arch Neurol 64:1706–7611.
52. Togashi K, Hara Y, Tominaga T, Higashi T, Konishi Y, Mori Y, Tominaga M. 2006. TRPM2 activation by cyclic ADP-ribose at body temperature is involved in insulin secretion. EMBO J 25:1804–1815.
53. Watanabe H, Murakami M, Ohba T, Takahashi Y, Ito H. 2008. TRP Channel and cardiovascular disease. Pharmacol Therap 118:337–351.
54. Wehage E, Eisfeld J, Heiner I, Jüngling E, Zitt C, Lückhoff A. 2002. Activation of the cation channel long transient receptor potential channel 2 (LTRPC2) by hydrogen peroxide. A splice variant reveals a mode of activation independent of ADP-ribose. J Biol Chem 277:23150–23156.

55. Xu C, Macciardi F, Li PP, Yoon IS, Cooke RG, Hughes B, Parikh SV, McIntyre RS, Kennedy JL, Warsh JJ. 2006. Association of the putative susceptibility gene, transient receptor potential protein melastatin type 2, with bipolar disorder. Am J Med Genet B Neuropsychiatr Genet 141:36–43.
56. Yamamoto S, Shimizu S, Kiyonaka S, Takahashi N, Wajima T, Hara Y, Negoro T, Hiroi T, Kiuchi Y, Okada T, Kaneko S, Lange I, Fleig A, Penner R, Nishi M, Takeshima H, Mori Y. 2008. TRPM2-mediated Ca^{2+} influx induces chemokine production in monocytes that aggravates inflammatory neutrophil infiltration. Nat Med 14:738–747.
57. Yanagihara R, Garruto RM, Gajdusek DC, Tomita A, Uchikawa T, Konagaya Y, Chen KM, Sobue I, Plato CC, Gibbs CJ Jr. 1984. Calcium and vitamin D metabolism in Guamanian Chamorros with amyotrophic lateral sclerosis and parkinsonism-dementia. Ann Neurol 15:42–48.
58. Yase Y. 1972. The pathogenesis of amyotrophic lateral sclerosis. Lancet 2:292–296.
59. Yuan JP, Kiselyov K, Shin DM, Chen J, Shcheynikov N, Kang SH, Dehoff MH, Schwarz MK, Seeburg PH, Muallem S, Worley PF. 2003. Homer binds TRPC family channels and is required for gating of TRPC1 by IP3 receptors. Cell 114:777–789.
60. Nazıroğlu M. 2010. TRPM2 cation channels, oxidative stress and neurological diseases: Where are we now? Neurochem Res. [Epub ahead of print] DOI: 10.1007/s11064-010-0347-4.

Oxidative Stress in Diabetes Mellitus

Stefano Comazzi

Abstract Diabetes mellitus of either human, or domestic animals is commonly associated to oxidative stress which is mainly involved in the onset of many complications such as cataract, neuropathy, vasculopathies, nephropathy and ketoacidosis. The pathogenesis of diabetic oxidative stress is a multifactorial process involving glucose auto-oxidation, formation of advanced glycation endproducts and activation of polyol, and protein-kinase pathways. Although the presence of oxidative stress in some specific diabetic complications is widely recognized its role is far to be completed elucidated as well as the use of antioxidants to prevent or manage complicated DM.

In fact, although many studies on laboratory animals and cell cultures indicate a possible role of antioxidants to prevent diabetic complications, convincing clinical studies are still lacking at the moment.

On the other side, oxidative stress can provide some useful information to monitor diabetic status. The role of oxidative stress in canine and feline diabetes mellitus is briefly discussed with particular regards to the link between activation of polyol patway and cataract in dogs and between diabetic ketoacidosys and oxidative stress in cats.

Keywords Oxidative stress • Diabetes Mellitus • Diabetic complications • Cataract • Sorbitol • AGE products • Superoxide anion

Diabetes mellitus (DM) is a common metabolic disorder both in dogs and cats, often associated with complications by which cataract, neuropathy and ketoacidosis are the most common in veterinary medicine. Pathogenesis of DM is mainly related

S. Comazzi (✉)
University of Milan, Department of Veterinary Pathology, Hygiene, and Health, Section of Veterinary General Pathology and Parasitology, Via Celoria 10, Milan 20133, Italy
e-mail: stefano.comazzi@unimi.it

L. Mandelker and P. Vajdovich (eds.), *Studies on Veterinary Medicine*, Oxidative Stress in Applied Basic Research and Clinical Practice 5,
DOI 10.1007/978-1-61779-071-3_5,

to decreased secretion (type 1) or action (type 2) of endogenous insulin, thus causing fasting hyperglycaemia, glycosuria and polyuria/polydipsia are major specific signs. Therapeutic approach is mainly centred to a well tailored exogenous insulin administration, even though prevention of long-term complications has always played a central role in management of diabetic patients.

In human medicine, many studies on the linkage between oxidative stress and either development and progression of DM or its complications are reported [3, 7]. This opens new interesting perspectives not only on management of diabetic patient but also on prevention of long term complications.

Until 2009, more than 5000 articles on peer-reviewed journals concerning with the relationship between DM and the increased production of free radicals and/or the impaired antioxidant defences have been published. However, the source of hyperglycaemia-induced oxidative stress is far to be completely elucidated. Even though no definitive studies have been available at the moment, the comprehension of the complex interactions between hyperglycaemia and production of free radicals could be useful on a practical point of view since: (1) it can lead to better understand pathogenesis of both type 1 and type 2 DM and to understand the role of antioxidant integration in preventing the onset of the disease in patients with diabetic predisposition; (2) it allows to select possible biomarkers to check diabetic status and tailor therapy; (3) it can clarify the possible role of antioxidant integration in therapeutic protocols for DM, in order to prevent long-term complications.

Mechanisms for Free Radical Production in DM

The production of free radicals in diabetic patients has been reported as related to several different mechanisms (Fig. 1).

1. *Glucose autoxidation* is thought to be the main source of free radicals. Glucose, in its enediol form is oxidised to an enediol radical anion, then converted into ketoaldehydes and superoxide anion radical. Superoxide anion (SO) is the main source of other radicals such as hydrogen peroxide, singlet oxygen, or, in the presence of transitional metals, highly reactive hydroxyl radicals. On the other side, superoxide also reacts with nitric oxide (NO) to reactive peroxinitrite radicals. All these reactive oxygen species (ROS) promote lipid peroxidation, destruction of membranes, and cleavage of DNA, thus inducing cell destruction. Moreover, increased free radical concentration, together with secondary impaired antioxidant content, causes alterations of structural and enzyme proteins, membrane lipids, and nucleic acids, both nuclear and mitochondrial, leading to cell damage or death.

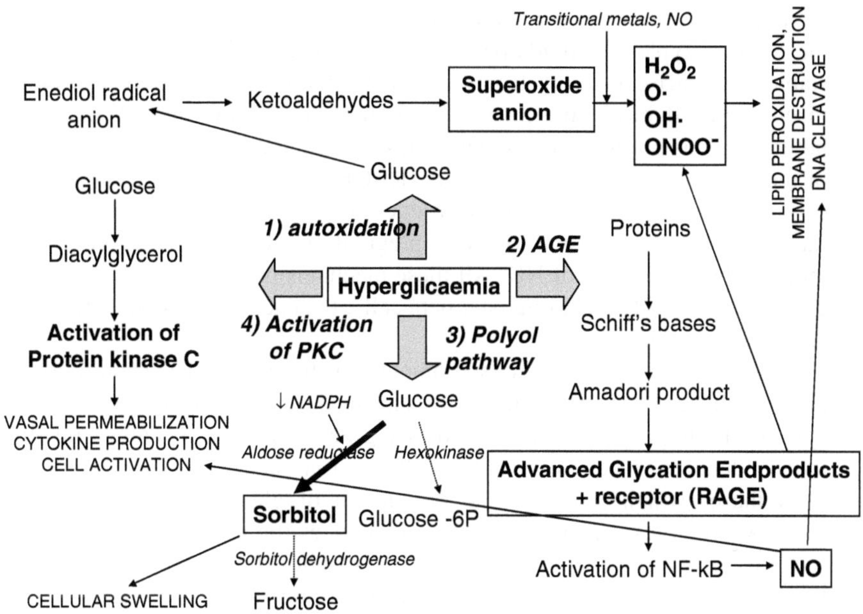

Fig. 1 Mechanisms for free radical production in DM: *1* glucose autoxidation, leading to direct production of ROS; *2* production of advanced glycation end-products, causing protein damages and production of NO; *3* activation of polyol pathway, with sorbitol accumulation and consumption of NADPH; and *4* activation of protein kinase C, activating vasal cells, and inducing vasal permeability

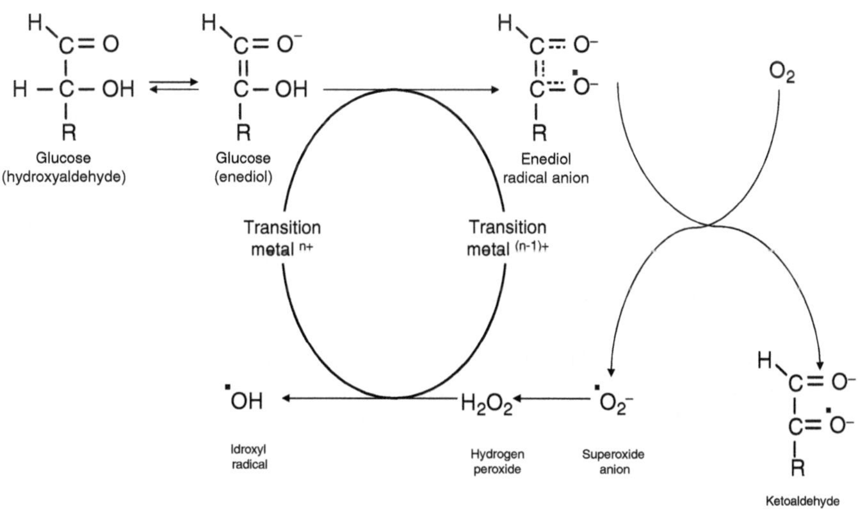

2. *Advanced glycation end-products (AGE) formation* is another important mechanism, typical of oxidative stress in Diabetes mellitus (DM). AGE are stable substances resulting from the nonenzymatic glycosilation of proteins through Schiff bases and an intermediate Amadori product. AGE are an heterogeneous group of molecules, whose accumulation in tissues, by means of the linkage with specific receptors (RAGE), leads to inactivation of intracellular enzymes, promotes ROS production, and activates the transcription factor NF-kB. NF-kB acts on the upregulation of many target genes and induces the production of NO. All these mechanisms contribute to aging and destruction of cells including islet beta cells, thus accelerating diabetes onset and complications.
3. *Activation of polyol pathway* is promoted by the elevated concentration of aldose reductase (AR) in some tissues such as the crystalline lens and nervous system. In hyperglycaemic conditions glucose passively enters insulin-independent cells (such as crystalline cells, neurons, and erythrocytes), saturates hexokinase (which promotes the production of glucose-6-phosphate), and strongly promotes activation of AR. AR induces the reduction of glucose to sorbitol, which is generally converted by sorbitol dehydrogenase (SDH) in fructose. However, in some cells AR concentration strongly exceeds SDH, thus sorbitol accumulates inducing osmotic cellular swelling and promoting some diabetic complications such as cataract. Moreover, increased activity of AR requires NADPH, thus leading to depletion of NADPH content and reducing the availability of NADPH for regeneration of reduced glutathione via glutathione reductase.
4. *Activation of protein kinase C (PKC) pathway* has been reported to be initiated by intracellular accumulation of glucose that causes de novo synthesis of diacylglycerol (DAG), which stimulates PKC. PKC plays a central role in activation of cells, finally leading to an increased production of extracellular matrix and cytokines and to enhanced contractility and permeability of vascular cells. The increased concentration of DAG is probably tissue-specific and may be involved in the pathogenesis of many complications of DM including cardiovascular and renal ones.

Increased production of ROS and depletion of antioxidant content is related both to the onset and progression of DM and to the development of diabetic complications.

Islet beta cells are particularly sensitive to ROS because their content of antioxidant substances is lower than that of other cell types. In type 1 diabetes the cause of insulin decrease is destruction of beta islet cells due to autoimmunitary insulitis. The death of islet beta cells is most likely mediated by the contact with activated macrophages and T lymphocytes and by the exposition to soluble mediators by cytokines, ROS and NO. In type 2 DM, insulin is not decreased, at least at the first phase of the disease, and hyperglycaemia is often associated with hyperlipidaemia. This causes an increased mitochondrial production of superoxide and the activation of inducible nitric oxide synthetase (iNOS), thus leading to the production of peroxinitrite, cellular damage, and impairment of response to endogenous insulin.

Recently great attention to mitochondrial dysfunction and production of ROS in type 2 has been focused even if the pathogenesis has not been completely elucidated.

Oxidative Stress and Diabetic Complications

High levels of free radicals cause damage to cellular proteins, membrane lipids, and nucleic acids and increase cellular death. Many ROS have been reported as correlated with different diabetic complications. However, the development of most diabetic complications have to be considered multifactorial, due to the interaction of different pathways leading to oxidative stress. Moreover, it must be considered that most complications are the result of the imbalance between production and abolition of ROS, that is, the increased ROS production and/or the decreased concentration of antioxidants. Oxidative stress has been widely demonstrated as related to diabetic complications in human medicine.

Cardiovascular Complications

Diabetes has been reported as an important risk factor for the development of cardiovascular diseases in humans such as coronary heart diseases, peripheral arterial diseases, hypertension, stroke, and cardiomyopathy. Cardiovascular complications are major causes of death in human diabetic patients [16]. The imbalance of pro-oxidants and antioxidants with excessive, destructive free radical chemistry is at the basis of most of these complications and plasma reduced glutathione (GSH) concentration can provide a good indicator of it.

AGE can also be related with coronary disease by interfering with LDL lipoprotein and facilitating deposition of cholesterol in atherosclerotic plaques in human diabetic patients with hyperlipidaemia.

Peripheral Microvascular Complications

Peripheral vascular diseases are common in human diabetic patients. Many studies show that ROS can play a role in vascular dysfunction and transforming, via oxidative damage, by decreasing the bioavailability of NO, by damaging endothelium, by accelerating endothelial cell migration, and by initiating inflammatory reaction. This leads to endothelial dysfunction which can evolve to hypertension and atherosclerosis [39]. Microvascular complications of diabetes carry a high morbidity and, when associated with macrovascular and cardiovascular complications, also high mortality in human medicine.

Neuropathy

The main pathogenesis of diabetic neuropathy is linked to increased apoptosis of neurons and glial cells due to two main mechanisms: peroxidation of lipids present in plasma, mitochondria, or endoplasmic reticulum membranes; and peroxidation of proteins by some antiapoptotic molecules (apoptosis inhibitory factor, complex 3, Bcl-2) and increasing the expression of some proapoptotic stress proteins.

In contrast with glial cells, neurons, as nondividing cells, are less prone to oxidative stress-induced DNA damages, yet mitochondrial DNA is quite sensitive to oxidative damage thus impairing energy regulation in high energy-requiring neurons. Both neuronal and glial damages may be at the basis of several neurodegenerative diseases [36].

Retinopathy

Retinopathy is one of the most important microvascular and long-term complications in diabetic human patients. It develops in stages and it leads to blindness and visual impairment. The central point leading to diabetic retinopathy is swelling of blood vessels, leaking fluids, neovascularisation, and detachment of the retina. Continued high plasma glucose concentration induces damages via repeated acute and also chronic cumulative changes. Hyperglycaemia damages the tight junction between endothelial cells and promotes endothelial apoptosis, thus increasing vascular permeability and leading to swelling of the retina. An attempt to compensate ischemia by means of new vasal production induces retinal detachment and blindness.

Compared with other tissues, the retina is particularly sensitive to oxidative stress because of the high content of polyunsaturated fatty acids. Another retinal peculiarity concerns the resistance to reverse retinal dysfunction after good glycaemic control is reinstituted. This is probably due to the accumulation of ROS and damaged molecules that are not easily removable after re-establishment of normoglycaemia [21].

Nephropathy

The principal event leading to diabetic nephropathy is the accumulation of extracellular matrix (ECM) protein in the glomerular mesangium and tubulointerstitium. Several different mechanisms are involved including an imbalance between the synthesis and degradation of ECM proteins that, because of their slow turnover, are especially susceptible to oxidative stress and glycation. Glycation of ECM components, leading to AGE formation, induces structural alterations and increased resistance to proteolytic digestion. The accumulation of ECM proteins is particularly apparent in the glomerular basal membrane and can induce microalbuminuria, a very sensitive early marker of diabetic nephropathy. AGE also interact with

RAGE influencing the expression of growth factors and cytokines regulating the proliferation of different renal cells. Finally it has also been demonstrated that AGE interact with the renin–angiotensin system and that the use of angiotensin-converting enzyme (ACE) inhibitors reduces the production of AGE and the formation of ROS in experimental DM [15].

Cataract

Cataract is another ocular complication of DM also in human diabetic patients and it is characterised by opacification of the crystalline lens correlated both to the duration and the severity of hyperglycaemia. Several mechanisms contribute to the onset of diabetic cataracts including swelling of the crystalline lens due to osmotic changes, glycosylation of lens proteins and decreased concentration of antioxidants. In particular great attention has been given to the increased concentration of sorbitol due to the elevated AR activity in the lens and to insulin-independency of crystalline cells. This causes increased water influx and thickening of fibres and subsequent cataract.

The concentration of sorbitol in crystalline has been found to be elevated in diabetic human patients and correlated to the one of the other insulin-independent tissues such as erythrocytes [17]. The use of AR inhibitors (tolrestat, epalrestat, ranirestat) to prevent cataract in experimental DM has been proposed in rats but at the moment these drugs are not suggested in management of patients with DM.

Another central point is the cross-linking of lens protein due to nonenzymatic glycosylation. The lens is characterised by an elevated content of long-lived proteins that are exceptionally stable and keep their transparency. The crystalline lens cells have virtually no turnover, thus they readily accumulate AGE which cause oxidation of thiol groups, cross-link formation, and aggregation of the crystalline proteins producing the high molecular weight insoluble molecules responsible for opacification. Moreover, glycation of membrane ATPase, such as Na-K pump, alters intracellular ion concentration and water movement via osmosis [1].

Ketoacidosis

Diabetic ketoacidosis (DKA) is probably the most common short-term complication occurring in diabetic patients. In contrast with other diabetic complications, in whose pathogenesis oxidative stress has been reported, little is known about the correlation between ketoacidosis and oxidative stress. Vantyghem et al. [35] measured malonildialdehyde (MDA) and total antioxidant status (TAS) in diabetic patients with DKA in poorly controlled diabetics without ketoacidosis, in well-compensated DM, and in control subjects, and found that MDA was increased and TAS decreased in DM with DKA but no statistical differences were found with poorly controlled DM. In 2002, Lee et al. [24] found that diabetic patients with DKA showed higher lipid peroxidation and lower vitamin A concentrations than normal and that after correction of DKA, lipid peroxidation increased and the concentrations of vitamins C and E decreased. These results suggest that oxidative stress can be linked both to ketoacidosis and poor compensation of DM even if no direct effect of ketone bodies on production of ROS has been demonstrated. However, the administration of antioxidant vitamins during treatment for DKA can be strongly recommended.

Use of Antioxidant to Prevent Diabetic Complications

Even though many studies suggested the linkage between oxidative stress and diabetic complications, at the moment no direct evidence that ROS production can directly cause or initiate diabetic complications is available as much as hyperglycaemia, hyperketonaemia, and high free fatty acid (FFA) concentrations could act to increase oxidative stress. On the other hand, an important issue is the possible use of antioxidants to prevent, stop, or delay the progression of DM or its complications.

Many different antioxidants, vitamins, coenzymes, and minerals, singularly or in combination, have been proposed and in some nonclinical trials, both on cultured cells and diabetic animal models, positive results in preventing ROS production or diabetic complications have been found [20, 22, 33]. However, large clinical trials on human patients have not been successful, perhaps because of the multiple pathways used by hyperglycaemia to mediate its adverse effects. The introduction of new multifunctional antioxidants might help to reach more clinical results in the near future.

Markers Useful in Monitoring Oxidant Stress in DM

Another important issue is the possible use of markers of oxidative stress to evaluate the level of compensation of DM, to monitor the effect of therapy, and to predict complications. Plasma markers are the simplest to use, however, in some cases, specific cells, such as erythrocytes (RBC), can be used as a good model of what can be found in other tissues in which it is difficult to test oxidative stress in vivo.

Evaluation of Oxidative Stress

Many different biomarkers of oxidative stress have been evaluated in DM, similarly to those already reported for other diseases. The same laboratory tests used in DM are validated for evaluation of oxidative stress in other conditions, thus the principle of the assays is not discussed in the present chapter.

Lipid peroxidation has been evaluated by means of fluorometric or spectrophotometric assays for thiobarbituric acid reactive substances (TBARS), total radical-trapping potential (TRAP), and MDA. They resulted in an increase in both natural and experimentally induced DM in plasma, RBC, and different organs such as the pancreas, lens, retina, nerve, and kidney. The increase of TBARS has been shown to be prevented by treatment with various antioxidants.

The concentration of the major intracellular redox substance, reduced glutathione (GSH), is another frequent test, based on a colorimetric assay that allows us to detect intracellular GSH concentration easily in several cells. In DM glutathione has been found to be decreased in the liver, kidney, pancreas, RBC, nerve, and precataractous lens. On the contrary, plasma concentration of GSH is usually below the lower concentration of linearity of the test, thus it is rarely evaluable.

Enzyme activities such as glutathione peroxidase (GPX) and reductase (GR) that are involved in the detoxification from hydrogen peroxide and regeneration of GSH can be assessed in cells by means of spectrophotometric kinetic tests. Even if there is not complete agreement on the activities of these enzymes in DM, GPX is found to be elevated in most tissues whereas it is decreased in the heart and retina and GR is reduced in the retina and plasma and increased in the heart.

Activities of other scavenging enzymes such as catalase and superoxide dismutase (SOD) can be easily tested in different organs. Catalase is reported to be elevated in the heart, liver, brain, and kidney of diabetic laboratory animals, however, these results are not consistent in all the studies. The same ambiguous results are available for SOD.

On the other hand, vitamins A, C, and E have also been evaluated in many studies due to their antioxidant role, using colorimetric, fluorimetric, or HPLC tests, but many discrepancies among different studies have been reported.

Advanced Glycation Endproducts

Nonenzymatic glycosilation of proteins leads to the production of several different AGE. Fructosamine and glycated haemoglobin are often used as common markers of protein glycation and are useful to monitor diabetic compensation. Even if specific immunological tests (mainly ELISA) have been developed for total or selected AGE, nonspecific fluorometric assays to test the total concentration are available too. The simplest test has been described in plasma and urine samples by Münch et al. [27] and directly measured autofluorescence of AGE by means of a fluorometer at 445-nm emission using an excitation wavelength of 370 nm. Increased AGE have been found in diabetic patients with nephropathy whereas AGE are usually normal in well-compensated diabetic patients [38].

Sorbitol

Evaluation of intracellular sorbitol concentration is another interesting issue: the cell membrane is not permeable to sorbitol thus its production, due to elevated glucose influx in noninsulin-dependent cells, leads to an increase in cytoplasmic concentration and contributes to cellular swelling. In vivo sorbitol concentration can be easily evaluated in erythrocytes using a fluorometric test as described by Shinohara et al. [32]. Briefly, RBC are lysed in distilled water and the concentration of haemoglobin in the solution is evaluated. After deproteinisation supernatant is separated and put in a cuvette in buffer (pH 8,6), containing EDTA, NAD+, and SDH. The fluorescence due to NADH produced by the transformation of sorbitol in fructose is then evaluated after 30 min incubation using a spectrofluorometer at 452-nm emission at an excitation wavelength of 366 nm. Sorbitol in erythrocytes and lens has been found to be increased in patients with DM but is normalised after administration of AR inhibitors or ascorbate.

Canine DM

DM is a common metabolic disorder in middle-aged and older dogs. From a pathogenetic point of view canine DM is, in most cases, similar to type 1 DM of human beings and is characterised by an impaired production of insulin due to beta islet destruction. Autoimmune diseases or pancreatitis are generally linked to development of islet destruction. Complications are frequent and include short-term complications such as DKA and long-term ones, including cataract [2].

Ketoacidosis with ketonuria is a common concurrent disease in noncompensated diabetic dogs. The production of ketone bodies is due to increased lipomobilisation followed by beta-oxidation of FFA. Ketonuria is usually associated with osmotic diuresis with loss of Na and K, hypovolaemia, and dehydration.

Within long-term complications bilateral cataract is the most diffuse disease occurring in 40–75% of patients [2, 4].Other complications, often reported in human beings, have been sporadically reported in dog, such as retinopathy, although subclinical in most cases, and other ocular complications [37], hypertension [34], dermatological disorders [30], nephropathy [26], neuropathy [23], and infectious diseases [28].

Oxidative Stress and Canine DM

To our knowledge, few extensive papers on oxidative stress and canine DM have been published, mainly regarding cataract, the most diffuse complication of canine DM, and erythrocytic abnormalities.

Cataract in diabetic dogs has been suggested to be due to both sorbitol accumulation in the lens and to glycosilation of lens proteins, similarly to the human counterpart. However, sorbitol accumulation is difficult to demonstrate in crystalline lens from diabetic dogs and, to our knowledge, at the moment no studies on sorbitol concentration in canine diabetic crystalline lens have been performed. On the other hand, the effect of AGE has been recently evaluated by means of immunohistochemistry in lens epithelial cells from dogs with diabetic and inherited cataract [5].

The authors did not find a statistical increase in RAGE expression in diabetic vs. inherited cataract lens, but an increase of p21 and PCNA expression in diabetic lens suggests a cell cycle and proliferation dysregulation. These results suggested that the effect of AGE cannot be the only triggering factor in development of canine diabetic cataract.

Another interesting issue regards haematological alterations due to DM-induced oxidative stress. DM can induce anaemia and other haematological complications and the study of oxidative stress on very simple cells such as the erythrocytes can be used as a good model of what happens in other insulin-independent organs.

DM has been shown to induce anaemia in about than 25% of diabetic dogs [18] and eccentrocytosis has been reported in many cases of canine DM with or without ketoacidosis [6]. In vivo these alterations are not related to altered osmotic fragility or to decreased content of GSH in RBC [11] but an increase in potassium and TBARS intraerythrocytic concentration has been found, mainly in poorly compensated diabetic dogs [8]. Regarding enzymes, erythrocytes from diabetic dogs showed increased activities of glucose 6-phosphate dehydrogenase (G6PD, the key enzyme of the penthose phosphate pathway; [11]), and of scavenging enzyme catalase [8], suggesting an attempt to compensate oxidative damage.

Effects of oxidative stress on erythrocytes are much more evident in vitro [12] in which erythrocytes incubated with a high concentration of glucose, simulating a severe hyperglycaemia, exhibited high osmotic fragility and 2,3 diphosphoglycerate intracellular concentrations in comparison with controls, whereas more evident effects are found when ketone bodies are present. In particular, an increase in mean corpuscular volume (MCV) and a higher consumption of glucose not associated with an adequate increase in lactate production suggest the activation of other metabolic pathways such as the polyol pathway leading to sorbitol accumulation and erythrocyte swelling. In diabetic dogs in vivo MCV is generally found to be normal but the higher content of osmotic products is probably compensated by the increased plasma osmolarity. These results suggest that oxidative stress can be found in erythrocytes from diabetic dogs and that ketoacidosis strongly increases oxidative damage, but many compensatory mechanisms may balance erythrocyte oxidation, thus avoiding severe haematological complications in most cases.

We recently evaluated the use of some haematological glycosilation markers to check diabetic status and compensation in diabetic dogs [13]. In particular sorbitol concentration in erythrocytes was found to be higher in diabetic dogs in comparison with controls and in diabetic, poorly compensated dogs and in dogs with ketoacidosis. Moreover, it tends to decrease after improvement of clinical conditions. This parameter can be useful not only to confirm the activation of the polyol pathway, leading to oxidative stress, in RBC as well as in other insulin-independent tissues, but could also permit the monitoring of diabetic status and response to therapy.

Regarding AGE, diabetic dogs showed an higher AGE concentration in plasma than control dogs, without any correlation with diabetic status, presence of long-term complications, or clinical conditions. This confirms that, also in dogs, production of AGE must be considered as a possible mechanism leading to diabetic complications but that this parameter is probably of little clinical use to manage diabetic patients.

Feline DM

DM is quite common in cats occurring mainly in animals over 7 years old, with a higher prevalence in males and neutered cats [29, 31]. From a pathogenetic point of view, feline DM is, in more than 85% of cases, similar to type 2 DM (noninsulin-dependent DM) of human beings and characterised by an impaired response to insulin production leading to hyperglycaemia, glycosuria, polyuria, and polydipsia. However, the progression of the disease frequently leads to secondary beta cell destruction, due to excessive deposition of amyloid and, subsequently, insulin therapy is often required in the late phase of the disease.

Ketoacidosis is considered less frequent in feline DM than in the canine counterpart and, when present, it is generally linked to a bad prognosis. This is probably due to the residual production of insulin in type 2 DM, the most common form in cats, which could be enough to reduce ketone bodies production in most cases.

Long-term complications, by which neuropathy, urinary and cardiac diseases, and haematological complications, are quite frequent. Neuropathy is the most common and severe long-term complication, leading to sensorimotor neuropathy in both pelvic and thoracic limbs. Symptoms vary from mild to extremely severe signs with plantigrade posture and/or secondary muscle atrophy [25].

Other important complications are the elevated percentage of infectious diseases, mainly from the urinary tract, due to impaired immune response, and haematological changes, by which a mild anaemia may occur.

On the other side, in contrast with dogs, cataract is a very rare complication in cats (less than 5% of cases), most likely because the lower levels of glucose in blood lead to a slow increase of the concentration of glucose in ocular fluid, thus inducing minimal glycoxidation and intracellular accumulation of sorbitol in the lens cells [31].

Oxidative Stress and Feline DM

Few studies on the relation among oxidative stress, diabetic status, and development of complications are available. In particular, attention has been paid on the correlation between DM and Heinz body formation [9]. Heinz bodies are clumps of denaturated globin, resulting from oxidation of haemoglobin, that precipitates in the inner membrane of RBC. If present in a high number, they can induce haemolytic anemia. Feline haemoglobin is very sensitive to oxidative damage due to the higher number of reactive sulfhydril groups in the globin chain. Intracellular glutathione is the main antioxidant substance in RBC. Heinz bodies have been reported to be increased in cats with DKA but not in DM alone, and the percentage of HB is found to be correlated with plasma beta-OH butyrate concentration. On the other hand, ketoacidotic cats showed reduced GSH concentration in RBC and TBARS were found to be slightly (although not significantly) increased in both ketotic and nonketotic diabetic cats in comparison with controls [10]. These results suggest an effect of ketoacidosis on haemoglobin oxidation and depletion of GSH even though in vitro studies failed to demonstrate a direct effect of ketones and hyperglycaemia on HB formation or GSH concentration in cultured RBC [14]. The mechanism for in vivo oxidation of haemoglobin in cats with DKA is still to be elucidated.

On the other hand, effects of hyperglycaemia on RBC membrane have also been investigated [19]. RBC from diabetic cats were found to be more rigid, probably due to alterations on some integral membrane proteins. This could be partly linked to some vascular complications of DM.

Regarding diabetic complications few studies are available on feline DM. An extensive study on neurological complications of feline DM was reported in 2002 [25]. Several symptoms and clinical signs have been reported associated to different severity. In this study the authors investigated the content of sorbitol, sugars, and water in nerves from diabetic cats with neuropathy. Diabetes was associated with a eightfold increase in nerve glucose and 12-fold in fructose, and with a statistical increase of water content. However, only a minimal increase in sorbitol concentrations was found in contrast with similar studies in rodents and humans. This is probably due to an elevated activity of SDH in feline nerves in comparison with other species, leading to transformation of most sorbitol in fructose. The authors suggested that the flux through the polyol pathway and not polyol accumulation per se, can cause neuropathy in cats as well as in humans and rodents. This could lead to Schwann cell injury, demyelinisation, and axonal injury leading to different clinical signs.

Conclusion

The study of oxidative stress in DM and in development of diabetic complications is a very interesting issue even though far from being completely clarified. In particular, dogs and cats seems to share a similar pathogenesis with human type 1 and 2 DM, respectively, even if extensive studies on the relation between DM and oxidative stress are still lacking in veterinary medicine. Further investigations and clinical trials could contribute to a better understanding not only of the onset of the disease and its complications but also of the possible role of integration with antioxidant drugs during therapy for DM in order to minimise the progression of the disease and complications.

References

1. Ahmed N. (2005) Advanced glycation endproducts-role in pathology of diabetic complications. Diabetes Res Clin Pract. 67(1):3–21
2. Basher AW, Roberts SM. (1995) Ocular manifestations of diabetes mellitus: diabetic cataracts in dogs. Vet Clin North Am Small Anim Pract. 25(3):661–676
3. Baynes JW, Thorpe SR. (1999) Role of oxidative stress in diabetic complications: a new perspective on an old paradigm. Diabetes. 48(1):1–9

4. Beam S, Correa MT, Davidson MG. (1999) A retrospective-cohort study on the development of cataracts in dogs with diabetes mellitus: 200 cases. Vet Ophthalmol 2(3):169–172
5. Bras ID, Colitz CM, Kusewitt DF, Chandler H, Lu P, Gemensky-Metzler AJ, Wilkie DA. (2007) Evaluation of advanced glycation end-products in diabetic and inherited canine cataracts. Graefes Arch Clin Exp Ophthalmol. 245(2):249–257
6. Caldin M, Carli E, Furlanello T, Solano-Gallego L, Tasca S, Patron C, Lubas G. (2005) A retrospective study of 60 cases of eccentrocytosis in the dog. Vet Clin Pathol. 34(3):224–231
7. Ceriello A, Morocutti A, Mercuri F, Quagliaro L, Moro M, Damante G, Viberti GC. (2000) Defective intracellular antioxidant enzyme production in type 1 diabetic patients with nephropathy. Diabetes 49(12):2170–2177
8. Chansaisakorn W, Sriphavatsarakorn P, Sopakdittapong P, Trisiriroj M, Pondeenana S, Buranakarl C. (2009) Oxidative stress and intraerythrocytic concentrations of sodium and potassium in diabetic dogs. Vet Res Commun 33(1):67–75
9. Christopher MM. (1989) Relation of endogenous Heinz bodies to disease and anemia in cats: 120 cases (1978–1987). J Am Vet Med Assoc. 194(8):1089–1095
10. Christopher MM, Broussard JD, Peterson ME. (1995) Heinz body formation associated with ketoacidosis in diabetic cats. J Vet Intern Med. 9(1):24–31
11. Comazzi S, Paltrinieri S, Spagnolo V, Sartorelli P. (2002) Some aspects of erythrocyte metabolism in insulin-treated diabetic dogs. Res Vet Sci. 72(1):23–27
12. Comazzi S, Spagnolo V, Bonfanti U. (2004) Erythroicyte changes in canine diabetes mellitus: in vitro effects of hyperglycaemia and ketoacidosis. Comp Clin Pathol. 12:199–205
13. Comazzi S, Bertazzolo W, Bonfanti U, Spagnolo V, Sartorelli P. (2008) Advanced glycation end products and sorbitol in blood from differently compensated diabetic dogs. Res Vet Sci. 84(3):341–346
14. Fallin CW, Christopher MM. (1996) In vitro effect of ketones and hyperglycemia on feline hemoglobin oxidation and D- and L-lactate production. Am J Vet Res. 57(4):463–467
15. Forbes JM, Cooper ME, Oldfield MD, Thomas MC. (2003) Role of advanced glycation end products in diabetic nephropathy. J Am Soc Nephrol. 14(8 Suppl 3):S254–S258
16. Haidara MA, Yassin HZ, Rateb M, Ammar H, Zorkani MA. (2006) Role of oxidative stress in development of cardiovascular complications in diabetes mellitus. Curr Vasc Pharmacol. 4(3):215–227
17. Hayashi R, Hayakawa N, Makino M, Nagata M, Kakizawa H, Uchimura K, Hamada M, Aono T, Fujita T, Shinohara R, Nagasaka A, Itoh M. (1998) Changes in erythrocyte sorbitol concentrations measured using an improved assay system in patients with diabetic complications and treated with aldose reductase inhibitor. Diabetes Care. 21(4):672–673
18. Hess RS, Saunders HM, Van Winkle TJ, Ward CR. (2000) Concurrent disorders in dogs with diabetes mellitus: 221 cases (1993–1998). J Am Vet Med Assoc. 217(8):1166–1173
19. Kaymaz AA, Tamer S, Albeniz I, Cefle K, Palanduz S, Ozturk S, Salmayenli N. (2005) Alterations in rheological properties and erythrocyte membrane proteins in cats with diabetes mellitus. Clin Hemorheol Microcirc. 33(2):81–88
20. Kowluru RA, Kennedy A. (2001) Therapeutic potential of anti-oxidants and diabetic retinopathy. Expert Opin Investig Drugs. 10(9):1665–1676
21. Kowluru RA, Chan PS. (2007) Oxidative stress and diabetic retinopathy. Exp Diabetes Res. 2007:43603
22. Koya D, Lee IK, Ishii H, Kanoh H, King GL. (1997) Prevention of glomerular dysfunction in diabetic rats by treatment with d-alpha-tocopherol. J Am Soc Nephrol. 8(3):426–435
23. Katherman AE, Braund KG (1983) Polyneuropathy associated with diabetes mellitus in a dog. J Am Vet Med Assoc. 182(5):522–524
24. Lee DM, Hoffman WH, Carl GF, Khichi M, Cornwell PE. (2002) Lipid peroxidation and antioxidant vitamins prior to, during, and after correction of diabetic ketoacidosis. J Diabetes Complicat. 16(4):294–300
25. Mizisin AP, Shelton GD, Burgers ML, Powell HC, Cuddon PA. (2002) Neurological complications associated with spontaneously occurring feline diabetes mellitus. J Neuropathol Exp Neurol. 61(10):872–884

26. Muñana KR. (1995) Long-term complications of diabetes mellitus, Part I: Retinopathy, nephropathy, neuropathy. Vet Clin North Am Small Anim Pract. 25(3):715–730
27. Münch G, Keis R, Wessels A, Riederer P, Bahner U, Heidland A, Niwa T, Lemke HD, Schinzel R. (1997) Determination of advanced glycation end products in serum by fluorescence spectroscopy and competitive ELISA. Eur J Clin Chem Clin Biochem. 35(9):669–677
28. Nichols R, Crenshaw KL. (1995) Complications and concurrent disease associated with diabetic ketoacidosis and other severe forms of diabetes mellitus. Vet Clin North Am Small Anim Pract. 25(3):617–624
29. Panciera DL, Thomas CB, Eicker SW, Atkins CE. (1990) Epizootiologic patterns of diabetes mellitus in cats: 333 cases (1980–1986). J Am Vet Med Assoc. 197(11):1504–1508
30. Peikes H, Morris DO, Hess RS. (2001) Dermatologic disorders in dogs with diabetes mellitus: 45 cases (1986–2000). J Am Vet Med Assoc. 219(2):203–208
31. Salgado D, Reusch C, Spiess B. (2000) Diabetic cataracts: different incidence between dogs and cats. Schweiz Arch Tierheilkd. 142(6):349–353
32. Shinohara R, Ohta Y, Yamauchi M, Ishiguro I. (1998) Improved fluorometric enzymatic sorbitol assay in human blood. Clin Chim Acta. 273(2):171–184
33. Stevens MJ, Obrosova I, Cao X, Van Huysen C, Greene DA. (2000) Effects of DL-alpha-lipoic acid on peripheral nerve conduction, blood flow, energy metabolism, and oxidative stress in experimental diabetic neuropathy. Diabetes. 49(6):1006–1015
34. Struble AL, Feldman EC, Nelson RW, Kass PH. (1998) Systemic hypertension and proteinuria in dogs with diabetes mellitus. J Am Vet Med Assoc. 213(6):822–825
35. Vantyghem MC, Balduyck M, Zerimech F, Martin A, Douillard C, Bans S, Degand PM, Lefebvre J. (2000) Oxidative markers in diabetic ketoacidosis. J Endocrinol Invest. 23(11):732–736
36. Vincent AM, Russell JW, Low P, Feldman EL. (2004) Oxidative stress in the pathogenesis of diabetic neuropathy. Endocr Rev. 25(4):612–628
37. Wyman M, Sato S, Akagi Y, Terubayashi H, Datiles M, Kador PF. (1988) The dog as a model for ocular manifestations of high concentrations of blood sugars. J Am Vet Med Assoc. 193(9):1153–1156
38. Yanagisawa K, Makita Z, Shiroshita K, Ueda T, Fusegawa T, Kuwajima S, Takeuchi M, Koike T. (1998) Specific fluorescence assay for advanced glycation end products in blood and urine of diabetic patients. Metabolism. 47(11):1348–1353
39. Yung LM, Leung FP, Yao X, Chen ZY, Huang Y. (2006) Reactive oxygen species in vascular wall. Cardiovasc Hematol Disord Drug Targets. 6(1):1–19

Oxidative Stress in the Spinal Cord of Dogs and Cats

Wendy Baltzer

Abstract Oxidative stress in spinal cord disease is considered a secondary mechanism of injury following a primary traumatic event such as vertebral fracture or intervertebral disk protrusion (intervertebral disk disease). The primary spinal cord injury often results in decreased perfusion of the spinal cord due to compression from bone, disk, hematoma, or granuloma. This decreased blood flow to the cord, or ischemia, causes neutrophils, macrophages, and eosinophils to release reactive oxygen species among other inflammatory mediators. Oxidative enzymes are upregulated in the spinal cord following any injury and are a major source of oxidative damage for weeks by stimulating inflammatory cells to release more reactive oxygen species, perpetuating the cycle of oxidative damage. A multiarmed therapeutic approach may be needed in the treatment of spinal cord injury that encompasses all or at least several of the major mechanisms of injury, namely, glutamate-mediated excitotoxicity, increased intracellular Ca^{2+}, and oxidative stress. These mechanisms lead to cell death through a loss of energy production, damage to nucleic acids and proteins, and apoptosis. Experimentally and in clinical trials, corticosteroids and N-acetyl-cysteine (NAC) have had limited success. Other free radical scavengers such as derivatives of acetylsalicylic acid and sulfasalazine, or aldehyde free radical scavengers are currently being investigated.

Keywords Canine • Feline • Dog • Cat • Spinal cord trauma • Reactive oxygen species • Isoprostanes • Ischemia reperfusion • Secondary spinal cord injury • Acreolin • Hydralazine • N-acetyl-cysteine

Oxidative stress is defined as an excess of reactive oxygen species (ROS) or a relative insufficiency of antioxidant defense mechanisms. ROS are generated from electron leakage from electron transport chains, inflammatory cells, UV light,

W. Baltzer (✉)
Small Animal Surgery, College of Veterinary Medicine, Oregon State University, Corvallis, OR 97331-4801, USA
e-mail: wendy.baltzer@oregonstate.edu

L. Mandelker and P. Vajdovich (eds.), *Studies on Veterinary Medicine*, Oxidative Stress in Applied Basic Research and Clinical Practice 5,
DOI 10.1007/978-1-61779-071-3_6,

drugs including chemotherapeutics, cigarette smoke, ozone and others. Oxidative stress in spinal cord disease is considered a secondary mechanism of injury following a primary traumatic event such as vertebral fracture or intervertebral disk protrusion (intervertebral disk disease, IVDD). The primary spinal cord injury often results in decreased perfusion of the spinal cord due to compression from bone, disk, hematoma or granuloma. This ischemia causes neuronal intracellular concentrations of ATP to decline, which then stimulates neurons to release the excitatory neurotransmitter glutamate [1]. Glutamate initiates a pathogenic cascade that results in cellular overload of calcium and increased production of ROS.

Neuronal ischemia also incites an inflammatory response with recruitment of neutrophils and macrophages and upregulation of cellular adhesion molecules. Neutrophils, macrophages and eosinophils release ROS among other inflammatory mediators. The free radical O_2^- is produced mainly by the enzyme nicotinamide adenine dinucleotide phosphate (NADPH) oxidase whose subunits are upregulated to form the active enzyme in the presence of proinflammatory cytokines [2, 3]. NADPH oxidase is upregulated in the spinal cord following injury and is thought to be a major source of oxidative damage for weeks following the primary injury [4]. ROS cause many secondary effects in the spinal cord following primary injury, including recruitment of additional inflammatory cells, which release more ROS, and the cycle of ROS-mediated neuronal damage is perpetuated [5].

Ischemia-reperfusion injury occurs as part of secondary injury in the spinal cord or with management of the injury such as surgical decompression. Reperfusion following ischemia in the spinal cord results in increased ROS production by mitochondria and stimulation of the enzyme xanthene oxidase to produce more ROS [1, 6, 7]. Peroxidation of lipids generates more reactive species including free radicals and reactive aldehydes such as malondialdehyde, 4-hydroxynonenal, and acrolein which bind to glutathione and result in additional oxidative stress [8–12].

The spinal cord is especially sensitive to oxidative stress because there are high levels of polyunsaturated fatty acids (PUFAs) present in the membranes of neurons and oligodendrocytes such as linoleic acid and arachidonic acid. The PUFAs react with ROS in cellular membranes resulting in propagation of new radicals. Lipid peroxidation will continue until all PUFAs are depleted or the free radicals interact with an antioxidant within the membrane, such as vitamin E [13]. Because lipid peroxidation is self-perpetuating, severe damage to neuron membranes results in alterations in ion channels, including calcium with an increase in intracellular calcium, eventually leading to apoptosis [14].

F2 isoprostanes (15F2t isoprostane or 8-isoprostaglandin F2α) are a class of free radical catalyzed products of the arachidonic acid pathway produced independently of cyclo-oxygenase enzymes in vivo. They have a high sensitivity and specificity for assessing oxidative stress in disease states and can be measured in the urine to assess lipid peroxidation [15]. In dogs with spinal cord injury from acute disk herniation, F2 isoprostanes significantly increase, indicating lipid peroxidation and oxidative stress occur in dogs with spinal cord injury [16].

Following the primary injury, secondary injury is also mediated by the interaction of ROS with nitric oxide (NO). NO production increases following spinal cord injury due to a rise in nitric oxide synthase enzymes iNOS and eNOS [4]. NO generation continues to be elevated for at least 14 days following spinal cord injury [4, 17]. NO and ROS react together to form highly reactive and cytotoxic molecules including peroxynitrite ($ONOO^-$) and peroxynitrous acid (ONOOH) which damage lipids, proteins and nucleic acids [18]. Nitrotyrosine, a marker of superoxide–NO interaction remains elevated for at least 14 days following spinal cord injury, indicating that nitrosative and oxidative stress continue for at least weeks following the primary injury [4, 17].

Reperfusion injury following an ischemic event is particularly damaging to the spinal cord because the spinal cord is subject to severe lipid peroxidation with high levels of PUFAs and because the neurons contain high numbers of mitochondria, which are susceptible to ROS and produce ROS following ischemia. Up to 33% of humans develop paraplegia following thoracoabdominal aorta surgery due to ischemia and reperfusion injury following temporary aortic cross-clamping [19]. Indeed, in some patients, there is delayed onset of paraplegia one to five days following aortic cross-clamping and this neurological dysfunction is thought to be due, in large part, to ROS-mediated damage and lipid peroxidation [20, 21]. In dogs following acute spinal cord injury from intervertebral disk herniation, secondary mechanisms of injury including lipid peroxidation begin within four h of the primary injury from the disk and peak four days later [22]. During this period, demyelination, axonal degeneration and myelomalacia occur in regions inside and outside the primary zone of trauma [23–25]. ROS-induced damage to neurons is not limited to lipid peroxidation-mediated inflammation and cell death. ROS also damage proteins and DNA resulting in cellular dysfunction and potentially carcinogenesis [26].

Because oxidative stress can decrease myelin production and cause axonal degeneration, its role in degenerative myelopathy has been investigated in dogs [27, 28]. Degenerative myelopathy or canine degenerative radiculomyelopathy is a progressive degenerative disease of the thoracolumbar spinal cord of dogs for which no treatment exists and is believed to be hereditary in nature [29, 30]. Production of the metabolite of lipid peroxidation, 8-isoprostane, in the cerebral spinal fluid of Pembroke Welsh corgis with degenerative myelopathy was not different from control dogs [29]. In a study investigating German shepherd dogs with degenerative myelopathy, no defect in alpha-tocopherol transfer protein gene or expression was found, indicating that vitamin E metabolism in these dogs is unlikely to be a part of the pathogenesis of this disease [31]. In addition, vitamin E levels are not altered nor does supplementation with vitamin E have any effect on dogs with degenerative myelopathy [32]. At this time there is no evidence for a role of oxidative stress in degenerative radiculomyelopathy of dogs.

Endogenous antioxidant defense mechanisms such as antioxidant proteins, small molecule antioxidants and antioxidant enzymes react with ROS to donate a single electron to the free radical resulting in a more stable compound [33]. The spinal cord contains relatively low amounts of the antioxidant glutathione [24, 34]. Following trauma, the spinal cord is quickly depleted of its antioxidants including glutathione, vitamin E, and ascorbic acid [7, 24, 34]. The free radical scavenging enzyme superoxide dismutase increased within 24 h following the primary spinal cord injury and remained elevated for at least 14 days to counteract the increased production of ROS. Unlike superoxide dismutase, catalase and glutathione peroxidase remain unchanged in the first 24 h following injury, however, by 14 days they too have responded and are present in increased amounts compared to normal spinal cord [4]. The delayed response of some of the antioxidant enzymes coupled with the rise in free radical production may account for some of the continued secondary injury and deteriorating neurological function seen during the four days following the primary injury. In addition, continued oxidant stress may play a role in the final functional outcome of many dogs following spinal cord injury.

An important target of therapy for spinal cord injury is oxidative stress, however, whereas in vivo studies in animals have shown successful treatment with antioxidants, studies in human and dog clinical trials have failed to produce similar results [21, 35–37]. Methylprednisolone has been used widely for its antioxidant properties in the treatment of spinal cord injury. Improved outcome has been demonstrated in some studies, however, serious side effects can occur in dogs including gastrointestinal bleeding, wound infections and pneumonia [38–43]. In blinded placebo-controlled clinical studies in humans with spinal cord injury, methylprednisolone sodium succinate can improve motor outcome if administered within eight h of the trauma [38, 44, 45]. In a recent retrospective study on the use of glucocorticoids for thoracolumbar intervertebral disk herniation, dogs treated with no glucocorticoids, methylprednisolone, prenisolone or dexamethazone had no difference in short-term outcome based on their neurological scores [46]. In addition, dogs treated with dexamethazone were over three times as likely to have a clinical complication compared with dogs given no glucocorticoids or those treated with methylprednisolone or prednisolone [46]. Complications included urinary tract infections and diarrhea [46].

In the CNS, where levels of other endogenous antioxidants (i.e., catalase) are low, the cells rely more heavily on glutathione for protection and N-acetyl-cysteine (NAC) is its precursor [24, 47]. NAC also directly scavenges ROS and numerous laboratory and clinical studies have demonstrated a beneficial effect from NAC administration [48–53]. In a blinded randomized, placebo-controlled clinical trial NAC was administered intravenously immediately prior to decompressive hemilaminectomy in dogs with acute thoracolumbar disk herniation. NAC was unable to ameliorate the neurological deterioration that occurred from the spinal cord injury in dogs with intervertebral disk disease [37]. A recent study of spinal cord ischemia in rabbits identified that, although NAC treatment improved neurological status, neuronal cell damage was still significant [54]. In addition, in closed brain trauma although neuron morphology is somewhat protected [55], brain perfusion and edema are unaffected by NAC treatment [56]. In the dogs with acute thoracolumbar disk herniation, spinal cord edema may have been severe enough that outcome was not affected by NAC administration. Although NAC did not alter urinary $15F_{2t}$ isoprostane excretion in dogs with IVDD-mediated spinal cord injury, it is possible that another antioxidant may have been more effective. The prostaglandin I_2 analogue, iloprost, when combined with NAC, improves neurological outcome in rabbits subjected to spinal cord ischemia more effectively than NAC alone [54].

A multiarmed therapeutic approach may be needed in the treatment of spinal cord injury that encompasses all or at least several of the major mechanisms of injury, namely, glutamate-mediated excitotoxicity, increased intracellular Ca^{2+}, and oxidative stress. These mechanisms lead to cell death through a loss of energy production, damage to nucleic acids and proteins, and apoptosis [57–65]. Neu200, a derivative of acetylsalicylic acid and sulfasalazine, inhibits NMDA receptor-mediated excitotoxicity and scavenges free radicals. Experimentally in rats, this compound improves functional recovery following acute spinal cord injury [57]. Another promising avenue of investigation for the treatment of spinal cord injury is through inhibition of free radical-generated aldehydes such as acreolin [8]. The acrolein scavenger hydralazine ameliorates secondary injury in the spinal cord, at least experimentally [66, 67].

Further understanding of the specific mechanism of oxidative stress in spinal cord injury and investigation into the best method for scavenging ROS may provide a better outcome in the future. Much work still needs to be done to understand how the various mechanisms of injury including oxidative and nitrosidative stress interact to produce the devastating effects of secondary injury in spinal cord trauma.

References

1. Juurlink BH, Paterson PG: Review of oxidative stress in brain and spinal cord injury: suggestions for pharmacological and nutritional management strategies. J Spinal Cord Med 21:309–334, 1998.
2. Bianca VD, Dusi S, Bianchini E, et al: beta-amyloid activates the O-2 forming NADPH oxidase in microglia, monocytes, and neutrophils. A possible inflammatory mechanism of neuronal damage in Alzheimer's disease. J Biol Chem 274:15493–15499, 1999.
3. Gao HM, Jiang J, Wilson B, et al: Microglial activation-mediated delayed and progressive degeneration of rat nigral dopaminergic neurons: relevance to Parkinson's disease. J Neurochem 81:1285–1297, 2002.
4. Vaziri ND, Lee YS, Lin CY, et al: NAD(P)H oxidase, superoxide dismutase, catalase, glutathione peroxidase and nitric oxide synthase expression in subacute spinal cord injury. Brain Res 995:76–83, 2004.
5. Schnell L, Fearn S, Klassen H, et al: Acute inflammatory responses to mechanical lesions in the CNS: differences between brain and spinal cord. Eur J Neurosci 11:3648–3658, 1999.
6. Braughler JM, Hall ED: Central nervous system trauma and stroke. I. Biochemical considerations for oxygen radical formation and lipid peroxidation. Free Radic Biol Med 6:289–301, 1989.
7. Hall ED: Lipid peroxidation. Adv Neurol 71:247–257; discussion 257–248, 1996.
8. Hamann K, Shi R: Acrolein scavenging: a potential novel mechanism of attenuating oxidative stress following spinal cord injury. J Neurochem 111(6):1348–1356, 2009.
9. Uchida K: Current status of acrolein as a lipid peroxidation product. Trends Cardiovasc Med 9:109–113, 1999.
10. Uchida K, Kanematsu M, Morimitsu Y, et al: Acrolein is a product of lipid peroxidation reaction. Formation of free acrolein and its conjugate with lysine residues in oxidized low density lipoproteins. J Biol Chem 273:16058–16066, 1998.
11. Esterbauer H, Schaur RJ, Zollner H: Chemistry and biochemistry of 4-hydroxynonenal, malonaldehyde and related aldehydes. Free Radic Biol Med 11:81–128, 1991.
12. O'Brien PJ, Siraki AG, Shangari N: Aldehyde sources, metabolism, molecular toxicity mechanisms, and possible effects on human health. Crit Rev Toxicol 35:609–662, 2005.

13. Radi R, Beckman JS, Bush KM, et al: Peroxynitrite-induced membrane lipid peroxidation: the cytotoxic potential of superoxide and nitric oxide. Arch Biochem Biophys 288: 481–487, 1991.
14. Montucschi P, Barnes P, Roberts LJ: Insights into oxidative stress: the isoprostanes. Curr Med Chem 14:703–717, 2007.
15. Moore K: Isoprostanes and the liver. Chem Phys Lipids 128:125–133, 2004.
16. McMichael MA, Ruaux CG, Baltzer WI, et al: Concentrations of 15F2t isoprostane in urine of dogs with intervertebral disk disease. Am J Vet Res 67:1226–1231, 2006.
17. Xu J, Kim GM, Chen S, et al: iNOS and nitrotyrosine expression after spinal cord injury. J Neurotrauma 18:523–532, 2001.
18. Dawson TM, Dawson VL, Snyder SH: Molecular mechanisms of nitric oxide actions in the brain. Ann N Y Acad Sci 738:76–85, 1994.
19. Mackey ME, Wu Y, Hu R, et al: Cell death suggestive of apoptosis after spinal cord ischemia in rabbits. Stroke 28:2012–2017, 1997.
20. Moore WM, Jr., Hollier LH: The influence of severity of spinal cord ischemia in the etiology of delayed-onset paraplegia. Ann Surg 213:427–431; discussion 431–422, 1991.
21. Chronidou F, Apostolakis E, Papapostolou I, et al: Beneficial effect of the oxygen free radical scavenger amifostine (WR-2721) on spinal cord ischemia/reperfusion injury in rabbits. J Cardiothorac Surg 4:50, 2009.
22. Olby N: Current concepts in the management of acute spinal cord injury. J Vet Intern Med 13:399–407, 1999.
23. Braughler JM, Hall ED: Involvement of lipid peroxidation in CNS injury. J Neurotrauma 9(Suppl 1):S1–S7, 1992.
24. Lucas JH, Wheeler DG, Guan Z, et al: Effect of glutathione augmentation on lipid peroxidation after spinal cord injury. J Neurotrauma 19:763–775, 2002.
25. Slatter D, Jerran RM, Dewey CW: Acute thoracolumbar disc extrusion in dogs-Part 1. Compend Contin Educ Prac Vet 21:922, 1999.
26. Karihtala P, Soini Y: Reactive oxygen species and antioxidant mechanisms in human tissues and their relation to malignancies. APMIS 115:81–103, 2007.
27. Tan SV, Guiloff RJ: Hypothesis on the pathogenesis of vacuolar myelopathy, dementia, and peripheral neuropathy in AIDS. J Neurol Neurosurg Psychiatry 65:23–28, 1998.
28. Ischiropoulos H, Beckman JS: Oxidative stress and nitration in neurodegeneration: Cause, effect, and association. J Clin Invest 111:163–169, 2003.
29. Coates JR, March PA, Oglesbee M, et al: Clinical characterization of a familial degenerative myelopathy in Pembroke Welsh Corgi dogs. J Vet Intern Med 21:1323–1331, 2007.
30. March P, Coates J, Abyad R, et al: Degenerative Myelopathy in 18 Pembroke Welsh Corgi Dogs. Vet Pathol, 2009.
31. Fechner H, Johnston PE, Sharp NJ, et al: Molecular genetic and expression analysis of alpha-tocopherol transfer protein mRNA in German shepherd dogs with degenerative myelopathy. Berl Munch Tierarztl Wochenschr 116:31–36, 2003.
32. Johnston PEJ: Chronic degenerative radiculomyelopathy: a study of the pathology and pathogenesis. Ph.D.: University of Glasgow, 1998.
33. Hermans N, Cos P, Maes L, et al: Challenges and pitfalls in antioxidant research. Curr Med Chem 14:417–430, 2007.
34. Romero FJ, Monsalve E, Hermenegildo C, et al: Oxygen toxicity in the nervous tissue: comparison of the antioxidant defense of rat brain and sciatic nerve. Neurochem Res 16:157–161, 1991.
35. Ohta S, Iwashita Y, Takada H, et al: Neuroprotection and enhanced recovery with edaravone after acute spinal cord injury in rats. Spine (Phila Pa 1976) 30:1154–1158, 2005.
36. Tator CH: Review of treatment trials in human spinal cord injury: issues, difficulties, and recommendations. Neurosurgery 59:957–982; discussion 982–957, 2006.
37. Baltzer WI, McMichael MA, Hosgood GL, et al: Randomized, blinded, placebo-controlled clinical trial of N-acetylcysteine in dogs with spinal cord trauma from acute intervertebral disc disease. Spine (Phila Pa 1976) 33:1397–1402, 2008.

38. Bracken MB, Shepard MJ, Holford TR, et al: Administration of methylprednisolone for 24 or 48 hours or tirilazad mesylate for 48 hours in the treatment of acute spinal cord injury. Results of the Third National Acute Spinal Cord Injury Randomized Controlled Trial. National Acute Spinal Cord Injury Study. JAMA 277:1597–1604, 1997.
39. Bracken MB, Shepard MJ, Holford TR, et al: Methylprednisolone or tirilazad mesylate administration after acute spinal cord injury: 1-year follow up. Results of the third National Acute Spinal Cord Injury randomized controlled trial. J Neurosurg 89:699–706, 1998.
40. Bracken MB: Methylprednisolone and spinal cord injury. J Neurosurg 96:140–141; author reply 142, 2002.
41. Amar AP, Levy ML: Pathogenesis and pharmacological strategies for mitigating secondary damage in acute spinal cord injury. Neurosurgery 44:1027–1039; discussion 1039–1040, 1999.
42. Hanson SM, Bostwick DR, Twedt DC, et al: Clinical evaluation of cimetidine, sucralfate, and misoprostol for prevention of gastrointestinal tract bleeding in dogs undergoing spinal surgery. Am J Vet Res 58:1320–1323, 1997.
43. Blight AR, Zimber MP: Acute spinal cord injury: pharmacotherapy and drug development perspectives. Curr Opin Investig Drugs 2:801–808, 2001.
44. Bracken MB, Shepard MJ, Collins WF, Jr., et al: Methylprednisolone or naloxone treatment after acute spinal cord injury: 1-year follow-up data. Results of the second National Acute Spinal Cord Injury Study. J Neurosurg 76:23–31, 1992.
45. Bracken MB, Shepard MJ, Collins WF, et al: A randomized, controlled trial of methylprednisolone or naloxone in the treatment of acute spinal-cord injury. Results of the Second National Acute Spinal Cord Injury Study. N Engl J Med 322:1405–1411, 1990.
46. Levine JM, Levine GJ, Boozer L, et al: Adverse effects and outcome associated with dexamethasone administration in dogs with acute thoracolumbar intervertebral disk herniation: 161 cases (2000-2006). J Am Vet Med Assoc 232:411–417, 2008.
47. Meister A: Glutathione, ascorbate and cellular protection. Cancer Res 54:1969s–1975s, 1994.
48. Sochman J: N-acetylcysteine in acute cardiology: 10 years later: what do we know and what would we like to know?! J Am Coll Cardiol 39:1422–1428, 2002.
49. Sevillano S, de Dios I, de la Mano AM, et al: N-acetylcysteine induces beneficial changes in the acinar cell cycle progression in the course of acute pancreatitis. Cell Prolif 36:279–289, 2003.
50. Tepel M, van der Giet M, Statz M, et al: The antioxidant acetylcysteine reduces cardiovascular events in patients with end-stage renal failure: a randomized, controlled trial. Circulation 107:992–995, 2003.
51. Birck R, Krzossok S, Markowetz F, et al: Acetylcysteine for prevention of contrast nephropathy: meta-analysis. Lancet 362:598–603, 2003.
52. Arakawa M, Ishimura A, Arai Y, et al: N-Acetylcysteine and ebselen but not nifedipine protected cerebellar granule neurons against 4-hydroxynonenal-induced neuronal death. Neurosci Res 57:220–229, 2007.
53. Sury MD, Frese-Schaper M, Muhlemann MK, et al: Evidence that N-acetylcysteine inhibits TNF-alpha-induced cerebrovascular endothelin-1 upregulation via inhibition of mitogen- and stress-activated protein kinase. Free Radic Biol Med 41:1372–1383, 2006.
54. Boga M, Discigil B, Ozkisacik EA, et al: The combined effect of iloprost and N-acetylcysteine in preventing spinal cord ischemia in rabbits. Eur J Vasc Endovasc Surg 31:366–372, 2006.
55. Hicdonmez T, Kanter M, Tiryaki M, et al: Neuroprotective effects of N-acetylcysteine on experimental closed head trauma in rats. Neurochem Res 31:473–481, 2006.
56. Thomale UW, Griebenow M, Kroppenstedt SN, et al: The effect of N-acetylcysteine on posttraumatic changes after controlled cortical impact in rats. Intensive Care Med 32: 149–155, 2006.
57. Springer JE, Rao RR, Lim HR, et al: The functional and neuroprotective actions of Neu2000, a dual acting pharmacological agent, in the treatment of acute spinal cord injury. J Neurotrauma 22:1–40, 2009.
58. Baines CP, Kaiser RA, Purcell NH, et al: Loss of cyclophilin D reveals a critical role for mitochondrial permeability transition in cell death. Nature 434:658–662, 2005.

59. Basso E, Fante L, Fowlkes J, et al: Properties of the permeability transition pore in mitochondria devoid of Cyclophilin D. J Biol Chem 280:18558–18561, 2005.
60. Schinzel AC, Takeuchi O, Huang Z, et al: Cyclophilin D is a component of mitochondrial permeability transition and mediates neuronal cell death after focal cerebral ischemia. Proc Natl Acad Sci U S A 102:12005–12010, 2005.
61. Nakagawa T, Shimizu S, Watanabe T, et al: Cyclophilin D-dependent mitochondrial permeability transition regulates some necrotic but not apoptotic cell death. Nature 434:652–658, 2005.
62. Brustovetsky N, Brustovetsky T, Purl KJ, et al: Increased susceptibility of striatal mitochondria to calcium-induced permeability transition. J Neurosci 23:4858–4867, 2003.
63. Cande C, Cohen I, Daugas E, et al: Apoptosis-inducing factor (AIF): a novel caspase-independent death effector released from mitochondria. Biochimie 84:215–222, 2002.
64. Chai J, Du C, Wu JW, et al: Structural and biochemical basis of apoptotic activation by Smac/DIABLO. Nature 406:855–862, 2000.
65. Petronilli V, Penzo D, Scorrano L, et al: The mitochondrial permeability transition, release of cytochrome c and cell death. Correlation with the duration of pore openings in situ. J Biol Chem 276:12030–12034, 2001.
66. Hamann K, Durkes A, Ouyang H, et al: Critical role of acrolein in secondary injury following ex vivo spinal cord trauma. J Neurochem 107:712–721, 2008.
67. Hamann K, Nehrt G, Ouyang H, et al: Hydralazine inhibits compression and acrolein-mediated injuries in ex vivo spinal cord. J Neurochem 104:708–718, 2008.

Oxidative Stress, Cognitive Dysfunction, and Brain Aging*

Elizabeth Head and Steven C. Zicker

Abstract Oxidative damage is a consistent feature of brain aging in all species studied. Decline in cognitive functions, which accompanies aging, may have a biological basis, and many of the disorders associated with aging may be preventable through dietary modifications that incorporate specific nutrient. Based on previous research and results of both laboratory and clinical studies in the canine model of human aging and disease, antioxidants may be one class of nutrient that may be beneficial. Brains of aged dogs accumulate oxidative damage to proteins and lipids, and mitochondrial dysfunction that may lead to impaired neuronal function. The production of free radicals and lack of increase in compensatory antioxidant enzymes may lead to increased damage to macromolecules within neurons. Reducing oxidative damage and mitochondrial dysfunction through a diet rich in antioxidants and mitochondrial cofactors significantly improves, or slows the decline of, learning and memory in aged dogs. Furthermore, there are clear links between the reduction of brain oxidative damage and mitochondrial impairments and improved or maintained cognitive function. However, determining which compounds, which combinations and dosage range, when to initiate intervention, and long-term effects constitute critical gaps in knowledge.

Keywords Cognitive dysfunction syndrome • Aging • Antioxidants • Carotenoids • Oxidative stress • Vitamins E and C

* This work was supported by Grant No. AG12694 from the National Institutes of Aging, U.S. Department of the Army, Contract No. DAMD17-98-1-8622. The content of the information does not necessarily reflect the position or the policy of the government, and no official endorsement should be inferred. Hill's Pet Nutrition, Topeka, KS, provided additional funding.

E. Head (✉)
Sanders-Brown Center on Aging, University of Kentucky, 800 S. Limestone St, Lexington, KY 40536, USA
e-mail: Elizabeth.head@uky.edu

L. Mandelker and P. Vajdovich (eds.), *Studies on Veterinary Medicine*, Oxidative Stress in Applied Basic Research and Clinical Practice 5,
DOI 10.1007/978-1-61779-071-3_7,

Introduction

Aging, in most species studied to date, is accompanied by the progressive accumulation of oxidative damage in many tissues [1]. The brain, in particular, is vulnerable to oxidative damage as it consumes approximately 20% of the body's total oxygen, has a high content of polyunsaturated fatty acids, and lower levels of endogenous antioxidant activity relative to other tissue [66]. Normal metabolic processes result in the release of reactive oxygen species (ROS), which in turn can lead to oxidative damage to proteins, lipids, DNA, and RNA [59]. ROS are produced primarily from mitochondria [56], intracellular organelles that are themselves vulnerable to oxidative damage [33]. Thus, mitochondrial dysfunction and production of ROS may be key contributors to the deleterious effects of aging on the brain.

Studies in both normal and pathological human brain aging provide correlative evidence in support of a role of oxidative damage in age-associated cognitive losses. Studies of human autopsy tissue show higher content of oxidative damage to DNA/RNA [49], proteins [71], and lipids [40] in aged as compared to young brain. Oxidative damage may also play a role in age-associated neurodegenerative diseases such as Alzheimer disease (AD) [39, 61, 73]. AD is associated with further increases in protein [9], lipid [40, 49, 60], DNA [36], and RNA/DNA oxidative damage [37, 53]. In addition, endogenous antioxidant activity in the AD brain is reduced relative to age-matched controls [69]. Proteins particularly vulnerable to oxidative damage have been identified by proteomics and a subset of these proteins may be directly or indirectly involved in the production and accumulation of AD neuropathology [8].

Mitochondrial function also appears compromised with age and disease in the human brain [1]. In normal aging, mitochondrial respiratory chain activity declines [54], mitochondrial metabolism-associated enzymes such as aconitase decrease [79], and the rate of somatic mitochondrial DNA mutations increases [43, 78]. In AD, similar types of losses in respiratory chain activity [6, 19, 26] and increases in mitochondrial DNA mutations [13] are observed but are higher when compared to age-matched controls. Furthermore, in AD, decreased cytochrome oxidase activity in the posterior cingulate cortex is correlated with hypometabolism seen by positron emission studies [77]. A gene array study in the cingulate cortex shows that energy-metabolism related genes, and specifically a 65% reduction in expression of mitochondrial electron transport chain genes, occurs in AD [32]. Thus, mitochondrial dysfunction and the production of ROS, combined with lower endogenous antioxidant activity may lead to increasing oxidative damage to molecules critically important to neuronal function.

Antioxidants, Human Brain Aging, and Disease

Based on correlative human neuropathology studies, antioxidants would be predicted to be associated with healthy aging, might serve to reduce the risk of developing AD, and may improve cognitive function in AD patients. Studies in humans have shown either a positive effect of antioxidant use on cognition and risk reduction for developing AD [17, 42, 51], or no significant effects [18, 30, 38, 41]. There have been few systematic and controlled clinical trials evaluating the effects of antioxidants on cognition in aged individuals or patients with AD. Vitamin E delays institutionalization in AD patients, suggesting some beneficial effects [65]. However, vitamin E alone did not improve cognition in patients with mild cognitive impairment, which is thought to be a precursor to AD [57]. Furthermore, in nondemented elderly women, vitamin E treatment was associated with minimal improvements in cognition [30].

In addition to studying the effects of cellular antioxidants on cognition and risk of AD, there are several studies of the effects of targeted cofactors that improve mitochondrial function, including acetylcarnitine (ALCAR) and lipoic acid (LA) [34]. ALCAR and LA may improve mitochondrial function and reduce the production of ROS, thus also reducing oxidative damage to proteins, lipids, and DNA/RNA [35]. In studies where ALCAR was administered to patients with moderate to severe AD, cognition either improved and/or slower deterioration was observed [5, 58, 62, 70]. In early-onset AD patients (less than 65 years of age), only small cognitive improvements were noted [75] but in other studies of younger patients with AD (less than 61 years) there was evidence for slowed disease progression [7, 76]. When the results of all these studies were combined in a meta-analysis, there were clear benefits of ALCAR administration in patients with AD, particularly with respect to slowing cognitive decline [48]. Moreover, combining ALCAR with acetylcholinesterase therapy in AD may provide additional benefits [3]. Similar evidence of maintenance of function was observed in a study of nine patients with AD or related dementias receiving 600 mg/day of LA for an average of 337 days [21]. In a follow-up study of 48 patients for a longer 48-month treatment period, maintenance of function similar to the smaller study was observed [20].

When taken together, however, studies of dietary or supplemental antioxidant intake in humans reveal variable results and appear far less robustly associated with positive functional outcomes than those reported in aging rodents [4, 28, 29]. Variability in outcomes of human antioxidant clinical trials may reflect inconsistencies in amounts of supplements administered, their form and source, compliance, as well as assessment and documentation of exact background of dietary intake of antioxidants. Interestingly, combinations of antioxidants may be superior to single compound supplementation [80] and dietary intake of antioxidants has been shown to be superior to supplements in human studies on cognition and risk of developing AD [2, 50]. In elderly women, supplementation with a combination of vitamins E and C can lead to improved memory [11]. Thus, single antioxidant administration (e.g., vitamin E alone) may prove to be more efficacious if administered in combination with other antioxidants (e.g., vitamin C, which helps to recycle vitamin E) and administered through diet rather than a tablet supplement. As described in later sections, the combination of antioxidants administered by way of a fortified food proved to be a potent intervention for improvement of cognition and reducing brain pathology when tested in a canine model of human brain aging.

Aging and Cognitive Dysfunction in Dogs

Advanced age in dogs is frequently associated with severe behavioral and cognitive deficits [52]. Age-dependent cognitive deficits in canines can be observed on many different measures of learning and memory. Deficits in complex learning tasks such as oddity discrimination learning [15, 47], size discrimination learning [22, 74], and spatial learning [12] occurs with age in dogs. Tasks sensitive to prefrontal cortex function, including reversal learning and visuospatial working memory, also deteriorate with age [22, 72]. In addition, egocentric spatial learning and reversal, measuring the ability of animals to select a correct object based on their own body orientation is age-sensitive [12]. Interestingly, on simple learning tasks and procedural learning measures, aged dogs performed equally as well as younger animals [45], suggesting that a subset of cognitive functions remains intact with age as it does in aging humans.

Memory also declines with age in dogs both for information about objects and location in space (spatial) [10, 24, 45]. Furthermore, studies of the time course of the development of cognitive decline demonstrate that deterioration in spatial ability occurs early in the aging process in canines, between 6 and 7 years of age [72] and provides researchers with guidelines for ages at which to start a treatment study. The neurobiological basis for cognitive decline in aging dogs may depend in part upon the progressive accumulation of oxidative damage to proteins, lipids, and DNA/RNA.

Oxidative Damage in Aging Dog Brain

In dog brain, the accumulation of carbonyl groups, which is a measure of oxidative damage to proteins, increases with age [23, 68] and is associated with reduced endogenous antioxidant enzyme activity or protein levels such as in glutamine synthetase and superoxide dismutase (SOD) [23, 27, 31, 55]. In several studies, a relation between age and increased oxidative damage has been inferred by measuring the amount of endproducts of lipid peroxidation (oxidative damage to lipids) including the extent of 4-hydroxynonenal (4HNE) [63, 64], or malondialdehyde [23]. Also, evidence of increased oxidative damage to DNA or RNA (8OHdG) in aged dog brain has been reported [14, 64]. If oxidative damage leads to progressive age-associated neuropathology and cognitive decline, then one could hypothesize that dietary antioxidants may prove beneficial.

Nutritional Antioxidants

A variety of antioxidant or antioxidant defense-associated molecules are derived from food sources. Vitamin E is found in high concentrations in nuts and oils, vitamin C is found in high concentrations in fruits, and beta-carotene is found in certain vegetables.

In addition trace minerals such as selenium, copper, zinc, and manganese, which are important to enzymes that specifically detoxify free radicals (Cu/Zn SOD) or help recycle antioxidants that detoxify free radicals (glutathione peroxidase), may be acquired from different food sources.

Recent research has shown that some molecules classified as mitochondrial cofactors (lipoic acid, l-carnitine) may enhance function of aged mitochondrion such that fewer ROS are produced during aerobic respiration. Chronic oxidative damage to enzymes and cell membranes may reduce the capability to bind mitochondrial enzyme cofactors thus reducing metabolic capacity [35]. Supplementation of foods with these mitochondrial cofactors increases the concentration within cells and restores binding to the enzymes that require them, which restores mitochondrial efficiency [35] and reduces oxidative damage to RNA [34].

Can Antioxidants Reduce Cognitive Impairments in Aged Dogs?

If brain aging in dogs is attributable to progressive accumulation of oxidative damage, which results in cognitive dysfunction, then reduction of oxidative damage via dietary fortification of antioxidants appears as a viable intervention option. A longitudinal investigation of the effects of dietary fortification of antioxidants on cognitive function of beagle dogs was thus completed, which included 48 aged (10–13 years of age) and 17 young beagles (3–5 years old). Each animal was assigned into one of two food groups using a counterbalanced design based on extensive baseline cognitive testing. No differences existed between cognitive ability of groups prior to dietary intervention.

An antioxidant-enriched food for maintenance of adult dogs was formulated to include a broad spectrum of antioxidants and two mitochondrial cofactors. The control and test foods had the following differences in formulation on an as-fed basis, respectively: dl-alpha-tocopherol acetate (120 vs. 1050 ppm), ascorbic acid as Stay-C (30 vs. 80 ppm), l-carnitine (20 vs. 260 ppm), and dl-alpha-lipoic acid (20 vs. 128 ppm). Based on an average weight of 10 kg per animal, the daily doses for each compound were 800 IU or 210 mg/day (21 mg/kg/day) of vitamin E, 16 mg/day (1.6 mg/kg/day) of vitamin C, 52 mg/day (5.2 mg/kg/day) of carnitine, and 26 mg/day (2.6 mg/kg/day) of lipoic acid. Fruits and vegetables were also incorporated at a 1-to-1 exchange ratio for corn, resulting in 1% inclusions of each of the following: spinach flakes, tomato pomace, grape pomace, carrot granules, and citrus pulp. This was equivalent to raising fruit and vegetable servings from 3 to 5–6/day. Serum vitamin E was increased ~75% by the antioxidant food in treated dogs [44].

Treatment with the antioxidant fortified food led to improvements in spatial attention (landmark task) learning as early as within 2 weeks of beginning the food intervention [44]. Subsequent testing of animals with a more difficult complex learning task, oddity discrimination, also revealed benefits of the fortified food [15].

Improved learning ability was maintained over time with the antioxidant treatment whereas untreated animals showed a progressive decline [46]. Interestingly, cognitive improvements were initially limited to aged animals as young dogs treated with the antioxidant fortified food were not different from control-fed dogs [67]. However, when initiated at a young age (2–4 years of age) it was found that prolonged administration of antioxidant fortified food resulted in significant improvement on a visual discrimination task compared to age-matched controls fed a nonfortified food over the same time period [16]. These results suggest that a relatively short administration period of a food fortified with antioxidants in aged dogs resulted in a slowing of the rate of decline in cognitive abilities, possibly attributable to either repair of age-associated oxidative damage or improved cellular function through improved mitochondrial function [25]. As a corollary to this finding the results in young dogs might suggest that early administration of an antioxidant fortified food acts in a way to delay onset of cognitive decline by either prevention of oxidative damage or continuance of optimal cellular function.

The improved cognitive outcomes were hypothesized to be attributable to enhanced mitochondrial function resulting in decreased oxidative damage in brain tissue from aged dogs administered antioxidant fortified food. Mitochondrial function was measured in aged dogs and revealed that antioxidant fortified food reduced age-associated mitochondrial dysfunction by reducing ROS production [25]. These results suggest that one mechanism by which the antioxidant fortified food improved cognition was by maintaining mitochondrial homeostasis by either the antioxidant fortification or the mitochondrial nutrient (l-carnitine, lipoic acid) targeted fortification. Oxidative damage to proteins from the parietal cortex was measured by derivatization with 2,4-dinitrophenylhydrazine (DNPH) which revealed reduced damage in dogs fed antioxidant fortified food [55]. These results suggest that oxidative damage to proteins was reduced in the parietal cortex, thought to be involved with landmark discrimination learning and other visual learning tasks that were improved in the cognitive assessments.

It is important to note that reduced oxidative damage to proteins and increased endogenous antioxidant activity were associated with improved cognition in the aging dogs on antioxidant fortified foods. Higher levels of one measure of protein oxidation (3-nitrotyrosine) and lower levels of antioxidant activity (glutathione-S-transferase) were subsequently correlated with higher error scores on a reversal learning problem and on a visuospatial task (i.e., impaired function) [55]. These results strongly suggest that oxidative damage, particularly to vulnerable proteins involved with energy metabolism, neuronal integrity, and antioxidant systems are key contributors to cognitive decline associated with aging in dogs. The most important aspect of this work is the discovery that cognitive performance may be improved relatively quickly in aged canines by dietary manipulation as well as slowing the onset of cognitive decline in younger dogs administered the fortified food at an early age for a prolonged period. Antioxidants may potentially act, therefore, to mitigate development of age-associated behavioral changes, and possibly even neuropathology, by counteracting oxidative stress.

Summary

Oxidative damage is a consistent feature of brain aging in all species studied. Decline in cognitive functions that accompanies aging may have a biological basis, and many of the disorders associated with aging may be preventable through dietary modifications that incorporate specific nutrients. Based on previous research and results of both laboratory and clinical studies in the canine model of human aging and disease, antioxidants may be one class of nutrient that may be beneficial. Brains of aged dogs accumulate oxidative damage to proteins and lipids, as well as mitochondrial dysfunction that may lead to dysfunction of neuronal cells. The production of free radicals and lack of increase in compensatory antioxidant enzymes may lead to detrimental modifications to important macromolecules within neurons. Reducing oxidative damage and mitochondrial dysfunction through food ingredients rich in a broad spectrum of antioxidants and mitochondrial cofactors significantly improves, or slows the decline of, learning and memory in aged dogs as well as delaying the onset of decline in younger dogs. Furthermore, there are clear links between the reduction of brain oxidative damage and mitochondrial impairments and improved or maintained cognitive function. However, determining which compounds, which combinations and dosage range, when to initiate intervention, and long-term effects constitute critical gaps in knowledge.

References

1. Ames, B.N., and Shigenaga, M.K. (1992). Oxidants are a major contributor to aging. Ann N Y Acad Sci 663:85–96.
2. Barberger-Gateau, P., Raffaitin, C., Letenneur, L., Berr, C., Tzourio, C., Dartigues, J.F., and Alperovitch, A. (2007). Dietary patterns and risk of dementia: the Three-City cohort study. Neurology 69:1921–1930.
3. Bianchetti, A., Rozzini, R., and Trabucchi, M. (2003). Effects of acetyl-L-carnitine in Alzheimer's disease patients unresponsive to acetylcholinesterase inhibitors. Curr Med Res Opin 19:350–353.
4. Bickford, P.C., Gould, T., Briederick, L., Chadman, K., Pollock, A., Young, D., Shukitt-Hale, B., and Joseph, J. (2000). Antioxidant-rich diets improve cerebellar physiology and motor learning in aged rats. Brain Res 866:211–217.
5. Bonavita, E. (1986). Study of the efficacy and tolerability of L-acetylcarnitine therapy in the senile brain. Int J Clin Pharm Ther Toxicol 24:511–516.
6. Bosetti, F., Brizzi, F., Barogi, S., Mancuso, M., Siciliano, G., Tendi, E.A., Murri, L., Rapoport, S.I., and Solaini, G. (2002). Cytochrome c oxidase and mitochondrial F1F0-ATPase (ATP synthase) activities in platelets and brain from patients with Alzheimer's disease. Neurobiol Aging 23:371–376.
7. Brooks, J.O., 3rd, Yesavage, J.A., Carta, A., and Bravi, D. (1998). Acetyl L-carnitine slows decline in younger patients with Alzheimer's disease: a reanalysis of a double-blind, placebo-controlled study using the trilinear approach. Int Psychogeriatr 10:193–203.
8. Butterfield, D.A. (2004). Proteomics: a new approach to investigate oxidative stress in Alzheimer's disease brain. Brain Res 1000:1–7.

9. Butterfield, D.A., and Kanski, J. (2001). Brain protein oxidation in age-related neurodegenerative disorders that are associated with aggregated proteins. Mech Ageing Dev 122:945–962.
10. Chan, A.D., Nippak, P.M., Murphey, H., Ikeda-Douglas, C.J., Muggenburg, B., Head, E., Cotman, C.W., and Milgram, N.W. (2002). Visuospatial impairments in aged canines (*Canis familiaris*): the role of cognitive-behavioral flexibility. Behav Neurosci 116:443–454.
11. Chandra, R.K. (2001). Effect of vitamin and trace-element supplementation on cognitive function in elderly subjects. Nutrition 17:709–712.
12. Christie, L.A., Studzinski, C.M., Araujo, J.A., Leung, C.S., Ikeda-Douglas, C.J., Head, E., Cotman, C.W., and Milgram, N.W. (2005). A comparison of egocentric and allocentric age-dependent spatial learning in the beagle dog. Prog Neuropsychopharmacol Biol Psychiatry 29:361–369.
13. Coskun, P.E., Beal, M.F., and Wallace, D.C. (2004). Alzheimer's brains harbor somatic mtDNA control-region mutations that suppress mitochondrial transcription and replication. Proc Natl Acad Sci U S A 101:10726–10731.
14. Cotman, C.W., and Head, E. (2008). The canine (dog) model of human aging and disease: dietary, environmental and immunotherapy approaches. J Alzheimers Dis 15:685–707.
15. Cotman, C.W., Head, E., Muggenburg, B.A., Zicker, S., and Milgram, N.W. (2002). Brain aging in the canine: a diet enriched in antioxidants reduces cognitive dysfunction. Neurobiol Aging 23:809–818.
16. de Rivera, C., Boutet, I., Zicker, S.C., and Milgram, N.W. (2005). A novel method for assessing contrast sensitivity in the beagle dog is sensitive to age and an antioxidant enriched food. Prog Neuropsychopharmacol Biol Psychiatry 29:379–387.
17. Engelhart, M.J., Geerlings, M.I., Ruitenberg, A., van Swieten, J.C., Hofman, A., Witteman, J.C., and Breteler, M.M. (2002). Dietary intake of antioxidants and risk of Alzheimer disease. JAMA 287:3223–3229.
18. Fillenbaum, G.G., Kuchibhatla, M.N., Hanlon, J.T., Artz, M.B., Pieper, C.F., Schmader, K.E., Dysken, M.W., and Gray, S.L. (2005). Dementia and Alzheimer's disease in community-dwelling elders taking vitamin C and/or vitamin E. Ann Pharmacother 39:2009–2014.
19. Gibson, G.E., Sheu, K.F., and Blass, J.P. (1998). Abnormalities of mitochondrial enzymes in Alzheimer disease. J Neural Transm 105:855–870.
20. Hager, K., Kenklies, M., McAfoose, J., Engel, J., and Munch, G. (2007). Alpha-lipoic acid as a new treatment option for Alzheimer's disease – a 48 months follow-up analysis. J Neural Transm 72:189–193.
21. Hager, K., Marahrens, A., Kenklies, M., Riederer, P., and Munch, G. (2001). Alpha-lipoic acid as a new treatment option for Azheimer type dementia. Arch Gerontol Geriatr 32:275–282.
22. Head, E., Callahan, H., Muggenburg, B.A., Cotman, C.W., and Milgram, N.W. (1998). Visual-discrimination learning ability and beta-amyloid accumulation in the dog. Neurobiol Aging 19:415–425.
23. Head, E., Liu, J., Hagen, T.M., Muggenburg, B.A., Milgram, N.W., Ames, B.N., and Cotman, C.W. (2002). Oxidative damage increases with age in a canine model of human brain aging. J Neurochem 82:375–381.
24. Head, E., Mehta, R., Hartley, J., Kameka, A.M., Cummings, B.J., Cotman, C.W., Ruehl, W.W., and Milgram, N.W. (1995). Spatial learning and memory as a function of age in the dog. Behav Neurosci 109:851–858.
25. Head, E., Nukala, V.N., Fenoglio, K.A., Muggenburg, B.A., Cotman, C.W., and Sullivan, P.G. (2009). Effects of age, dietary, and behavioral enrichment on brain mitochondria in a canine model of human aging. Exp Neurol 220:171–176.
26. Hirai, K., Aliev, G., Nunomura, A., Fujioka, H., Russell, R.L., Atwood, C.S., Johnson, A.B., Kress, Y., Vinters, H.V., Tabaton, M., et al. (2001). Mitochondrial abnormalities in Alzheimer's disease. J Neurosci 21:3017–3023.
27. Hwang, I.K., Yoon, Y.S., Yoo, K.Y., Li, H., Choi, J.H., Kim, D.W., Yi, S.S., Seong, J.K., Lee, I.S., and Won, M.H. (2008). Differences in lipid peroxidation and Cu,Zn-superoxide dismutase in the hippocampal CA1 region between adult and aged dogs. J Vet Med Sci 70:273–277.

28. Joseph, J.A., Denisova, N.A., Bielinski, D., Fisher, D.R., and Shukitt-Hale, B. (2000). Oxidative stress protection and vulnerability in aging: putative nutritional implications for intervention. Mech Ageing Dev 116:141–153.
29. Joseph, J.A., Denisova, N., Villalobos-Molina, R., Erat, S., and Strain, J. (1996). Oxidative stress and age-related neuronal deficits. Mol Chem Neuropathol 28:35–40.
30. Kang, J.H., Cook, N., Manson, J., Buring, J.E., and Grodstein, F. (2006). A randomized trial of vitamin E supplementation and cognitive function in women. Arch Intern Med 166:2462–2468.
31. Kiatipattanasakul, W., Nakamura, S., Kuroki, K., Nakayama, H., and Doi, K. (1997). Immunohistochemical detection of anti-oxidative stress enzymes in the dog brain. Neuropathology 17:307–312.
32. Liang, W.S., Reiman, E.M., Valla, J., Dunckley, T., Beach, T.G., Grover, A., Niedzielko, T.L., Schneider, L.E., Mastroeni, D., Caselli, R., et al. (2008). Alzheimer's disease is associated with reduced expression of energy metabolism genes in posterior cingulate neurons. Proc Natl Acad Sci U S A 105:4441–4446.
33. Liu, J., Atamna, H., Kuratsune, H., and Ames, B.N. (2002). Delaying brain mitochondrial decay and aging with mitochondrial antioxidants and metabolites. Ann N Y Acad Sci 959:133–166.
34. Liu, J., Head, E., Gharib, A.M., Yuan, W., Ingersoll, R.T., Hagen, T.M., Cotman, C.W., and Ames, B.N. (2002). Memory loss in old rats is associated with brain mitochondrial decay and RNA/DNA oxidation: Partial reversal by feeding acetyl-L-carnitine and/or R-alpha-lipoic acid. Proc Natl Acad Sci U S A 99:2356–2361.
35. Liu, J., Killilea, D.W., and Ames, B.N. (2002). Age-associated mitochondrial oxidative decay: improvement of carnitine acetyltransferase substrate-binding affinity and activity in brain by feeding old rats acetyl-L-carnitine and/or R-alpha-lipoic acid. Proc Natl Acad Sci U S A 99:1876–1881.
36. Lovell, M.A., and Markesbery, W.R. (2007). Oxidative DNA damage in mild cognitive impairment and late-stage Alzheimer's disease. Nucleic Acids Res 35:7497–7504.
37. Lovell, M.A., and Markesbery, W.R. (2008). Oxidatively modified RNA in mild cognitive impairment. Neurobiol Dis 29:169–175.
38. Luchsinger, J.A., Tang, M.X., Shea, S., and Mayeux, R. (2003). Antioxidant vitamin intake and risk of Alzheimer disease. Arch Neurol 60:203–208.
39. Markesbery, W.R. (1997). Oxidative stress hypothesis in Alzheimer's disease. Free Radic Biol Med 23:134–147.
40. Markesbery, W.R., and Lovell, M.A. (1998). Four-hydroxynonenal, a product of lipid peroxidation, is increased in the brain in Alzheimer's disease. Neurobiol Aging 19:33–36.
41. Masaki, K.H., Losonczy, K.G., Izmirlian, G., Foley, D.J., Ross, G.W., Petrovitch, H., Havlik, R., and White, L.R. (2000). Association of vitamin E and C supplement use with cogntive function and dementia in elderly men. Neurology 54:1265–1272.
42. Maxwell, C.J., Hicks, M.S., Hogan, D.B., Basran, J., and Ebly, E.M. (2005). Supplemental use of antioxidant vitamins and subsequent risk of cognitive decline and dementia. Dement Geriatr Cogn Disord 20:45–51.
43. Michikawa, Y., Mazzucchelli, F., Bresolin, N., Scarlato, G., and Attardi, G. (1999). Aging-dependent large accumulation of point mutations in the human mtDNA control region for replication. Science 286:774–779.
44. Milgram, N.W., Head, E., Muggenburg, B.A., Holowachuk, D., Murphey, H., Estrada, J., Ikeda-Douglas, C.J., Zicker, S.C., and Cotman, C.W. (2002). Landmark discrimination learning in the dog: effects of age, an antioxidant fortified diet, and cognitive strategy. Neurosci Biobehav Rev 26:679–695.
45. Milgram, N.W., Head, E., Weiner, E., and Thomas, E. (1994). Cognitive functions and aging in the dog: Acquisition of nonspatial visual tasks. Behav Neurosci 108:57–68.
46. Milgram, N.W., Head, E., Zicker, S.C., Ikeda-Douglas, C.J., Murphey, H., Muggenburg, B., Siwak, C., Tapp, D., and Cotman, C.W. (2005). Learning ability in aged beagle dogs is preserved by behavioral enrichment and dietary fortification: a two-year longitudinal study. Neurobiol Aging 26:77–90.

47. Milgram, N.W., Zicker, S.C., Head, E., Muggenburg, B.A., Murphey, H., Ikeda-Douglas, C., and Cotman, C.W. (2002). Dietary enrichment counteracts age-associated cognitive dysfunction in canines. Neurobiol Aging 23:737–745.
48. Montgomery, S.A., Thal, L.J., and Amrein, R. (2003). Meta-analysis of double blind randomized controlled clinical trials of acetyl-L-carnitine versus placebo in the treatment of mild cognitive impairment and mild Alzheimer's disease. Int Clin Psychopharmacol 18:61–71.
49. Montine, T.J., Neely, M.D., Quinn, J.F., Beal, M.F., Markesbery, W.R., Roberts, L.J., and Morrow, J.D. (2002). Lipid peroxidation in aging brain and Alzheimer's disease. Free Radic Biol Med 33:620–626.
50. Morris, M.C., Evans, D.A., Bienias, J.L., Tangney, C.C., Bennett, D.A., Aggarwal, N., Wilson, R.S., and Scherr, P.A. (2002). Dietary intake of antioxidant nutrients and the risk of incident Alzheimer disease in a biracial community study. JAMA 287:3230–3237.
51. Morris, M.C., Evans, D.A., Bienias, J.L., Tangney, C.C., and Wilson, R.S. (2002). Vitamin E and cognitive decline in older persons. Arch Neurol 59:1125–1132.
52. Neilson, J.C., Hart, B.L., Cliff, K.D., and Ruehl, W.W. (2001). Prevalence of behavioral changes associated with age-related cognitive impairment in dogs. J Am Vet Med Assoc 218:1787–1791.
53. Nunomura, A., Perry, G., Pappolla, M.A., Wade, R., Hirai, K., Chiba, S., and Smith, M.A. (1999). RNA oxidation is a prominent feature of vulnerable neurons in Alzheimer's disease. J Neurosci 19:1959–1964.
54. Ojaimi, J., Masters, C.L., Opeskin, K., McKelvie, P., and Byrne, E. (1999). Mitochondrial respiratory chain activity in the human brain as a function of age. Mech Ageing Dev 111:39–47.
55. Opii, W.O., Joshi, G., Head, E., Milgram, N.W., Muggenburg, B.A., Klein, J.B., Pierce, W.M., Cotman, C.W., and Butterfield, D.A. (2008). Proteomic identification of brain proteins in the canine model of human aging following a long-term treatment with antioxidants and a program of behavioral enrichment: relevance to Alzheimer's disease. Neurobiol Aging 29:51–70.
56. Perez-Campo, R., Lopez-Torres, M., Cadenas, S., Rojas, C., and Barja, G. (1998). The rate of free radical production as a determinant of the rate of aging: evidence from the comparative approach. J Comp Physiol [B] 168:149–158.
57. Petersen, R.C., Thomas, R.G., Grundman, M., Bennett, D., Doody, R., Ferris, S., Galasko, D., Jin, S., Kaye, J., Levey, A., et al. (2005). Vitamin E and donepezil for the treatment of mild cognitive impairment. N Engl J Med 352:2379–2388.
58. Pettegrew, J.W., Klunk, W.E., Panchalingam, K., Kanfer, J.N., and McClure, R.J. (1995). Clinical and neurochemical effects of acetyl-L-carnitine in Alzheimer's disease. Neurobiol Aging 16:1–4.
59. Poon, H.F., Calabrese, V., Scapagnini, G., and Butterfield, D.A. (2004). Free radicals and brain aging. Clin Geriatr Med 20:329–359.
60. Pratico, D., Clark, C.M., Lee, V.M., Trojanowski, J.Q., Rokach, J., and FitzGerald, G.A. (2000). Increased 8,12-iso-iPF2alpha-VI in Alzheimer's disease: correlation of a noninvasive index of lipid peroxidation with disease severity. Ann Neurol 48:809–812.
61. Pratico, D., and Delanty, N. (2000). Oxidative injury in diseases of the central nervous system: focus on Alzheimer's disease. Am J Med 109:577–585.
62. Rai, G., Wright, G., Scott, L., Beston, B., Rest, J., and Exton-Smith, A.N. (1990). Double-blind, placebo controlled study of acetyl-l-carnitine in patients with Alzheimer's dementia. Curr Med Res Opin 11:638–647.
63. Rofina, J.E., Singh, K., Skoumalova-Vesela, A., van Ederen, A.M., van Asten, A.J., Wilhelm, J., and Gruys, E. (2004). Histochemical accumulation of oxidative damage products is associated with Alzheimer-like pathology in the canine. Amyloid 11:90–100.
64. Rofina, J.E., van Ederen, A.M., Toussaint, M.J., Secreve, M., van der Spek, A., van der Meer, I., Van Eerdenburg, F.J., and Gruys, E. (2006). Cognitive disturbances in old dogs suffering from the canine counterpart of Alzheimer's disease. Brain Res 1069:216–226.
65. Sano, M., Ernesto, C., Thomas, R.G., Klauber, M.R., Schafer, K., Grundman, M., Woodbury, P., Growdon, J., Cotman, C.W., Pfeiffer, E., Schneider, L.S., and Thal, L.J. (1997). A controlled trial of selegiline, alpha-tocopherol, or both as treatment for Alzheimer's disease. N Engl J Med 336:1216–1222.

66. Shulman, R.G., Rothman, D.L., Behar, K.L., and Hyder, F. (2004). Energetic basis of brain activity: implications for neuroimaging. Trends Neurosci 27:489–495.
67. Siwak, C.T., Tapp, P.D., Head, E., Zicker, S.C., Murphey, H.L., Muggenburg, B.A., Ikeda-Douglas, C.J., Cotman, C.W., and Milgram, N.W. (2005). Chronic antioxidant and mitochondrial cofactor administration improves discrimination learning in aged but not young dogs. Prog Neuropsychopharmacol Biol Psychiatry 29:461–469.
68. Skoumalova, A., Rofina, J., Schwippelova, Z., Gruys, E., and Wilhelm, J. (2003). The role of free radicals in canine counterpart of senile dementia of the Alzheimer type. Exp Gerontol 38:711–719.
69. Smith, C.D., Carney, J.M., Starke-Reed, P.E., Oliver, C.N., Stadtman, E.R., Floyd, R.A., and Markesbery, W.R. (1991). Excess brain protein oxidation and enzyme dysfunction in normal aging and in Alzheimer disease. Proc Natl Acad Sci U S A 88:10540–10543.
70. Spagnoli, A., Lucca, U., Menasce, G., Bandera, L., Cizza, G., Forloni, G., Tettamanti, M., Frattura, L., Tiraboschi, P., Comelli, M., et al. (1991). Long-term acetyl-L-carnitine treatment in Alzheimer's disease. Neurology 41:1726–1732.
71. Stadtman, E.R. (1992). Protein oxidation and aging. Science 257:1220–1224.
72. Studzinski, C.M., Christie, L.A., Araujo, J.A., Burnham, W.M., Head, E., Cotman, C.W., and Milgram, N.W. (2006). Visuospatial function in the beagle dog: an early marker of cognitive decline in a model of human aging and dementia. Neurobiol Learn Mem 86:197–204.
73. Sultana, R., Perluigi, M., and Butterfield, D.A. (2009). Oxidatively modified proteins in Alzheimer's disease (AD), mild cognitive impairment and animal models of AD: role of Abeta in pathogenesis. Acta Neuropathol 118:131–150.
74. Tapp, P.D., Siwak, C.T., Estrada, J., Muggenburg, B.A., Head, E., Cotman, C.W., and Milgram, N.W. (2003). Size and reversal learning in the beagle dog as a measure of executive function and inhibitory control in aging. Learn Mem 10:64–73.
75. Thal, L.J., Calvani, M., Amato, A., and Carta, A. (2000). A 1-year controlled trial of acetyl-l-carnitine in early-onset AD. Neurology 55:805–810.
76. Thal, L.J., Carta, A., Clarke, W.R., Ferris, S.H., Friedland, R.P., Petersen, R.C., Pettegrew, J.W., Pfeiffer, E., Raskind, M.A., Sano, M., et al. (1996). A 1-year multicenter placebo-controlled study of acetyl-L-carnitine in patients with Alzheimer's disease. Neurology 47:705–711.
77. Valla, J., Berndt, J.D., and Gonzalez-Lima, F. (2001). Energy hypometabolism in posterior cingulate cortex of Alzheimer's patients: superficial laminar cytochrome oxidase associated with disease duration. J Neurosci 21:4923–4930.
78. Wallace, D.C. (1992). Mitochondrial genetics: a paradigm for aging and degenerative diseases? Science 256:628–632.
79. Yan, L.-J., Levine, R.L., and Sohal, R.S. (1997). Oxidative damage during aging targets mitochondrial aconitase. Proc Natl Acad Sci U S A 94:11168–11172.
80. Zandi, P.P., Anthony, J.C., Khachaturian, A.S., Stone, S.V., Gustafson, D., Tschanz, J.T., Norton, C., Welsh-Bohmer, K.A., Breitner, J.C., and Cache County Study Group (2004). Reduced risk of Alzheimer disease in users of antioxidant vitamin supplements: the Cache County Study. Arch Neurol 61:82–88.

The Role of Oxidative Stress in Ocular Disease

Gustavo L. Zapata

Abstract The eye is an organ particularly sensitive to oxidative stress because it is formed by different highly susceptible tissues that play different related roles aimed at preserving the visual function. This sensitivity due to the variety of tissues leads to a large number of extraocular and ocular mechanisms that produce an oxidative imbalance affecting the ocular apparatus. A change in either one or several of these tissues can produce an alteration in the visual function. In addition, the eyes are under a particular risk of oxidative stress due to their high exposure to different extraocular agents, such as oxygen (Cejková et al., Histol Histopatol 23:1477–1483, 2008), the high content of polyunsaturated fatty acids in the retina (Stone et al., Exp Eye Res 28:387–397, 1979), their high exposure to light and environmental contaminants (Novaes et al., Environ Health Perspect, 115:1753–1756, 2007), and their exposure to ultraviolet radiation (Young, Prog Biophys Mol Biol 92:80–85, 2006). Among the ocular processes that generate oxidative imbalance, we can mention inflammatory diseases on the ocular surface, the uveal tract, and the retina, the formation of cataracts in the lens, degenerative diseases in the retina, and glaucoma. The aim of the present work is to discuss the role of oxidative stress in the pathophysiology of diverse ocular disorders that affect the ocular surface, the uveal tract, the lens, and the nervous tissue.

Keywords Oxidative stress • Eye • Uveitis • Glaucoma • Cataract • Retinal pigment epithelium • Nitric oxide

---

G.L. Zapata (✉)
Servicio Central de Laboratorio, Facultad de Ciencias Veterinarias, Universidad Nacional de La Plata, La Plata, Buenos Aires, Argentina
e-mail: zapatagu@fcv.unlp.edu.ar

L. Mandelker and P. Vajdovich (eds.), *Studies on Veterinary Medicine*, Oxidative Stress in Applied Basic Research and Clinical Practice 5,
DOI 10.1007/978-1-61779-071-3_8,

Ocular Surface

The tear is composed mainly of water, and, to a lesser extent, of lipids and proteins. It has different functions: it lubricates the ocular surface, provides the cornea with oxygen and nutrients for its metabolism, keeps the substance related to the immune defense in suspension, and regulates oxidative stress [1]. This last function is carried out by scavengers such as cysteine, ascorbic acid, GSH (reduced glutathione), uric acid, tyrosine, catalase, and peroxidase, which are normally present in the tear [2–4].

Lactoferrin is known to be a tear-specific protein secreted by the acinar epithelial cells of the main and accessory lacrimal glands. Several authors have reported different activities of this protein. Shimmura et al. showed its antioxidant capacity by inhibiting the formation of some free radicals (hydroxils) [5], demonstrated the antimicrobial capacity of the tear [6], Arnold et al. showed that some of the antimicrobial activity of the tears was the result of the activity of lactoferrin [7]. Malmquist et al. suggested its anti-inflammatory capacity [8] and Fushijara et al. demonstrated the protective effect against ultra violet (UV) rays [9].

A frequent disease involving the ocular surface (conjunctiva, cornea, and tear) in veterinary ophthalmology is known to be the keratoconjunctivitis sicca. This disease is characterized by both a qualitative and quantitative deficit of tear production which triggers a series of events that result in the alteration of the physiology of the ocular surface. Among the etiological agents that cause a decrease in the production of tears (quantitative deficit) we can mention the canine distemper virus (which is the cause of glandular adenitis), chronic blepharoconjunctivitis, congenital acinar hypoplasia, drug toxicity, and trauma. In addition, keratoconjunctivitis sicca can be induced by drugs (i.e. sulfa drugs) or be mediated by the immune system [10].

On the other hand, among the alterations that cause qualitative deficit we can mention meibomian gland dysfunction, chronic blepharoconjunctivitis, and vitamin A deficiency [11]. All these factors favor both the generation of reactive oxygen species (ROS) and the presence of various types of free radicals in the tear; that is, they favor oxidative stress. These, in turn, destroy proteins, mucous components, and lipids contained in the tear leading to the ultimately loss of its stability [12]. If the antioxidative mechanisms normally present in the precorneal tear are inadequate, the tear becomes unstable as its components are destroyed. Different processes, such as inflammation of the cornea and the conjunctiva, corneal neovascularization, and metaplasia of the epithelium, which are associated with the loss of corneal transparency and ocular secretion, are triggered at this point (Fig. 1).

The cornea is a highly differentiated tissue that allows the refraction and transmission of light. It is basically a concave–convex lens with an anterior face in close contact with the precorneal tear and a posterior face covered with aqueous humor. Because this liquid is mainly responsible for maintaining the physiological requirements of the cornea, the cornea lacks vascularization. It is composed of a stratified squamous nonkeratinized epithelium, a stroma of connective tissue, and an endothelial cell monolayer. Although this nonvascular tissue is apparently simple in its composition, the great regularity and uniformity of its structure allows the

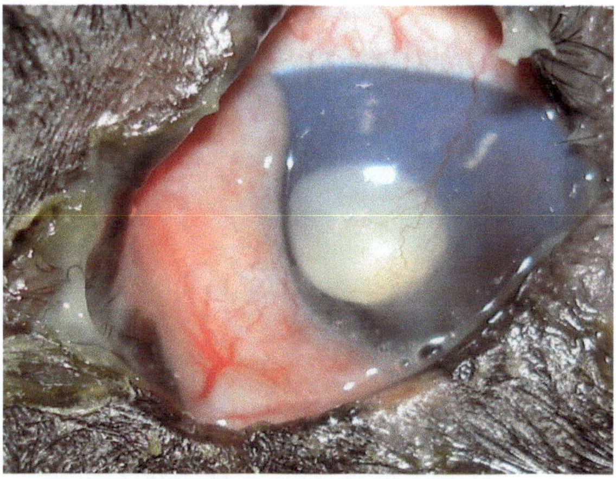

Fig. 1 Left eye of the dog with chronic quantitative tear deficiency. Note hyperemia and thick discharges, vascularized and opaque cornea (chronic keratitis)

accurate transmission and refraction of light [13]. It possesses effective low molecular weight antioxidants (such as ascorbic acid, glutathione, and alpha tocopherol) [14–18] and high molecular weight antioxidant enzymes (such as superoxide dismutase, glutathione peroxidase and reductase, and catalase) [19–21]. Works carried out by Kovaceva et al. have shown differences between the concentration of antioxidants in the corneal epithelium of animals with daily activity and that of animals with nocturnal behavior; these differences are considered due to the exposure of diurnal animals to UV rays [22].

Due to its location, the cornea is constantly exposed to the effects of radiation, atmospheric oxygen, environmental chemicals or pollutants, and physical factors. These external agents can cause an oxidative imbalance in the cornea. Corneal inflammatory processes (keratitis), which have a high diagnostic frequency (Fig. 2), can also produce this alteration.

Cells in the epithelium and corneal stroma are capable of producing superoxide via NADPH oxidase [23]. The superoxide thus produced potentially contributes to the processes that occur during corneal inflammation. This process involves neutrophil infiltration, which may obviate ROS production by the membrane-bound NADPH oxidase system and cytoplasmic granule myeloperoxidases [12, 24]. As suggested by Chandler et al. this process can be increased by exposure to UV, which causes overexpression of matrix metalloproteinases that are observed with changes in the cornea that allow an influx of inflammatory cells and vascularization [25].

Oxidative stress can alter the natural equilibrium between proteinases and proteinase inhibitors in favor of the former, engendering pathological degradation of collagen and proteoglycans within the corneal stromal extracellular matrix [26].

Independently of the cause, the deficit of antioxidant agents in the ocular surface, both in tears as in the cornea, is extremely important because the ocular surface is

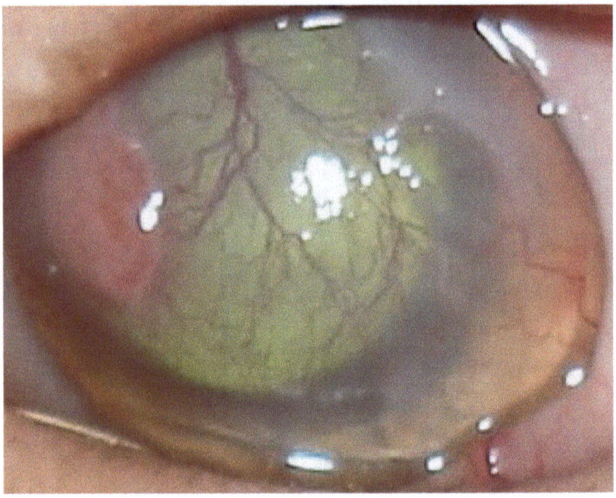

Fig. 2 Keratitis affected left eye. The superficial cornea is both infiltrated and vascularized

the mucosal surface most environmentally exposed to the effects of external agents [27–29] which exacerbate the production of ROS and free radicals, thus increasing the oxidative injury [24].

Therefore, it would be interesting to find the association between the treatment of the primary cause of the disease and antioxidant agents in some lesions of the ocular surface, in order to minimize the alterations of the oxidative disbalance and thus maintain corneal transparency.

Uveitis

The inflammation of the uveal tract (iris, ciliary body, and choroid), known as uveitis, is the most frequently diagnosed disease in this tissue, in both dogs and cats. The process can affect different structures that form the uveal tract or the uvea as a whole. In veterinary medicine, there are numerous causes that can trigger uveitis, ranging from local injuries to systemic diseases (Table 1) [30]. The alteration in visual function in patients with uveitis is caused by different factors, one of which is the damage caused by infiltration of inflammatory cells. The disruption of the blood barrier resulting in protein extravasation and cellular infiltration into the aqueous humor and anterior chamber is clinically called Tynall Effect (scattering of particles in acolloid medium), and if the process allows an excessive increase in cells deposited in the anterior chamber it is called hypopyon, and changes in the metabolism of corneal endothelium that are observed with loss of corneal transparency as a result of edema (Fig. 3) [31]. The inflammatory cells accumulated in the uveal tract, mainly mononuclear and polymorphonuclear phagocytes (PMNs), generate ROS. All these processes are related to an increase in cytokines, metabolites of arachidonic acid (eicosatetraenoic acid), nitric oxide (NO), including radicals,

Table 1 Uveitis triggers

Exogenous	Endogenous
Trauma	Infections (viruses, fungi, and bacteria)
Corneal ulcers	Parasites
Deep wounds	Cataracts
	Neoplasia
	Immune mediation

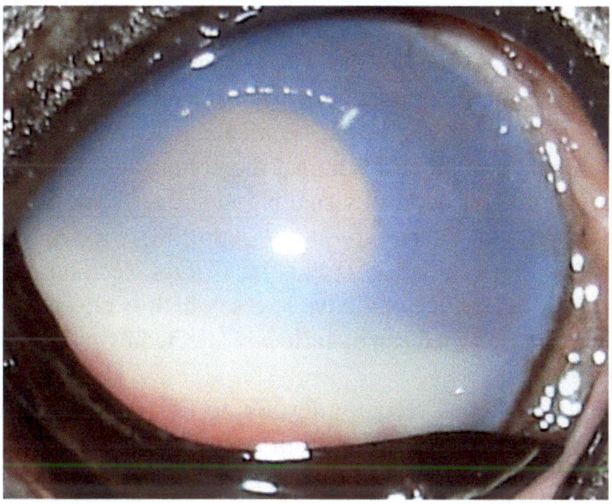

Fig. 3 Anterior acute uveitis in a dog with lymphosarcoma. Note severe corneal edema, hipopyon is limited to the lower anterior chamber and miosis

such as superoxide anion ($O_2 \cdot^-$) and hydroxyl radical (HO·) and nonradicals, such as hydrogen peroxide (H_2O_2) and singlet oxygen ($^1O_2$). Different studies showed that the cytokines, tumor necrosis factor alpha (TNFα), and several types of interleukins enhance the inflammatory response in both the aqueous humor and uveal tissue [32, 33]. ROS play a key role in the development of uveitis, because these can increase vascular permeability, damage to extracellular matrix proteins, enhance cytokine expression, activate endothelial cells, increase adhesion molecule expression, and increase the formation of chemotactic factors [34]. Superoxide and hydrogen peroxide are among the most important primary species generated by these phagocytes, This process, and other free radicals are formed and are associated with tissue damage. Although H_2O_2 is a strong oxidant, its reaction with organic compounds is low. In the presence of H_2O_2, the myeloperoxidase released by phagocytes catalyzes the formation of hypochlorous acid. This acid reacts with other metabolites derived from O_2^- to form HO·, these hydroxyl radicals have been implicated as the ultimate cytotoxic agent because of their high reactivity, including peroxidation of membrane lipids [35].

On the other hand, it has been shown that activated phagocytes synthesize considerable amounts of NO, which is the oxidation product of one of the guanidino-nitrogens of L-arginine. The reaction is catalyzed by two different forms of the enzyme NO synthase (NOS) [32]. Park et al. showed that the IL-1, TNF-ά, and IFN-γ are directly implicated in the production of NO [36]. The reaction between superoxide and nitric oxide proceeds with near diffusion-limited rates, yielding peroxynitrite ($ONOO^-$), and is the most potent biological oxidant obtained from NO and O_2 [33]. Physiologically, the uvea possesses a high metabolic aerobic activity with a consequent production of O_2^- and H_2O_2. However, diverse intracellular enzymatic systems prevent the accumulation of metabolic waste highly toxic for the eye.

The inflammatory process of the choroid (choroiditis) usually involves the retina, producing chorioretinitis, due to the close relationship between both tissues (Fig. 4). The retina is a structure particularly sensitive to oxidative damage because of its high consumption of O_2 and its high percentage of polyunsaturated fatty acids (PUFAs) [37] which are required by retinal membranes to build up a highly fluid microenvironment for the rhodopsin visual pigment that is responsible for the light signal transduction and its direct exposure to light [38]. Free radicals may trigger the initiation of lipid peroxidation (LPO) of PUFAs, leading to the release of lipoperoxyl radicals that induce oxidative stress, which is closely related to photoreceptor membrane lysis. Vision can be seriously affected by oxidative stress damage to the retinal membranes.

Special attention has been recently devoted to ROS as factors that initiate the retinal degeneration induced by inflammation. In fact, in experimental uveitis, all the products of LPO, as well as a high concentration of peroxynitrites ($ONOO^-$), have been identified [39].

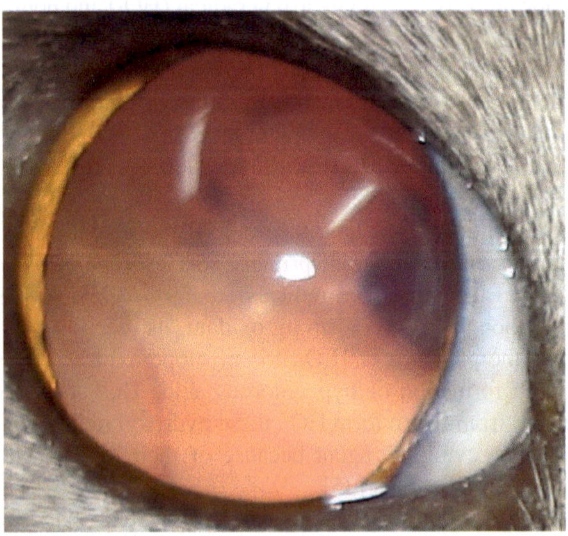

Fig. 4 Complete bullous retinal detachment in this domestic shorthair with advanced chorioretinitis

The aim of the therapy of the inflammatory processes of the uveal tract is based on preventing the development of the inflammation with steroid and nonsteroid anti-inflammatory drugs. A large number of research works have shown the use of antioxidants and their positive effects in the anti-inflammatory processes.

Glaucoma

The regulation of the intraocular pressure (IOP) depends on the balance of the complex mechanisms involved in the production and drainage of the aqueous humor. The aqueous humor is generated and evacuates constantly and in equilibrium in the structures of the anterior segment of the eye, for its own nutrition and the maintenance of the ocular tone [40]. Glaucoma is an optic neuropathy, characterized by the progressive degeneration of retinal ganglion cells (RGCs) associated with an elevated IOP and visual field damage, representing the final stage caused by a number of different conditions affecting the eye. In dogs, glaucoma has an estimated prevalence of 1.8% [41], whereas in cats it is often secondary to other ocular diseases [42]. In dogs, the most frequently diagnosed glaucoma is closed-angle glaucoma, which is characterized by a degeneration of the optic nerve head and a progressive deterioration of the visual function, in spite of the decrease observed in the IOP, which suggests the existence of other factors producing optic neuropathy (Fig. 5) [43]. Although it is considered that the increase in the IOP is the most important risk factor (neurodegeneration) of the glaucoma, there are several associated factors such as increases in the levels of glutamate [44], the metabolic alteration of nitric oxide [45], vascular factors (ischemia) [46], and oxidative damage caused by overproduction of ROS [47].

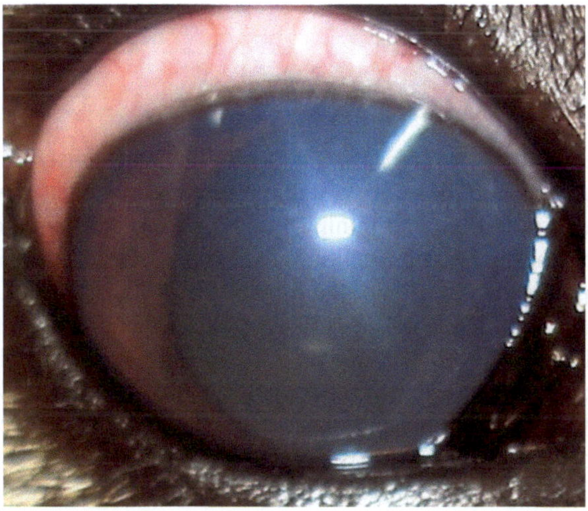

Fig. 5 Early state primary glaucoma. Both corneal edema and mydriasis are clinicals signs

Several works have shown that oxidative DNA damage in the cells that compose the trabecular meshwork is associated with an alteration in the structure of the extracellular matrix, which produces a decrease in the drainage of the aqueous humor and an increase in the IOP because of its constant production [48, 49]. Weinstein et al. have demonstrated an increase in active matrix metalloproteinases in the iridocorneal angle tissue from glaucomatous eyes when compared to normal eyes [50]. Alterations in the irrigation (ischemia) of the retina and the optic nerve increase the levels of glutamate in the vitreous, which in turn produces cellular degeneration of the retina caused by excitotoxic damage to the RGCs through activation of ionotropic and metabotropic glutamate receptors [51–53]. Activation of glutamate receptors is thought to initiate damage in the retina by a cascade of biochemical effects such as activation of neuronal nitric oxide synthase (nNOS) and increase in intracellular Ca^{2+}. Glutamate has been described as a major contributing factor to the loss of RGCs.

Several works have associated the increase in endothelin-1 with the etiology of glaucoma. An increase in this endogenous vasoconstrictor is related to ischemia, which can exacerbate the neurodegeneration by increasing the levels of glutamate [54]. Nitric oxide plays a significant role in the pathophysiology of glaucoma, because an increase in its synthesis has been associated with a decrease in the number of RGCs. The uncontrolled increase in NO may lead to significant oxidative stress, as NO couples with the superoxide anion radical ($O_2\cdot^-$) to generate the highly reactive and destructive peroxynitrite anion ($ONOO\cdot^-$). Indeed, there is certainly ample evidence documenting the toxicity of NO in neural tissue [45]. Gionfriddo et al. have shown that the intracellular antioxidant glutathione is decreased in the glaucomatous eye [55]. Finally, several authors have shown the processes by which oxidative stress plays its role in the pathogenesis of glaucoma and have allowed us to gain insights into the molecular mechanisms involved in the physiopathology of glaucoma and the later degeneration of the retina.

Cataract

The crystalline is a transparent, soft, intraocular biconvex lens. The fiber cells are filled with high concentrations of crystallines, the proteins that provide a uniform refractive index to the fiber and minimize light scatter. All these components are inside an elastic capsule, which conditions the molecular transport with the aqueous humor and supports the insertion of the zonular fibers [56]. In its equatorial region, the lens has a layer of epithelial cells that terminally differentiate into fiber cells. These cells elongate toward both the anterior and the posterior poles. The newly formed fiber cells develop over the older fibers and displace them toward the center of the tissue. Cataract is an opacity of the lens of the eye. There are many distinct types of cataract defined on the basis of the section of the lens that has become opaque (Fig. 6) [57]. This alteration affects mainly canines, especially certain breeds [58], and is one of the main causes of blindness in this species. Park et al.

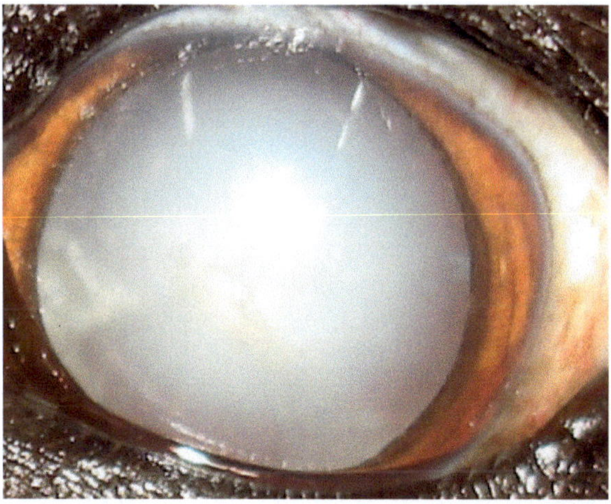

Fig. 6 Typical appearance of a mature or complete cataract

have demonstrated that the age of onset and gender distribution of cataracts in small-breed dogs are different depending on the breed [59]. Cataracts in different dog breeds can be of genetic, metabolic, toxic, or traumatic origin, or due to senility, environmental factors and other causes (Figs. 7 and 8) [60–62].

Cataracts originate from modifications in the uniformity of the protein density of the crystalline, which produces changes in its refraction index and is manifested by its opacity. Although the biochemical events that lead to these modifications are not well understood yet, several researchers have found that in this disorder insoluble proteins are increased and that the insolubilization process is accelerated. It seems that this insolubilization is produced by the formation of covalent bonds as a result of protein oxidation. The sulfur-containing amino acids present in the proteins are usually very susceptible to oxidative damage, generating disulfides that can cause the establishment of double bridges between proteins or subunits and the formation of aggregates [63]. The crystalline has a high concentration of reduced glutathione, which maintains the thiol groups reduced, which in turn contributes to its complete transparency.

Ultraviolet (UV) radiation reaches the crystalline lens directly, where it generates free radicals that damage essential molecules. This situation provides favorable conditions for the induction of ocular destruction by ROS and reactive nitrogen species [64].

The exact mechanism leading to cataract formation is still largely unknown [65]. The lens epithelium plays an important role in the regulation of nutrient and ion transport to all cells in the lens, and actively participates in adaptive responses to environmental and chemical stressors [66]. Oxidative stress is currently no longer the hypothesis for the development of cataracts, but it is known that it not only causes protein damage and precipitation in lens fibers, but also contributes to the dysfunction and apoptotic death of lens epithelial cells [67–69].

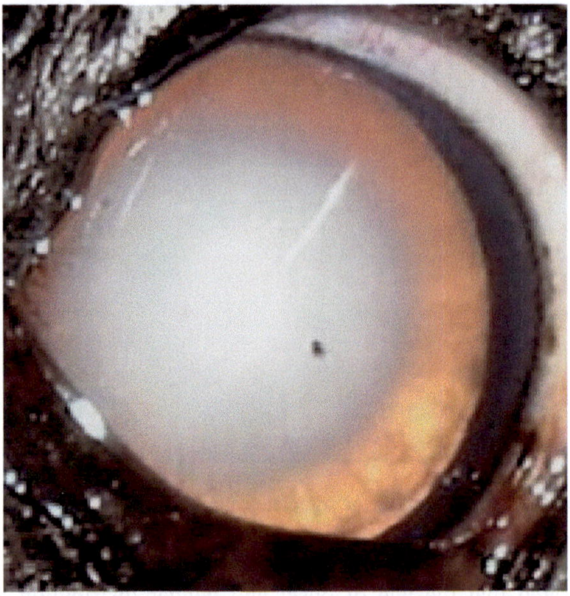

Fig. 7 Typical age-related cataract in a 13-years-old English Cocker Spaniel. Note the dense nuclear sclerosis

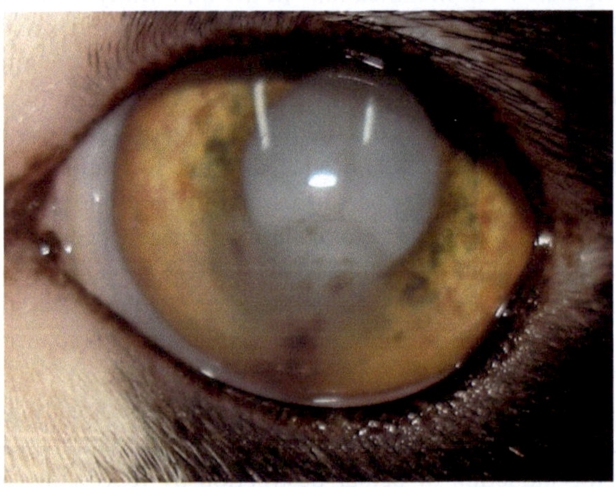

Fig. 8 Toxoplasma-induced anterior uveitis, as in this domestic shorthair, is characterized by iridal nodules, keratic precipitates and cataract

McCarty et al. and Spector have demonstrated that the damage to the lens epithelium caused by UV radiation or oxidative stress (i.e., an increase in H_2O_2) represents an initiating event in cataract formation, emphasizing the importance of maintaining the integrity of the epithelial layer [70, 71]. Wu et al. have recently reported that sustained oxidative stress impairs the cell signaling involved in NF-κB activation, which increases the death of the lens epithelial cells and exacerbates cataract genesis [72].

The lens, as well as other tissues, has an antioxidant system and high levels of glutathione. In patients with cataracts, these levels are lower than normal [73]. This is in agreement with other studies that have reported an increase in ROS and other free radicals in the aqueous humor both in patients with cataracts and in cell cultures of lens epithelium [74]. The oxidative stress associated with increased levels of ROS is known to accelerate cataract formation in laboratory animal models [75]. Another antioxidant agent is superoxide dismutase. Dingbo et al. demonstrated an increase in the expression of this antioxidant agent in lens epithelial cells when stimulating them with H_2O_2 [76].

On the other hand, Babizhayen et al. showed that the initial stages of cataract are characterized by the accumulation of primary LPO products and a reliable increase in oxiproducts of fatty acyl content of lenticular lipids. The lens opacity degree correlates with the level of LPO fluorescent end-product accumulation in the lens, which is associated with sulfhydryl group oxidation of lens proteins due to a decrease in reduced glutathione concentration in the lens [77]. In most tissues of the body, including the lens, superoxide is converted into hydrogen peroxide by superoxide dismutase, but even hydrogen peroxide can become highly toxic because it produces the hydroxyl radical OH· This toxicity is prevented by catalase and glutathione peroxidase. These enzymes protect the lens by a system of antioxidant molecules, the lynchpin of which is glutathione, and the role of which has been expertly reviewed by Marjorie Lou [78].

Surgery (phacoemulsification and aspiration) is the main therapeutic tool against cataracts. However, adverse effects, such as corneal epithelium damage, mechanical and thermal damage due to the procedure, and formation of free radicals due to the effect of ultrasound (because ultrasound causes the disintegration of water molecules and the consequent formation of hydroxyl radicals) have been described. However, because sodium hyaluronidate is known for its capacity to remove free radicals, it is used for this surgery technique.

Nonsurgical alternatives based on antioxidants, for both the systemic and the topical treatment as well as for the prevention of cataracts, are currently being tested [79–81]. Finally, it is important to point out that some studies with dietary supplements with antioxidants in dogs and human patients with cataracts have shown a decrease in the development of cataracts [82–84].

Retina

The retina is the innermost layer of the eyeball and it is made up of cells with vastly different functions. The retina is composed of two types of photoreceptor cells, rods and cones, that contain a photosensitive pigment called rhodopsin, which allows them to perceive the light stimulus.

The functioning and maintenance of these types of cells require a high consumption of energy: three to four times more oxygen than other cells of the retina and neurons of the central nervous system (CNS). They are probably the cells of the organisms with the highest oxidative metabolism. In as much as they generate adenosine triphosphate mainly from glucose by oxidative metabolism and glycolysis. They are composed of a high percentage of PUFAs, which are required by retinal membranes to build up a highly fluid microenvironment for the rhodopsin visual pigment that is responsible for the light signal transduction. On the other hand, high unsaturation indices make these membranes highly vulnerable to oxidative damage [85].

In the retina, the main sources of ROS are exposure to light, mitochondrial respiration, cellular metabolic reactions such as phagocytosis of outer segment tips by retinal pigment epithelium (RPE) cells, lipofuscin phototoxicity, and protoporphirin photosensitization. From the physiological point of view, ROS are constantly generated in the course of photochemical reactions that convert light energy to chemoelectrical signal [86].

The RPE plays a key role in the functioning and metabolism of photoreceptors. This epithelium controls the volume and composition of fluids in the subretinal space by means of the transport of ions, fluids, and metabolites, and is responsible for the maintenance of the integrity of the RPE–photoreceptor interface. In addition, the RPE plays a key role in the phagocytosis of the external segment of photoreceptors, an important process in the regeneration of the photopigment and the activity of detoxification (antioxidants).

There are different mechanisms by which the retina can suffer alterations in the reduction–oxidation (redox) status. It is known, for instance, that certain wavelengths can cause phototoxicity or photochemical lesion [87]. Short wavelength radiations (the spectrum of rhodopsin) and blue light (absorption peak of 440 nm) have an impact on the function of the pigmentary epithelium and photoreceptors, producing photochemical damage and death by apopotosis [88]. This light energy causes the generation of free radicals in the eye tissues, thus promoting LPO (lipid perioxidation). The neuroretina is extremely sensitive to this process because of its high content of PUFAs, especially the external segment of the cones and rods, which contains a high concentration of 22:6 ω3 (DHA) [89].

The molecular and cellular pathological events include apoptosis of the photoreceptors and the accumulation of the pigment lipofuscin in the cells of the pigmentary epithelium. Lipofuscin contains several fluorophores that are responsible for its characteristic broadband fluorescence when excited with UV or blue light. Irradiation of lipofuscin with short wavelength light can generate superoxide anions, singlet oxygen, hydrogen peroxide, and lipid hydroperoxides, and the generation of these ROS increases as wavelength decreases These processes cause damage to the mitochondrial DNA and to the RPE cells and photoreceptors (Fig. 9).

In order to contrast the formation of oxidant agents, the retina presents enzymes such as superoxidase dismutase (SOD), catalase, heme oxygenase, glutathione peroxidase, phospholipase, and nonenzymatic antioxidants, such as vitamins E, A, and C, melanin, lutein, and zeaxanthin [20,21,90].

Fig. 9 Cat with retinal degeneration. Note the generalized change in reflectivity or hyperreflectivity and dilated pupils

The inflammatory process that involves the retina is another cause of oxidative imbalance such as the one developed in uveitis. There are numerous causes that develop inflammatory responses that involve the retina. Studies show that leukocytic infiltration of the retina and upregulation of tumor necrosis factor-alfa (TNF-alfa) lead to an increase in the iNOS level, which in turn leads to production of cytotoxic peroxynitrite ($ONOO\cdot^-$). Other studies carried out first by Ito et al. and then by Rajendram et al. show that $ONOO\cdot^-$ can cause photoreceptor mitocondrial oxidative stress that leads to photoreceptor apoptosis [91,92].

Reactive nitrogen species such as nitric oxide and its subproducts nitrate, nitrite, peroxynitrite, and peroxyl radical have a direct role in cellular signaling, vasodilation, and inflammation [72].

Different studies have demonstrated and associated the degenerative processes of the retina with an oxidative imbalance both in elder people and animal models [93]. In the last decades, the advances in veterinary medicine have been able to increase the expectancy of life of domestic animals, which makes the study of diseases related to aging interesting. As mentioned above, the RPE is key to the function of photoreceptors both because it provides them with the photopigment and because it removes oxidative material from them. The death of these cells by accumulation of oxidative material eventually results in the degeneration of photoreceptors, a disease known as age-related macular degeneration [94]. Several biochemical structures of oxidative damage of the RPE cells have been characterized. The best known of these is the accumulation of lipofuscin, which contributes to the oxidation of molecular substrates in the RPE cells through the production of ROS and also interferes with oxidative respiration in mitochondria [88,95].

Different lines of work on oxidative damage and the retina have shown that the physiopathological development of retinal degeneration as the cause of oxidative mitochondrial damage is shared by various diseases such as diabetes [96], glaucoma [97], blood–retinal barrier in hypoxic ischemia by different causes [92], hereditary diseases [98], and retinitis pigmentosa [99]. McLellan et al. studied the role of primary metabolic vitamin E deficiency in the development of retinal pigment epithelial dystrophy in cocker spaniel dogs [100]. Numerous works in different species, both in vitro and in vivo, have shown the protective efficacy on the retina or in the different cells that integrate it with different types of antioxidant agents [91,101–104].

Summary

There is growing evidence that a variety of intrinsic and extrinsic factors can alter redox homeostasis. ROS and reactive nitrogen species are products of normal cellular metabolism. It is also clear that an excessive production of free radicals causes damage to biological material and is an essential event in the etiopathogenesis of various diseases. The present chapter discusses the evidence regarding the role of oxidative damage in the pathogenesis of several eye diseases and the oxidative imbalance associated with aging.

The impact of antioxidants on prevention, manifestation, and progression of eye disease has become an important and controversial issue in the last years. In veterinary ophthalmology, it would be interesting to use antioxidants as a preventive therapy in certain breeds with a predisposition to present eye diseases that alter the oxidative balance such as glaucoma and cataracts. On the other hand, the therapeutic objectives for the different eye diseases could be widened by associating the classical protocols with antioxidant agents. Further studies should be carried out to evaluate the penetrating ability of different antioxidants in different eye tissues.

References

1. Crouch R.K., Goletz P., Snyder A. et al. Antioxidant enzymes in human tears. J Ocul Pharmacol 1991; 7: 253–258.
2. Gogia R., Richer S.P. and Rose R.C. Tear fluid content of electrochemically active components including water soluble antioxidants. Curr Eye Res 1998; 17: 257–263.
3. Choy C.K., Benzie I.F. and Cho P. Ascorbic acid concentration and total antioxidant activity of human tear fluid measured using the FRASC assay. Invest Ophthalmol Vis Sci 2000; 41: 3293–3298.

4. Gebherd R. Anti-oxidative capacity of various artificial tear preparations. Graefe's Arch Clin Exp Ophthalmol 2001; 239: 222–226.
5. Shimmura A., Suematsu M., Shimoyama M., Tsubota K., Oguchi Y. and Ishimura Y. Subthreshold UV radiation-induced peroxide formation in cultured corneal epithelial cells: The protective effects of lactoferrin. Exp Eye Res 1996; 63: 519–526.
6. Selinger O., Selinger R. and Reed W. Resistance to infection of the external eye: The role of tear. Surv Ophthamol 1979; 24: 33.
7. Arnold R.R., Cole M.F. and McGhee J.R. A bactericidal effect for human lactoferrin. Science 1977; 197: 263–265.
8. Malmquist J., Thovell J.I. and Wollheim F.A. 1977. Lactoferrin and lysozyme in arthritic exudative. Acta Med Scand 1977; 202: 313–318.
9. Fujihara T., Nagano T., Endo K. et al. Lactoferrin protects against UV-B irradiation-induced corneal epithelial damage in rats. Cornea 2000; 19: 207–211.
10. Kaswan R.L., Bounous D. and Hirsh S.G. Diagnosis and management of keratoconjunctivitis sicca. Vet Med 1995; 90: 539–560.
11. Williams DL. Immunopathogenesis of keratoconjunctivitis sicca in the dog. Vet Clin North Am Small Anim Pract. 2008; 38(2): 251–268.
12. Pauloin T., Dutot M., Joly F., Warnet J. and Rat P. High molecular weight hyaluronan decreases UVB-induced apoptosis and inflammation in human epithelial corneal cells. Mol Vis 2009; 15: 577–583.
13. Edelhauser H. and Ubels J. Cornea and sclera. In: Kaufman H. and Alm A. (Eds), Adler's Physiology of the Eye, 10th edn. CV Mosby, Madrid, Spain 2004; 47–114.
14. Bhuyan K. and Bhuyan D. Catalase in ocular tissue and its intracellular distribution in corneal epithelium. Am J Ophthalmol 1970; 69: 147–153.
15. Brubaker R., Bourne W., Bachman L. and Mclaren J. Ascorbic acid content of human corneal epithelium. Invest Ophthalmol Vis Sci 2000; 41: 1681–1683.
16. Jo S., Lee S., Chun H., Lee S., Kok H., Lee S., Chum J., Park J. and Huh T. Cellular defence against UVB-induced phototoxicity by cytosolic NADP(+)-dependent isocitrate dehydrogenase. Biochem Biophys Res Commun 2002; 292: 542–549.
17. Lee S., Koh H., Park D., Song B., Huh T. and Park J. Cytosolic NADP(+)-dependent isocitrate dehydrogenase status modulates oxidative damage to cells. Free Radic Biol Med 2002; 32: 1185–1196.
18. Ayala M. and Sodetberg P. Vitamin E can protect against ultraviolet radiation-induced cataract in albino rats. Ophthalmic Res 2004; 36: 264–269.
19. Čejková J., Tipek S., Crkovská J., Ardan T., Pláteník J., Čejka Č. and Midelfart A. UV rays, the prooxidant/antioxidant imbalance in the cornea and oxidative eye damage. Physiol Res 2004; 53: 1–10.
20. Atalla L., Fernandez M. and Rao N. Immunohistochemical localization of catalase in ocular tissue. Curr Eye Res 1987; 7: 1181–1187.
21. Atalla L., Seavanian A. and Rao N. Immunohistochemical localization of glutathione peroxidase in ocular tissues. Curr Eye Res 1988; 7: 1023–1027.
22. Kovaceva J., Plátenik J., Vejrazka M., Štipek S., Aradn T., Čejka Č., Midelfart A. and Čejkova J. Differences in activities of antioxidant superoxide dismutase, glutathione peroxidase and prooxidant xanthine oxidoreductase/xanthine oxidase in the normal corneal epithelium of various mammals. Physiol Res 2007; 56: 105–112.
23. Yoshizumi M., Nakamura T., Kato M., Ishioka T., Kozawa K., Wakamatsu K. and Kimura H. Release of cytokines/chemokines and cell death in UVB-irradiated human keratinocytes, HaCaT. Cell Biol Int 2008; 32: 1405–1411.
24. Shoham A., Hadziahmetovic M., Dunaief J., Mydlarski M. and Schipper H. Oxidative stress in diseases of the human cornea. Free Radic Biol Med 2008; 45: 1047–1055.
25. Chandler H., Kusewitt D. and Colitz C. Modulation of matrix metalloproteinases by ultraviolet radiation in the canine cornea. Vet Ophthalmol 2008; 11(3): 135–144.
26. Ollivier FJ., Gilger B.C., Barrie K.P., Kallberg M.E., Plummer C.E., O'Reilly S., Gelatt K.N. and Brooks D.E. Proteinases of the cornea and preocular tear film. Vet Ophthalmol 2007; 10: 199–206.

27. Carubelli R., Nordquist R. and Rowsey J. Role of reactive oxygen species in corneal ulceration. Effect of hydrogen peroxide generated in situ. Cornea 1990; 9: 161–169.
28. Gebherd R. Anti-oxidative capacity of various artificial tear preparations. Graefe's arch. Clin Exp Ophthalmol 2001; 239: 222–226.
29. Wakamatsu T., Dogru M. and Tsubota K. Tearful relations: Oxidative stress, inflammation and eye disease. Arq Bras Oftalmol 2008; 71(6): 72–79.
30. Peiffer R., Wilcock B., Dubielzig R., Render J. and Whiteley H. Fundamentals of veterinary ophthalmic pathology. In: Gelatt K.N. (Ed.), Veterinary Ophthalmology, 3rd ed. Lippincott Williams & Wilkins, Philadelphia, 1999; 355–425.
31. Collins B.K. and Moore C.P. Diseases and surgery of the canine anterior uvea. In: Gelatt K.N. (Ed.), Veterinary Ophthalmology, 3rd ed. Lippincott Williams & Wilkins, Philadelphia, 1999; 755–795.
32. Allen J., Keng T. and Privalle C. Nitric oxide and peroxynitrite production in ocular inflammation. Environ Health Perspect 1998; 106(5): 1145–1149.
33. Wu G.S., Zhang J. and Rao N.A. Peroxynitrite andoxidative damage in experimental autoimmune uveitis. Invest Ophthalmol Vis Sci 1997; 38(7): 1333–1339.
34. Shiels I.A., Sanderson S.D. and Taylor S.M. Arterially perfused eye model of uveitis. Aust Vet J 1999; 77(2): 100–104.
35. Rao N.A. and Wu G.S. Free radical mediated photoreceptor damage in uveitis. Prog Retin Eye Res 2000; 19(1): 41–68.
36. Parks D., Cheung M., Chan C. and Roberge F. The role of nitric oxide in uveitis. Arch Ophthalmol 1994; 112: 544–546.
37. Ofri R. Optics and physiology of vision. In: Gelatt K.N. (Ed.), Veterinary Ophthalmology, 3rd ed. Lippincott Williams & Wilkins, Philadelphia 1999; 183–216.
38. Fliesler S.J. and Anderson R.E. Chemistry and metabolism of lipids in the vertebrate retina. Prog Lipid Res 1983; 22: 79–131.
39. Rajendram R., Saraswathy S. and Rao N. Photoreceptor mitochondrial oxidative stress in early experimental autoimmune uveoretinitis. Br J Ophthalmol 2007; 91: 531–537.
40. Gabelt B. and Kaufman P. Hydrodynamics of aqueous humor. In: Kaufman H. and Alm A. (Eds), Adler's Physiology of the Eye, 10th ed. CV Mosby, Madrid, Spain 2004; 238–289.
41. Gelatt K.N. and MacKay E.O. Prevalence of the breed-related glaucomasin pure-bred dogs in North America. Vet Ophthalmol 2004; 7: 97–111.
42. Blocker T. and Van der Woerdt A. The feline glaucomas: 82 cases (1995–1999). Vet Ophthalmol 1999; 4: 81–85.
43. Gelatt K.N. and Brooks D.E. The canine glaucomas. In: Gelatt KN (Ed.), Veterinary Ophthalmology, 3rd ed. Lippincott Williams & Wilkins, Baltimore, 1999; 701–754.
44. Shen F., Chen B., Danias J., Lee K.C., Lee H., Su Y., Podos S.M. and Mittag T.W. Glutamate-induced glutamine synthase expression in retinal Muller cells after short-term ocular hypertension in the rat. Invest Ophthalmol Vis Sci 2004; 45: 3107–3112.
45. Wallace K. and Eng-Ang L. Hypoxia-ischemia and retinal ganglion cell damage. Clin Ophthalmol 2008; 2(4): 879–889.
46. Alyahya K., Chen C., Mangan B., Gionfriddo J., Legare M., Dubielzig R. and Madl J. Microvessel loss, vascular damage and glutamate redistribution in th retinas of dogs with primary glaucoma. Vet Ophthalmol 2007; 10(Supplement 1): 70–77.
47. Izzotti A., Bagnis A. and Sacca S. The role of oxidative stress in glaucoma. Mutat Res 2006; 612: 105–114.
48. Russell P. and Johnson D.H. Enzymes protective of oxidative damage present in all decades of life in the trabecular meshwork, as detected by two-dimensional gel electrophoresis protein maps. J Glaucoma 1996; 5: 317–324.
49. Sacca S.C., Pascotto A., Camicione P. et al. Oxidative DNA damage in the human trabecular meshwork: Clinical correlation in patients with primary open-angle glaucoma. Arch Ophthalmol 2005; 123: 458–463.
50. Weinstein W., Dietrich U., Sapienza J., Carmichael K., Moore P. and Krunkosky T. Identification of ocular matrix metalloproteinases present within the aqueous humor and

iridocorneal drainage angle tissue of normal and glaucomatous canine eyes. Vet Ophthalmol 2007; 10(1): 108–116.
51. Vorwerk C.K., Naskar R., Schuettauf F. et al. Depression of retinal glutamate transporter function leads to elevated intravitreal glutamate levels and ganglion cell death. Investig Ophthalmol Vis Sci 2000; 41: 3615–3621.
52. Brooks D.E., Garcia G.A., Dreyer E.B. et al. Vitreous body glutamate concentration in dogs with glaucoma. Am J Vet Res 1997; 58: 864–867.
53. Mozaffarieh M., Grieshaber M.C. and Flammer J. Oxygen and blood flow: Players in the pathogenesis of glaucoma. Mol Vis 2008; 14: 224–233.
54. Källberg M., Brooks D., Gelatt K., Garcia-Sanchez G., Szabo N. and Lambrou G. Endothelin-1, nitric oxide, and glutamate in the normal and glaucomatous dog eye. Vet Ophthalmol 2007; 10(1): 46–52.
55. Gionfriddo J., Freeman K., Groth A., Scofield V., Alyahya K. and Madl J. Alpha-Luminol prevents decreases in glutamate, glutathione, and glutamine synthase in the retinas of glaucomatous DBA/2J mice. Vet Ophthalmol 2009; 12(5): 325–332.
56. Zigler J.S., Jr. Lens proteins. In: Albert D.M. and Jakobiec F.A. (Eds), Principles and Practice of Ophthalmology, Basic Science. W. B. Saunders Co., Philadelphia, 1994; pp. 97–113.
57. Davidson M.G. and Nelms S.R. Diseases of the lens and cataract formation. In: Gelatt K.N. (Ed.), Veterinary Ophthalmology, 3rd edn. Lippincott Williams & Wilkins, Philadelphia, 1999; 797–825.
58. Engelhardt A., Stock K.F., Hamann H., Brahm R., Grußendorf H., Rosenhagen C.U. and Distl O. A retrospective study on the prevalence of primary cataracts in two pedigrees from the German population of English Cocker Spaniels. Vet Ophthalmol 2008; 11(4): 215–221.
59. Park S.A., Yi N.Y., Jeong M.B., Kim W.T., Kim S.E., Chae J.M. and Seo K.M. Clinical manifestations of cataracts in small breed dogs. Vet Ophthalmol 2009; 12(4): 205–210.
60. Baumworcel N., Soares A., Helms G., Rei P. and Castro M. Three hundred and three dogs with cataracts seen in Rio de Janeiro, Brazil. Vet Ophthalmol 2009; 12(5): 299–301.
61. Colitz C., Sugimoto Y., Lu P., Barden C., Thomas-Ahner J. and Chandler H. ERα increases expression and interacts with TERT in cataractous canine lens epithelial cells. Mol Vis 2009; 15: 2259–2267.
62. Mellersh C., McLaughlin B., Ahonen S., Pettitt L., Lohi H. and Barnett K. Mutation in HSF4 is associated with hereditary cataract in the Australian Shepherd. Vet Ophthalmol 2009; 12(6): 372–378.
63. Andley U. Effects of alpha-crystallin on lens cell function and cataract pathology. Curr Mol Med 2009; 9(7): 887–892.
64. Robman L. and Taylor H. External factors in the development of cataract. Eye 2005; 19(10): 1074–1082.
65. Harding JJ. Viewing molecular mechanisms of ageing through a lens. Ageing Res Rev 2002; 1: 465–479.
66. Mathias R. and Rae J. The lens: local transport and global transparency. Exp Eye Res 2004; 78: 689–698.
67. Li W.C., Kuszak J.R., Dunn K., Wang R.R., Ma W., Wang G.M., Spector A., Leib M., Cotliar A.M., Weiss M. et al. Lens epithelial cell apoptosis appears to be a common cellular basis for non-congenital cataract development in humans and animals. J Cell Biol 1995; 130: 169–181.
68. Williams D. Oxidation, antioxidants and cataract formation: A literature review. Vet Ophthalmol 2006; 9(5): 292–298.
69. Truscott RJ. Age-related nuclear cataract-oxidation is the key. Exp Eye Res 2005; 80(5): 709–725.
70. McCarty C. and Taylor H. Recent developments in vision research: Light damage in cataract. Investig Ophthalmol Vis Sci 1996; 37: 1720–1723.
71. Spector A. Oxidative stress-induced cataract: Mechanism of action. FASEB J 1995; 9: 1173–1182.

72. Wu M., Bian Q., Liu Y., Fernandes A., Taylor A., Pereira P. and Shang F. Sustained oxidative stress inhibits NF-κB activation partially via inactivating the proteasome. Free Radic Biol Med 2009; 46: 62–69.
73. Ganea E. and Harding JJ. Glutathione-related enzymes and the eye. Curr Eye Res 2006; 31(1): 1–11.
74. Goswami S., Sheets N., Zavadil J., Chauhan B, Bottinger E., Reddy V., Kantorow M. and Cvekl A. Spectrum and range of oxidative stress responses of human lens epithelial cells to H_2O_2 insult. Investig Ophthalmol Vis Sci 2003; 44(5): 2085–2093.
75. Bhuyan D.K., Huang X., Kuriakose G. et al. Menadione-induced oxidative stress accelerates onset of Emory mouse cataract in vivo. Curr Eye Res 1997; 16: 519–526.
76. Lin D., Barnett M., Grauer L., Robben J., Jewell A., Takemoto L. and Takemoto D. Expression of superoxide dismutase in whole lens prevents cataract formation. Mol Vis 2005; 11: 853–858.
77. Babizhayen M.A., Deyev A.I. and Yermakova U.N. Lipid peroxidation and cataracts: N-acetylcarnosine as a therapeutic tool to manage age-relate cataracts in human and in canine eyes. Drugs R D 2004; 5(3): 125–139.
78. Lou MF. Redox regulation in the lens. Prog Retin Eye Res 2003; 22: 657–682.
79. Elanchezhian R., Sakthivel M., Isai M., Geraldine P. and Thomas P.A. Evaluation of lenticular antioxidant and redox system components in the lenses of acetyl-L-carnitine treatment in BSO-induced glutathione deprivation. Mol Vis 2009; 15: 1485–1491.
80. Shang F., Lu M., Dudek E., Reddan J. and Taylor A. Vitamin C and vitamin E restore the resistance of GSH-depleted lens cells to H2O2. Free Rad Biol Med 2003; 34: 521–530.
81. Manikandan R., Thiagarajan R., Beulaja S., Chindhu S., Mariammal K., Sudhandiran G. and Arumugam M. Anti-cataractogenic effect of curcumin and aminoguanidine against selenium-induced oxidative stress in the eye lens of Wistar rat pups: An in vitro study using isolated lens. Chem-Biol Interact 2009; 181:202–209
82. Elanchezhian R., Sakthivel M., Isai M., Geraldine P., Thomas P.A., Christen W.G., Liu S., Glynn R.J., Gaziano J.M. and Buring J.E. A prospective study of dietary carotenoids, vitamins C and E, and risk of cataract in women. Arch Ophthalmol 2008; 126(1): 102–109.
83. Fernandez M.M. and Afshari N.A. Nutrition and the prevention of cataracts. Curr Opin Ophthalmol 2008; 19(1): 66–70.
84. Mares J.A. High-dose antioxidant supplementation and cataract risk. Nutr Rev 2004; 62(1): 28–32.
85. Narfström K. and Ekesten B. Diseases of the canine ocular fundus. In: Gelatt K.N. (Ed.), Veterinary Ophthalmology, 3rd edn. Lippincott Williams & Wilkins, Baltimore, 1999; 869–933.
86. Valko M., Leibfritz D., Moncola J., Cronin M., Mazur M. and Telser J. Free radicals and antioxidants in normal physiological functions and human disease. Int J Biochem Cell Biol 2007; 39: 44–84.
87. Demontis G.C., Longani B. and Marchiafara P.L. Molecular steps involved in light-induced oxidative damage to retinal rods. Invest Ophtal Sci 2002; 43: 2421–27.
88. Liang F.Q. and Godley B.F. Oxidative stress-induced mitochondrial DNA damage in human retinal pigment cells: a possible mechanism for RPE aging and age-related macular degeneration. Exp Eye Res 2003; 76(4): 397–403.
89. Zapata G.L., Guajardo M.H. and Terrasa A.M. The in vitro protective effect of alpha tocopherol on oxidative injury in the dog retina. Vet J 2008; 177: 266–272.
90. Behndig A., Svensson B., Marklund S. and Karlsson K. Superoxide dismutase isoenzymes in the human eye. Invest Ophthalmol Vis Sci 1998; 39: 471–475.
91. Ito S., Wu G.S., Kimoto T., Hisatomi T., Ishibashi T., Rao N.A. Peroxynitrite-induced apoptosis in photoreceptor cells. Curr Eye Res 2004; 28: 17–24.
92. Kaur C., Foulds W. and Ling E. Blood–retinal barrier in hypoxic ischaemic conditions: Basic concepts, clinical features and management. Prog Retin Eye Res 2008; 27: 622–647.
93. Xu H., Chen M. and Forrester J. Para-inflammation in the aging retina. Prog Retin Eye Res 2009; 28: 348–368.

94. Kang K., Lemke G. and Kim J. The PI3K-PTEN tug-of-war, oxidative stress and retinal degeneration. Trends Mol Med 2009; 15(5): 191–198.
95. Jarrett S.G., Lin H., Godley B.F. and Boulton M.E. Mitochondrial DNA damage and its potential role in retinal degeneration. Prog Retin Eye Res 2008; 27: 596–607.
96. Kanwar M., Chan P., Kern T. and Kowluru RA. Oxidative damage in the retinal mitochondria of diabetic mice: Possible protection by superoxide dismutase. Invest Ophthalmol Vis Sci 2007; 48 (8): 3805–3811.
97. Tezel G. Oxidative stress in glaucomatous neurodegeneration: Mechanisms and consequences. Prog Retin Eye Res 2006; 25(5): 490–513.
98. Carelli V., La Morgia C., Valentino M.L., Barboni P., Ross-Cisneros F.N. and Sadun A. Retinal ganglion cell neurodegeneration in mitochondrial inherited disorders. Biochim Biophys Acta 2009; 1787: 518–528.
99. Usui S., Oveson B., Lee S., Jo Y., Yoshida T., Miki A., Miki K., Iwase T., Lu L. and Campochiaro P. NADPH oxidase plays a central role in cone cell death in retinitis pigmentosa. J Neurochem 2009; 110: 1028–1037.
100. McLellan G.J., Cappello R., Mayhew I.G., Elks R., Lybaert P., Watte C. and Bedford P.G. Clinical and pathological observations in English cocker spaniels with primary metabolic vitamin E deficiency and retinal pigment epithelial dystrophy. Vet Rec 2003; 153: 287–292.
101. Organisciak D.T., Jiang Y.L., Wang H.M. et al. The protective effect of ascorbic acid in retinal light damage of rats exposed to intermittent light. Invest Ophthalmol Vis Sci 1990; 31: 1195–202.
102. Terrasa A.M., Guajardo M.H., Marra C.A. and Zapata G. Alpha Tocopherol protects against oxidative damage to lipids of the rod outer segments of the equine retina. Vet J 2009; 182 (3): 463–468.
103. Mukherjee P.K., Marcheselli V.L., Serhan C.N. and Bazan N.G. Neuroprotectin D1: A docosahexaenoic acid-derived docosatriene protects human retinal pigment epithelial cells from oxidative stress. PNAS 1 2004; 101(22): 8491–8496.
104. Hanneken A., Lin F., Johnson J. and Maher P. Flavonoids protect human retinal pigment epithelial cells from oxidative-stress–induced death. Invest Ophthalmol Vis Sci 2006; 47: 3164–3177.
105. Cejková J., Ardan T., Simonova Z., Malec J., Dotrelová D. and Brunová B. Decreased expression of antioxidant enzymes in the conjunctival epithelium of the dry eye and its possible contribution to the development of ocular surface oxidative injuries. Histol Histopatol 2008; 23(12): 1477–1483.
106. Stone W.L., Farnsworth C.C. and Dratz E.A. A reinvestigation of the fatty acid content of bovine, rat and frog retinal rod outer segments. Exp Eye Res 1979; 28: 387–397.
107. Novaes P., do Nascimento Saldiva P., Kara-José N., Macchione M., Matsuda M., Racca L. and Berra A. Ambient levels of fair pollution induce goblet-cell hyperplasia in human conjunctival epithelium. Environ Health Perspect 2007; 115(129): 1753–1756.
108. Young A.R. Acute effects of UVR on human eyes and skin. Prog Biophys Mol Biol 2006; 92: 80–85.

Oxidative Stress in Heart Failure

Gordon Moe

Abstract Heart failure (HF) is a complex clinical syndrome whose pathogenesis involves an interplay of neurohormonal activation and inflammatory processes at the cellular and molecular levels. Oxidative stress describes an imbalance between antioxidant defense and the production of reactive oxygen species (ROS), which at high levels cause cell damage but at lower levels induce subtle changes in intracellular signaling pathways. Substantial evidence to date suggests the involvement of oxidative stress in the pathophysiology of HF as well as its antecedent conditions such as cardiac hypertrophy and adverse remodeling following myocardial infarction. Oxidative stress may indeed represent one of the common pathways for cell death/apoptosis, mitochondrial dysfunction, cardiac remodeling, and dysfunction. There is increasing evidence implying that ROS play an important role in the development and progression of HF. However, although levels of ROS are elevated in HF, the relative contribution of the different intracellular sources of ROS, the precise mechanisms and their impact on the progression of HF remain unclear. Further delineation of the downstream signaling pathways involved in ROS accumulation is important in order to improve understanding of these processes and also for the development of new therapies. Drugs such as the angiotensin-converting enzyme (ACE) inhibitors and the statins may act in part through such mechanisms. Despite disappointing results in using antioxidants in human studies, it is likely that modulation of endogenous antioxidants in human HF will continue to hold potential for both the treatment and prevention of HF.

Keywords Myocardial ischemia • Heart failure • Coronary artery disease • Myocyte apoptosis • Uncouplers • PARP inhibitor

---

G. Moe (✉)
Li Ka Shing Knowledge Institute, St. Michael's Hospital, University of Toronto,
Toronto, ON, Canada M5B 1W8
e-mail: moeg@smh.toronto.on.ca

L. Mandelker and P. Vajdovich (eds.), *Studies on Veterinary Medicine*,
Oxidative Stress in Applied Basic Research and Clinical Practice 5,
DOI 10.1007/978-1-61779-071-3_9,

Reactive Oxygen Species in Myocardial Ischemia and Heart Failure

Generation of reactive oxygen species (ROS) together with impaired antioxidant defenses is observed in the failing heart as well as in myocardial ischemia and reperfusion. During the first moments of reperfusion and myocardial injury a burst of ROS occurs that is associated with changes in mitochondria, for example, permeability transition (PT) pore opening [1]. The source of ROS generation during early reperfusion has not been clearly determined; it may be of either mitochondrial or cytoplasmic origin. On the other hand, during ischemia, and likely during the early acute pathways of ischemic preconditioning, the source of ROS generated clearly involves the mitochondrial electronic transport chain (ETC), and this may be different than the source of ROS generated during early reperfusion [2, 3]. The mitochondria are central to the process of myocardial failure because ROS generation by mitochondria affects the expression of genes encoding key elements of the apoptotic pathway, including mitochondrial PT pore components such as adenine nucleotide translocator (ANT) and porin.

A number of studies have shown that in the failing heart cardioprotective interventions may prevent or slow ROS production via antioxidants such as manganese superoxide dismutase (Mn-SOD) and target apoptosis. Furthermore, when exposed to an insult, cardiomyocytes are able to mount a cardioprotective response. In addition, chemical, metabolic, and even physical stressors have been shown to generate cardioprotective responses that share elements of the ischemic preconditioning model and in these responses, the mitochondria are the key factors.

As well as regulating the cellular bioenergetic supply in the form of adenosine triphosphate (ATP), the organelle has many roles. A participant in the trigger and mediation of cardioprotective responses, mitochondria house and regulate pivotal early events in the apoptotic pathway. This plasticity enables the organelle to function both as a target and as a player in myocardial signal transduction events and in the generation of ROS in response to a variety of cellular insults providing an appropriate antioxidant response.

The time course of Mn-SOD induction, a mitochondrial-specific protective response to ROS mediated damage, correlates well with that of the appearance of ischemic tolerance. However, the activities of other cellular antioxidants such as catalase and copper-SOD (Cu-SOD) are not similarly affected. In the canine model of pacing-induced HF, our group has demonstrated a marked increase in left ventricular (LV) tissue aldehyde levels suggesting increased free radical-induced damage [4, 5]. The increased levels of oxidative stress correlated with the onset of reduced cardiac mitochondrial complex III and V activities. This has been corroborated by measurements of ROS levels in paced compared to unpaced dogs, revealing that both $O_2\cdot^-$ and OH· radicals generated from mitochondrial-produced H_2O_2 are increased in the failing heart and correlated with the severity of LV dysfunction [6, 7].

Oxidative stress also appears to be involved in the generation of large-scale myocardial mtDNA deletions that have been shown in pacing-induced HF [4], as well as in studies of ameroid constriction-mediated myocardial ischemia in the dog [8]. Furthermore, neonatal cardiac myocytes treated with TNF-α displayed a significant increase in ROS levels, accompanied by an overall decline in mtDNA copy number and decreased complex III activity [9]. The TNF-α mediated decline in mtDNA copy number might result from an increase in mtDNA deletions. ROS-induced mtDNA damage, resulting in respiratory complex enzyme dysfunction, contributes to the progression of LV remodeling and failure after myocardial infarction (MI). In a murine model of MI and remodeling created by ligation of the left anterior descending coronary artery, increased ROS production (e.g., OH· level) was found in association with decreased levels of mtDNA and ETC activities, suggesting impairment of mitochondrial function [10].

In addition, chronic release of ROS has been linked to the development of LV hypertrophy, extracellular matrix (ECM) remodeling, and HF. Thus, blocking/reducing ROS production may be an important target in the treatment of HF. Indeed, recent observations have suggested that overexpression of the genes for peroxiredoxin-3 (Prx-3), a mitochondrial antioxidant, or mitochondrial transcription factor A (TFAM), might prevent the decline in mtDNA copy number (likely secondary to mtDNA damage, i.e., increased deletions) and also the mitochondrial dysfunction observed in HF [11]. Thus, preventing oxidative stress and mtDNA damage may be an attractive approach in HF treatment.

In addition to chronic ROS generation from the mitochondrial organelle, ROS can also be derived from nonmitochondrial NADPH oxidase, which in endothelial cells is activated by cytokines, neurohormones, and growth factors (e.g., angiotensin II, norepinephrine, and TNF-α) [12, 13]. Furthermore, long-term alterations in cardiac phenotype can be driven by redox-sensitive gene expression, and in this way ROS may act as potent intracellular second messengers. In cardiac myocytes, NADPH oxidase plays a prominent role in the hypertrophic pathway [14, 15]. In addition, NADPH oxidase activity is significantly increased in the failing versus nonfailing myocardium [15]. It has been reported that the statins, by modulating the ROS-generating activity of NADPH oxidase, can inhibit cardiac hypertrophy by cholesterol-independent mechanisms [16, 17]. The statins have been shown to block the isoprenylation and activation of members of the Rho guanosine triphosphatase (GTPase) family such as Rac1, an essential component of NADPH oxidase. Taken together, it appears that blocking ROS production with statins may have some beneficial effects in patients with myocardial hypertrophy and chronic HF.

Myocardial ischemia, an important cause of heart failure in humans, can produce changes in the endogenous defense mechanisms against oxygen free radicals, mainly through a reduction in the activity of mitochondrial superoxide dismutase (SOD) and a decline in tissue content of reduced glutathione [18]. Increased ROS production in both the mitochondria and white blood cells (WBC) and toxic oxygen metabolite production are exacerbated by readmission of oxygen during postischemic reperfusion. Furthermore, oxidative stress resulting from both increased ROS generation and diminished antioxidant reserve, promotes the oxidation of thiol groups in proteins and lipid peroxidation leading first to reversible damage, and eventually to necrosis. Mitochondria of circulating WBC and platelets sense oxidative stress during capillary passage and react by producing ROS.

Although a number of observations have shown that severe HF is associated with oxidative stress, the mediator role of WBC and platelets of oxidative stress in HF has only been recently investigated. Ijsselmuiden et al. have studied a group of HF patients and healthy volunteers (control group) [19]. WBC and platelets of both arterial and venous blood samples were quantitatively analyzed with respect to the development of cytoplasmic and mitochondrial oxidative stress using fluorescent dyes. The increased number and percentage of cells with both cytosolic and mitochondrial fluoresence signified extensive presence of oxidative stress and were significantly greater than those found in control samples. Moreover, myocardial oxidative stress (gauged by coronary sinus sampling) exceeded systemic production (gauged by peripheral venous sampling). Furthermore, partial pulmonary clearance of ROS containing cells was identified in controls by the reduction in arterial compared to venous samples; in contrast, no significant difference (between arterial and venous samples) was found in ROS-containing cells in severe HF patients suggesting defective pulmonary clearance in these cases. Thus, in severe HF the proportion of WBC and platelets that are ROS-positive is raised, possibly because cytosolic ROS-positive WBC and platelets are normally cleared in the lungs and this function appears to be deficient in severe HF whereas mitochondrial ROS production is increased. The raised numbers of circulating ROS-positive WBC and platelets amplify OS in HF. A more rigorous approach could have been employed by first carrying out the study in an animal model that might eliminate the concerns raised by the drug treatments, and the potential heterogeneity of HF etiology, which may have an impact on these findings.

Coronary artery disease (CAD), with consequent myocardial ischemia and necrosis, is a well-recognized leading cause of HF, thus it is relevant to note here that ROS play an integral role in the genesis and progression of CAD and HF. In the vessel wall ROS contribute to the formation of oxidized low density lipoprotein (LDL), a major player in the pathogenesis of atherosclerosis and ROS-mediated activation of matrix metalloproteinases (MMPs) may play a contributory role in vessel plaque rupture, initiating coronary thrombosis and occlusion [20, 21]. Observations in vitro and animal studies have shown that in the failing heart, ROS influence several components of the cardiac phenotype and its remodeling, including contractile function, interstitial fibrosis, endothelial dysfunction, and myocyte hypertrophy. ROS contribute to the remodeling processes in a number of ways, including activating MMPs that participate in the reconfiguration of the ECM; acting as signaling molecules in the development of compensatory hypertrophy; and contributing to myocyte loss via apoptosis signaling.

Excessive production of nitric oxide (NO) has also been implicated in the pathogenesis of chronic HF [22]. Uncoupling of constitutive nitric oxide synthase (NOS) leads to the overproduction of superoxide ($O_2\cdot^-$) and peroxynitrite ($ONOO^-$), two extremely potent oxidants. Peroxynitrite produced from the reaction of highly reactive NO with the superoxide anion impairs cardiovascular function through multiple mechanisms, including activation of MMPs and nuclear enzyme poly

(ADP-ribose) polymerase (PARP). Increased oxidative stress resulting from the overproduction of superoxide also mediates the dysregulation of S-nitrosylation of proteins at specific cysteine residues by reactive nitrogen species (RNS), a more selective modification of proteins than found with protein oxidation. This redox mechanism has been demonstrated to lead to altered myocardial excitation-contractility and vascular reactivity [23, 24].

Myocyte Apoptosis

One of the mechanisms from which increased oxidative stress may contribute to HF progression is apoptosis. Both the generation of ROS and the onset of apoptotic cell death are important and often connected events in cell homeostasis. Increased oxidative stress and ROS accumulation can lead to myocyte hypertrophy, interstitial fibrosis (through matrix metalloproteinase activation), and apoptosis. Apoptosis can be blocked or delayed by treatment with antioxidants and thiol reductants [25]. Moreover, overexpression of antioxidant proteins (e.g., MnSOD, glutathione peroxidase [GPx], and metallothionein [MT]) can block the progression of apoptosis [26, 27]. Attenuation of apoptosis by overexpression of the antiapoptotic protein Bcl-2 was associated with protection against ROS and oxidative stress [28]. Cells from transgenic mice containing ablated genes encoding antioxidant proteins such as glutathione peroxidase (GPx) have both increased OS levels and increased apoptosis [29, 30]. Agents that induce apoptosis, for example, TNF-α, also promote high levels of mitochondrial-generated ROS [31]. In addition, ROS production can target an array of signal transducers (e.g., JNK, TNF-α, and MAPK) that interface with the apoptotic machinery. This includes both activation of the proapoptotic Bcl-2 family and mediators including TNF-α, as well as the modulation of their gene expression by activation of specific transcription factors such as NF-κB.

An important model system has emerged to study the induction of oxidative stress and apoptosis using H_2O_2 treated cultured cardiomyocytes [32]. This approach allowed molecular and biochemical appraisal of cellular and mitochondrial events presaging, accompanying, and following the induction of cardiomyocyte apoptosis. In addition, it allowed the study of short-lived signaling intermediates as well as the use of various treatments including antioxidants to stem the development of cardiomyocyte oxidative stress and apoptotic progression [33]. Not all cardiomyocyte apoptosis is triggered by oxidative stress and ROS generation. Gathered observations have shown no evidence of ROS or NO involvement in the palmitate-mediated induction of apoptosis in neonatal rat cardiomyocytes [34], illustrating the complex, multipathway, and parallel nature of signaling in the apoptotic pathway.

Cesselli and coworkers attempted to link common pathways involved in mediating oxidative stress and apoptosis in dogs with heart failure induced by right ventricular pacing [35]. Using immunohistochemistry and immunoblotting techniques, the investigators tracked the changes in multiple apoptotic pathways including the caspases, mitochondrial cytochrome *c* release, as well as proteins involved in DNA damage.

Interestingly, myocyte, endothelial cell, and fibroblast apoptosis were detected before severe impairment of cardiac function became apparent. Cell death increased with the duration of pacing, and myocyte death exceeded endothelial cell and fibroblast death throughout heart failure development. Nitrotyrosine formation and $p66^{shc}$ levels progressively increased with pacing and were associated with cell apoptosis. p50 (ΔN) fragments were augmented and paralleled the degree of cell death. Cytochrome *c* release and activation of caspase-9 and -3 increased from 1 to 4 weeks of rapid pacing.

These findings indicate that cardiac cell death precedes the development of LV dysfunction, providing support that apoptosis participates in the progression of LV remodeling. $p66^{shc}$ is a signaling molecule that appears to link oxidative stress and apoptosis [36]. It is an oxidant stress-induced, proapoptotic proto-oncogene known to be activated by phosphorylation in response to stimuli such as H_2O_2, UV radiation, or epidermal growth factor. In mice, transgenic disruption of $p66^{shc}$ or inactivation of its ability to be phosphorylated appears to confer resistance to oxidative damage and prolong life. In the study, cytochrome *c* release from the intermembrane mitochondrial compartment to the extramitochondrial space, translocation of the inactive procaspase-9 from the mitochondria to the cytosol, and nitrotyrosinylation of proteins in myocytes were identified here at the early and late stages of heart failure. These biochemical events increased with duration of pacing and the appearance of clinical heart failure. p53 fragments due to the loss of the C-terminus (p50 [ΔC]) or N-terminus (p50 [ΔN]) were detected in the paced myocytes, indicating an interaction between activated p53 and distinct forms of DNA damage. p53 fragments also increased with the duration of pacing. Taken together, these findings support the notion that oxidative damage may have at least played a partial role in myocyte death during the onset and evolution of heart failure.

Mitochondrial Dysfunction

Recent studies in this pacing model have also revealed that heart failure progression is commonly accompanied by marked changes in myocardial bioenergetic reserves, mitochondrial function, metabolic substrate utilization, as well as an overall depletion of myocardial high-energy phosphates [4, 37, 38]. Reduced levels of myocardial respiratory complex I [6], III, and V activities [4], and mitochondrial fatty acid β oxidation [38] have been observed in the pacing-induced heart failure model primarily in LV tissues. In the atrial tissues from paced hearts, Cha et al. reported severely reduced activities of the phosphotransfer enzymes creatine kinase and adenylate kinase, as well as depletion of high-energy phosphates (ATP and creatine phosphate), indicating that atrial bioenergetics was also affected in the paced failing [37]. Although the molecular and biochemical basis underlying the specific mitochondrial enzyme changes as well as the myocardial shift in substrate utilization have thus far been undetermined, a time course analysis revealed a parallel increase of oxidative stress markers, myocyte apoptosis, and respiratory enzymatic dysfunction in the left ventricle in early heart failure, thereby suggesting the existence of a link among these factors [39]. A similar association was found between the elevation of atrial levels of malondialdehyde, a marker of oxidative stress, and atrial bioenergetic deficit in dogs with pacing-induced heart failure [37].

An attractive hypothesis for the mitochondrial contribution as a mediator of heart failure relates to its pivotal role in the generation of ROS. ROS-mediated damage affects lipids, DNA, and proteins, and might be expected to most severely affect mitochondrial macromolecules. Effects of ROS damage on a wide spectrum of cardiomyocyte functions including contractility, calcium transport, and cycling, and on further mitochondrial ATP generation could explain several of the deficits seen in heart failure. Furthermore, mitochondrial ROS appear to play a role in activation of select nuclear gene expression by eliciting a novel transcriptional programming [40].

In an attempt to further delineate the molecular basis of a potential linkage between myocardial enzymatic dysfunction and apoptotic remodeling and establish their significance in the pacing-induced HF, we recently examined the regional distribution of these subcellular changes assessed by immunoblot and with enzymatic analysis applied to specific subpopulations of mitochondria from paced and unpaced animals [41]. Enzymatic dysfunction was examined for in mitochondrial subpopulations and immunoblot analysis was performed using homogenate proteins from the left atrium (LA) and LV of paced and control mongrel dogs. We found a greater range of enzymatic defects (complex I, III, and V) in mitochondria subpopulations from the LV as compared with the LA, where only complex V was defective. As shown in Fig. 1, analysis of paced LV tissue proteins demonstrated a downregulated expression of both mitochondrial genes such as cytochrome b and nuclear genes (e.g., ATP synthase β subunit, mitochondrial creatine kinase). Protease-activated products of both mitochondrial (e.g., apoptosis inducing factor) and cytosolic (e.g., caspase-3) apoptogenic proteins were increased in both the LA and LV. Nuclear-localized apoptotic markers such as p53 and p21were also significantly increased in the LV of the paced dogs. Thus, abnormal activity of several mitochondrial enzymes and increased apoptogenic pathway appear to be mediated, at least in part, by an orchestrated shift in expression (both nuclear and mitochondrial DNA) of respiratory chain subunits (e.g., cytochrome b, ATP-b), mitochondrial bioenergetic enzymes (e.g., mitochondrial creatine kinase), global transcription factor (e.g., PGC-1), as well as apoptotic proteins (e.g., p53, p21), with distinct differences in their regional distribution and in the subpopulations of affected mitochondria.

Oxidative Stress and Cardiac Function

Oxygen serves as the critical terminal electron acceptor in the mitochondrial ETC; in its absence the ETC shuts down and cardiac demands for ATP are not met. Molecular oxygen is also central in both the formation of NO, a primary determinant of both vascular tone and cardiac contractility, and in the generation of ROS

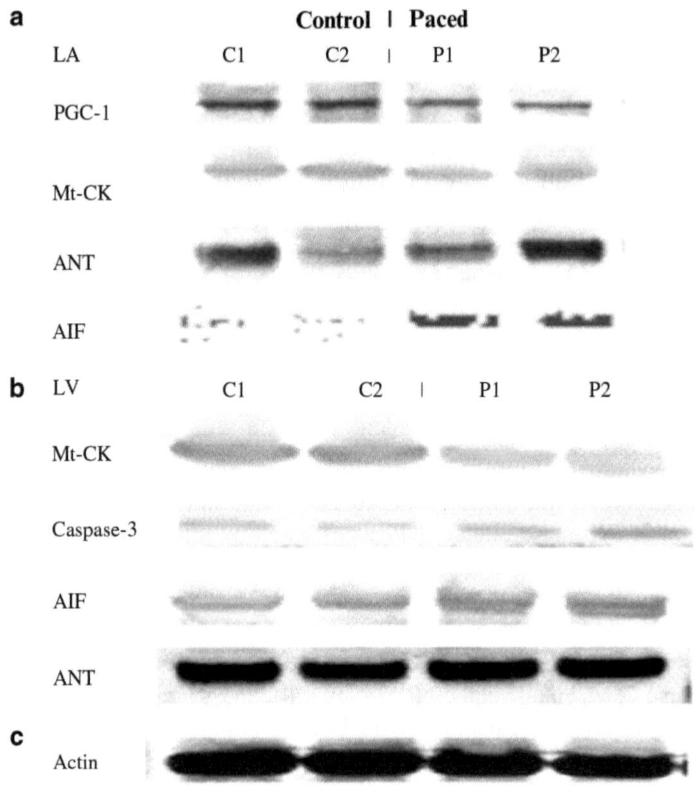

Fig. 1 Western immunoblot analysis of proteins from left ventricular and left atrial tissues. Representative Western immunoblots (**a**, **b**) showing specific peptide levels in homogenates of left atrium (LA) and left ventricle (LV) from control (C) and paced (P) dog hearts; (**c**) actin. Reproduced with permission from: [41]

(and subsequent oxidative stress) as a significant by-product of energy metabolism during its sequential acceptance of electrons in the mitochondrial ETC. The end results of NO/redox disequilibrium have implications for cardiac and vascular homeostasis and may result in the development of atherosclerosis, myocardial tissue remodeling, and hypertrophy. ROS/RNS generation is also attributed to the transit from hypertrophic to apoptotic phenotypes, a possible mechanism of myocardial failure [42].

Endothelial dysfunction is a critical component in the systemic vasoconstriction and reduced peripheral perfusion that characterizes patients with HF [43]. Endothelial regulation of vascular tone is mediated mainly by NO. Increased oxidative stress in patients with HF may be related in part to decreased bioavailability of NO secondary to reduced expression of endothelial NO synthase (eNOS) and increased generation of ROS. These react with NO in the setting of decreased antioxidant defenses that would normally clear these radicals, culminating in attenuated endothelium-dependent vasodilation particularly in patients with advanced HF. Moreover, abnormal production and/or distribution of ROS and reactive nitrosative species in blood creates oxidative and/or nitrosative stresses in the failing myocardium and endothelium.

As previously noted, the uncoupling of constitutive NOS leads to increased generation of superoxide ($O_2\cdot^-$) and peroxynitrite ($ONOO^-$). Peroxynitrite affects cardiovascular function and contributes to the pathogenesis of cardiac and endothelial dysfunction secondary to MI, chronic HF, diabetes, atherosclerosis, hypertension, aging, and various forms of shock. Moreover, pharmacological inhibition of xanthine oxidase (XO)-derived superoxide formation, as well as neutralization of peroxynitrite, or inhibition of PARP, may provide benefit in various forms of cardiovascular injury [23]. Increased oxidative stress resulting from an overproduction of superoxide mediates the dysregulation of S-nitrosylation of proteins at specific cysteine residues by RNS, a more selective modification of proteins than found with protein oxidation. This redox mechanism has been found to lead to altered myocardial excitation-contractility and vascular reactivity [23, 24].

In the vasculature, disease states such as hypertension and the activation of the renin–angiotensin–aldosterone system (RAAS) can result in reduced vascular NO, as NO synthase becomes uncoupled with the oxidative depletion of its cofactor tetrahydrobipterin (BH_4), leading to the production of superoxide ($O_2\cdot^-$) in lieu of NO [44]. Both NO and NOS have been implicated in the modulation of cardiac relaxation [45]. Using a series of carefully controlled invasive hemodynamic studies and noninvasive imaging methods, Silberman and colleagues recently studied a mouse model that developed diastolic dysfunction in the absence of systolic dysfunction or cardiac hypertrophy [46]. Cardiac tissues from these mice showed increased oxidation of BH4 as well as increased superoxide formation, reduced NO production, and dephosphorylated phospholamban. Feeding hypertensive mice BH_4 improved cardiac BH4 stores, phosphorylated phospholamban levels, and the diastolic dysfunction. The investigators concluded cardiac oxidation, independently of vascular changes, could lead to uncoupled cardiac NOS and diastolic dysfunction. The heart preparations analyzed in this study contained both the eNOS and nNOS isoforms that are expressed within cardiac myocytes, in addition to the eNOS that is robustly expressed in cardiac endothelial cells, which may raise a challenge to separating the roles of cell- and isoform-specific NOS pathways at the molecular level. Nevertheless, the observation that BH4 and related redox pathways play a key role in cardiac diastolic function, represents an important addition to the growing appreciation of the salutary biological effects of BH4 in overall cardiovascular homeostasis [47, 48].

Taken together, studies that have explored the pathways whereby BH4 ameliorates diastolic dysfunction [46] or endothelial dysfunction [49] have provided indirect evidence that implicates the RAAS in the pathophysiological response. Angiotensin II modulates ROS production in diverse biologic tissues and has been shown to stimulate eNOS-derived O_2^- production, which in turn is suppressed by BH4 [49]. The vascular dysfunction seen in models of diabetes, hypercholesterolemia, and hypertension is associated with enhanced oxidation of BH4 to BH2 and can be restored in vivo by BH4 supplementation [47, 48, 50]. Moens et al. [51] have reported that administration of BH4 can attenuate the cardiac hypertrophy and myocardial fibrosis seen in a model of pressure overload, in association with an improvement in oxidative stress. These investigators provided evidence indicating that the effects of BH4 were due to effects on NOS pathways in cardiac myocytes but did not involve activation of cGMP-dependent pathways in the heart.

To further assess the role of renin-angiotensin-aldosterone system in mediating oxidative stress in HF, we recently studied the impact of angiotensin II type 1 receptor (AT_1R) blockade on hemodynamics, LV remodeling, oxidative stress, and tissue expression of $AT(_1)R$ and angiotensin II type 2 receptors (AT_2R) in a canine model of pacing-induced HF [52]. Animals were randomized to rapid right ventricular-pacing for 3 weeks to severe heart failure and treated with candesartan or placebo from day 3 onwards, or no pacing. Candesartan significantly reduced pulmonary arterial and LV diastolic pressure, LV end-diastolic and end-systolic volume and ascites, increased cardiac output, dP/dt, and ejection fraction, while reversing the marked increase in aldehydes, a marker of oxidative stress, observed in the placebo group (Table 1). Although candesartan did not alter LV AT_1R protein expression compared to placebo or sham, it reversed the decrease in AT_2R protein observed in the placebo group. We therefore conclude that in the pacing model of heart failure, chronic AT_1R blockade attenuates hemodynamic deterioration and limits LV remodeling and dysfunction, in part by reversing oxidative stress and AT_2R downregulation.

Table 1 Effect of angiotensin receptor blockade on LV tissue aldehyde concentrations (Reproduced with permission from Moe et al. [52])

Aldehydes (ρmol/100 mg)	Sham	Paced/placebo	Paced/candesartan
Saturated			
Heptanal	148±26	297±54*	172±43***
Octanal	177±21	177±32	197±28
Decanal	72±3	77±8	61±9
Unsaturated			
t-2-Heptanal	3,447±420	3,557±810	2,387±485***
t-2-Octanal	149±18	205±19*	154±19***
t-2-Nonenal	210±18	275±36*	168±14***
t,t-2,4,-Nonadienal	17±4	24±3	18±4
4-OH-Hexanal	1,469±102	2,337±168**	1,728±268***
4-OH-Nonenal	693±53	1,200±175*	1,189±175
4-OH-Decanal	5.7±0.7	6.9±0.2	5.8±0.7
Malondialdehyde	2,009±108	14,125±1,002**	5,563±874***

*$P<0.05$ vs. sham
**$P<0.01$ vs. sham
***$P<0.05$ vs. paced/placebo dogs

Antioxidant Therapy

ROS and cellular redox states regulate an extensive number of vital pathways in the myocardium, including energy metabolism, survival and stress responses, apoptosis, inflammatory response, and oxygen sensing. Gathered observations have shown that the role ROS play in ischemia and reperfusion injury, as well as the role that antioxidant therapy may play, varies whether one use whole organ and animal models or isolated cell models; findings are often contradictory, thereby explaining in part why clinical trials of antioxidants frequently have shown mixed results. The powerful cell-damaging ROS oxidants can be neutralized by an array of protective antioxidant scavenger enzymes, as well as by various lipid and water-soluble compounds including ascorbic acid, glutathione, thioredoxin, and α-tocopherol. The antioxidant enzymes are located in a variety of cellular compartments including the mitochondria (e.g., MnSOD, glutathione peroxidase, thioredoxin reductase, peroxisomes, e.g., catalase, microsomes, e.g., cytochrome p450) and in the cytosol (e.g., CuSOD and cytosolic thioredoxin reductase). In general, there are significantly lower levels of antioxidants in myocardial mitochondria than in liver mitochondria, but the consensus opinion is that the antioxidant capacity of the heart is generally sufficient to handle normal levels of ROS production, but not the greater ROS accumulation that occurs in myocardial ischemia [53].

A mitochondrial isoform of catalase with low specific activity has been found in rats [54, 55]. This mitochondrial catalase activity was detected in the heart but not in the liver or skeletal muscle and it appeared to increase during caloric-restricted diets and in the diabetic heart [56, 57]. The role that this enzyme plays has not been fully determined although there is evidence of its participation in the prevention of excess lipid peroxidation in myocardial ischemia [58]. On the other hand, a mitochondria-specific catalase has not been found in the heart of transgenic mice even after overexpression of the catalase gene [59]. SODs catalyze the removal of superoxide radicals by the formation of H_2O_2, however, GPx catalyzes the breakdown of H_2O_2 to water and oxidized glutathione (GSSG) by using reduced glutathione (GSH). Because GPx is located in both the mitochondria and cytosol, H_2O_2 can be removed from either compartment depending on the availability of glutathione. A small fraction of the total cellular pool of GSH is sequestered in the mitochondria by the action of a carrier that transports GSH from cytosol to the mitochondrial matrix [60]. Upon exposure to increased exogenous ROS, isolated perfused rat hearts are rapidly depleted of their antioxidant reserves, including those of SOD and GSH, rendering them more vulnerable to the action of oxidative injury [61].

Another important mechanism in the antioxidant reactions is the sequestering of iron and copper ions to keep them from reacting with superoxide or H_2O_2. The antioxidant dexrazoxane prevents site-specific iron-based oxygen radical damage by chelating free and loosely bound iron. In addition, it has been used as a cardioprotective drug against doxorubicin-induced oxidative damage to myocardial mitochondria in both humans and animals [62]. The antioxidant metal-binding protein metallothionein (MT) also provides cardioprotection by reacting with ROS produced by ischemia/reperfusion and doxorubicin treatment as found in studies with a cardiac-specific MT-overexpressing transgenic mouse model [63]. MT expression is also inducible within the heart (and other tissues) by TNF-α, IL-6, doxorubicin, and metals such as cadmium and Zn [63, 64], although its cardioprotective role has yet to be determined.

The uncoupling of mitochondrial respiration from OXPHOS ATP production, by either artificial uncouplers such as 2,4-dinitrophenol (DNP) or natural uncouplers such as laurate, fatty acids, and mitochondrial uncoupling (UCP) proteins, strongly inhibits $O_2 \cdot^-$ and H_2O_2 formation in mitochondria [65, 66]. ROS production is favored when the mitochondrial membrane potential is above a specific threshold. Under conditions where the mitochondrial membrane potential is at its peak, for example, state 4 respiration, ROS production is increased. It is noteworthy that increased mitochondrial membrane potential slows electron transport through the respiratory chain, resulting in increased half-life of the ubiquinone free radical and the likelihood that electrons will interact with oxygen to form ROS [67]. Uncouplers prevent the transmembrane electrochemical H^+ potential difference ($\Delta \breve{u}_H$) from being above a threshold critical for ROS formation by respiratory complexes I and III. This has been corroborated in transgenic mice in which UCP3 protein is lacking, resulting in enhanced ROS production and increased OS in heart and skeletal muscle [68], in transgenic mice with UCP1 overexpression [69], and in cardiomyocytes with UCP2 overexpression, in which ROS is markedly attenuated [70].

Pharmacological inhibition of xanthine oxidase-derived superoxide formation and neutralization of peroxynitrite or inhibition of poly(ADP-ribose) polymerase (PARP) have been reported to provide significant benefit in various forms of cardiovascular injury [71]. Using rat and mouse models of HF, beneficial effects of a novel ultrapotent PARP inhibitor have also been demonstrated by Pacher et al. [72]. The effect of INO-1001 on the development of HF induced by permanent ligation of the left anterior descending coronary artery was assessed, as was HF induced by doxorubicin, and acute HF induced by bacterial endotoxin. In the coronary ligation model, significantly depressed LV performance and impaired vascular relaxation of aortic rings were found; PARP inhibition significantly improved both cardiac function and vascular relaxation. In the doxorubicin model, a single injection of doxorubicin induced high mortality and a significant decrease in LV systolic pressure, $+dP/dt$, $-dP/dt$, stroke volume, stroke work, ejection fraction, and cardiac output. Treatment with the PARP inhibitor reduced doxorubicin-induced mortality and markedly improved cardiac function; on the other hand, PARP inhibition did not interfere with doxorubicin's antitumor effect. In the endotoxin model of cardiac dysfunction, PARP inhibition attenuated the reduction in myocardial contractility elicited by endotoxin. Taken together, these data support the view that PARP inhibition may be an effective approach in the experimental treatment of various forms of acute and chronic HF.

Mitochondrial ROS and OS are implicated in the pathogenesis of the cardiac damage elicited by ischemia/reperfusion and the cardiomyopathies associated with Fridreich's ataxia (FRDA) and doxorubicin [73, 74]. Several oxygen radical scavengers including CoQ10, vitamin E, dexrazoxane, and idebenone have been used for treatment [75–77]. In doxorubicin-induced cardiomyopathy, the free-radical scavenger dexrazoxane has been shown to protect the heart from doxorubicin-associated oxidative damage, and has been recommended for clinical use to attenuate the myocardial damage that may occur in children with acute lymphoblastic leukemia treated with doxorubicin chemotherapy [78]. Both CoQ10 and idebenone have been found to markedly improve cardiac function and reduce cardiac hypertrophy in patients with FRDA [79–81]. Of interest, ataxia and other central nervous system symptoms are less affected by the administration of these antioxidants than the cardiac phenotype. In addition, idebenone seems to improve the cardiac dysfunction observed in mitochondrial cardiomyopathy [75].

CoQ10, in addition to its role as an antioxidant, also serves multiple cellular functions, including participation as an electron carrier in the respiratory chain and as an activating cofactor for mitochondrial uncoupling proteins. Furthermore, CoQ10 appears to have a beneficial effect in several neurological disorders with cardiac involvement, including MELAS and KSS syndromes [82]. Moreover, a significant reduction in the incidence of cardiac conduction abnormalities seen in patients with KSS or CPEO syndromes has been reported using CoQ10 at relatively high doses ranging from 60 to 150 mg/day [83]. Also, clinical improvement was observed in patients with advanced HF after CoQ10 supplementation to standard therapy [84]. Nonetheless, because the sample size and the design used in these studies raised concerns as to general clinical application of CoQ10 in treating HF, a large double-blind multisite clinical trial is currently underway to test its efficacy [85].

β-blocker and angiotensin-converting enzyme (ACE) inhibitor are routinely used in the management of HF. Whether carvedilol, a vasodilator β-blocker with antioxidant activity, indeed exerts beneficial antioxidant effects in patients with HF remains controversial. Whereas some investigators reported a reduction of oxidative stress in HF patients treated with carvedilol [86] others did not [87]. Using immunohistochemistry, Nakamura et al. evaluated the expression of 4-hydroxy-2-nonenal (HNE)-modified protein (a major lipid peroxidation product) in endomyocardial biopsy tissues from 23 patients with dilated cardiomyopathy (DCM) and 13 control subjects with normal cardiac function [88]. Expression of HNE-modified protein was found in all myocardial tissue samples from patients with DCM. Expression was distinctive in cardiomyocytes cytosol. Myocardial HNE-modified protein levels in patients with DCM were significantly increased compared with the levels in control subjects. In addition, biopsy samples from 11 patients with DCM were examined before and after treatment with carvedilol (5–30 mg/day; mean dosage, 22 ± 8 mg/day). Following treatment, myocardial HNE-modified protein levels decreased by 40% and with improvement in HF. Taken together, this study confirmed that oxidative stress is elevated in HF and the administration of carvedilol resulted in reduction of oxidative stress and improvement in cardiac function.

These same investigators have also examined whether levels of 8-hydroxy-2-deoxyguanosine (8-OHdG), a marker of oxidative DNA damage, were elevated in the serum and myocardium of patients with DCM, and whether carvedilol could decrease 8-OHdG [89]. Serum 8-OHdG was measured by enzyme immunoassay in

56 patients with DCM and in 20 control subjects. DCM patients had significantly increased levels of 8-OHdG compared to control subjects. Interestingly, immunohistochemically positive 8-OHdG staining was detected in the nuclei of cardiomyocytes from DCM endomyocardial tissue patients, but not in controls. After treatment with carvedilol, the serum levels of 8-OHdG in DCM patients decreased by 19%, together with improvement in HF. Thus, levels of 8-OHdG are elevated in serum and myocardium of HF patients and carvedilol seems to be an effective way to reduce oxidative DNA damage.

The antioxidative properties of β-blockers and ACE inhibitors in severe HF have been studied by Chin et al. [90]. From a group of 66 outpatients with HF, 46 patients who were on an established treatment with ACE inhibitors were started on β-blockers, and 20 patients not previously on ACE-inhibitors were treated with lisinopril. Baseline parameters were compared to 22 healthy control subjects. Serum lipid hydroperoxides (LHP) and total antioxidant capacity (TAC) were determined as indices of oxidative damage and antioxidant defense, and plasma von Willebrand factor (vWf) as an index of endothelial damage/dysfunction. The baseline indices for the measures of oxidative damage and endothelial function in the 66 HF patients were significantly higher than healthy control subjects. After 3 months of maintenance therapy with β-blockers, a significant reduction in LHP levels occurred, but not in TAC or vWf. ACE inhibitor therapy also significantly reduced vWf levels, but did not have any statistically significant effects on LHP or TAC. These findings suggest that oxidative stress in advanced severe HF may be due to increased free radical production or inefficient free radical clearance by scavengers and that β-blockers, but not ACE inhibitors, reduced lipid peroxidation (although no relation was found between a reduction in oxidative damage and endothelial damage/dysfunction).

Treatment with ACE inhibitors, angiotensin, aldosterone, and endothelin receptor blockers has been shown to beneficially modulate endothelial dysfunction in severe HF. As pointed out by Bauersachs and Widder [91], these therapies increase NO bioactivity by either modulation of ROS generation, thereby preventing the interaction of superoxide anions with NO, and/or by increasing eNOS expression/activity. In experiments in rats with large MI, treatment with AVE9488, a novel eNOS transcription enhancer, attenuates cardiac remodeling and endothelial dysfunction. Furthermore, antioxidants, L-arginine, cofactors of endothelial NO-synthase, and exercise training positively modulate endothelial function [92].

Conclusions

Molecular oxygen is central in both the formation of NO, a primary determinant of both vascular tone and cardiac contractility and in the generation of ROS, as a significant by-product of energy metabolism during its sequential acceptance of electrons in the mitochondrial ETC. These short-lived intermediates can either act as an important signaling molecule or induce irreversible oxidative damage to proteins, lipids, and nucleic acids, therefore ROS and oxygen exert both beneficial and deleterious effects. ROS play an integral role not only in the genesis of CAD but also in its progression. Several in vitro and animal studies have demonstrated that in the failing heart, ROS influence several components of the cardiac phenotype and cardiac remodeling, including contractile function, interstitial fibrosis, endothelial dysfunction, and myocyte hypertrophy. Furthermore, ROS contribute to the remodeling processes in a number of ways including activation of MMPs that participate in reconfiguration of the ECM, acting as signaling molecules in the development of compensatory hypertrophy, and contributing to myocyte loss via apoptosis signaling. ROS can be generated in the heart and endothelial tissues by nonmitochondrial reactions, including the involvement of xanthine oxidase (XO), NAD(P)H oxidases, and cytochrome p450. In addition, increased ROS and toxic oxygen metabolite production in both the myocardial mitochondrial organelle and leukocytes are exacerbated by readmission of oxygen during postischemic reperfusion.

Oxidative stress, resulting from both increased ROS generation and diminished antioxidant protection promotes the oxidation of thiol groups in proteins and lipid peroxidation leading initially to reversible damage, and finally to necrosis. Reduced bioavailability of NO and increased generation of ROS within the vascular wall are key determinants in endothelial dysfunction, and this imbalance between NO and ROS mainly results from neurohumoral activation associated with HF. Furthermore, the activation of the RAAS may play a central role. Although the use of scavengers to treat oxidative stress has been considered for quite some time, the success of this approach has been questionable at least. Nonetheless, there is increasing evidence that HF generation of ROS/RNS may be ameliorated by targeting ROS/RNS-generating enzymes and upstream mediators. β-blockers, but not necessarily ACE inhibitors, appear to reduce lipid peroxidation although so far no relation has been demonstrated between a reduction in oxidative damage and endothelial damage/dysfunction. Nevertheless, therapies that ameliorate endothelial function in heart failure have been shown to improve exercise tolerance and outcomes making endothelial dysfunction an important target to develop novel therapies based on the concepts reviewed in this chapter.

References

1. Hess ML, Manson NH. Molecular oxygen: friend and foe. The role of the oxygen free radical system in the calcium paradox, the oxygen paradox and ischemia/reperfusion injury. *J Mol Cell Cardiol* 1984;16:969–985.
2. Becker LB, vanden Hoek TL, Shao ZH, Li CQ, Schumacker PT. Generation of superoxide in cardiomyocytes during ischemia before reperfusion. *Am J Physiol* 1999;277:H2240–H2246.
3. Becker LB. New concepts in reactive oxygen species and cardiovascular reperfusion physiology. *Cardiovasc Res* 2004;61:461–470.
4. Marin-Garcia J, Goldenthal MJ, Moe GW. Abnormal cardiac and skeletal muscle mitochondrial function in pacing-induced cardiac failure. *Cardiovasc Res* 2001;52:103–110.

5. Moe GW, Marin-Garcia J, Konig A, Goldenthal M, Lu X, Feng Q. In vivo TNF-alpha inhibition ameliorates cardiac mitochondrial dysfunction, oxidative stress, and apoptosis in experimental heart failure. *Am J Physiol Heart Circ Physiol* 2004;287:H1813–H1820.
6. Ide T, Tsutsui H, Kinugawa S, Utsumi H, Kang D, Hattori N, Uchida K, Arimura K, Egashira K, Takeshita A. Mitochondrial electron transport complex I is a potential source of oxygen free radicals in the failing myocardium. *Circ Res* 1999;85:357–363.
7. Ide T, Tsutsui H, Kinugawa S, Suematsu N, Hayashidani S, Ichikawa K, Utsumi H, Machida Y, Egashira K, Takeshita A. Direct evidence for increased hydroxyl radicals originating from superoxide in the failing myocardium. *Circ Res* 2000;86:152–157.
8. Marin-Garcia J, Goldenthal MJ, Ananthakrishnan R, Mirvis D. Specific mitochondrial DNA deletions in canine myocardial ischemia. *Biochem Mol Biol Int* 1996;40:1057–1065.
9. Suematsu N, Tsutsui H, Wen J, Kang D, Ikeuchi M, Ide T, Hayashidani S, Shiomi T, Kubota T, Hamasaki N, Takeshita A. Oxidative stress mediates tumor necrosis factor-alpha-induced mitochondrial DNA damage and dysfunction in cardiac myocytes. *Circulation* 2003;107:1418–1423.
10. Ide T, Tsutsui H, Hayashidani S, Kang D, Suematsu N, Nakamura K, Utsumi H, Hamasaki N, Takeshita A. Mitochondrial DNA damage and dysfunction associated with oxidative stress in failing hearts after myocardial infarction. *Circ Res* 2001;88:529–535.
11. Tsutsui H, Kinugawa S, Matsushima S. Mitochondrial oxidative stress and dysfunction in myocardial remodelling. *Cardiovasc Res* 2009;81:449–456.
12. Sorescu D, Griendling KK. Reactive oxygen species, mitochondria, and NAD(P)H oxidases in the development and progression of heart failure. *Congest Heart Fail* 2002;8:132–140.
13. Griendling KK, Sorescu D, Ushio-Fukai M. NAD(P)H oxidase: role in cardiovascular biology and disease. *Circ Res* 2000;86:494–501.
14. Sabri A, Hughie HH, Lucchesi PA. Regulation of hypertrophic and apoptotic signaling pathways by reactive oxygen species in cardiac myocytes. *Antioxid Redox Signal* 2003;5:731–740.
15. Heymes C, Bendall JK, Ratajczak P, Cave AC, Samuel JL, Hasenfuss G, Shah AM. Increased myocardial NADPH oxidase activity in human heart failure. *J Am Coll Cardiol* 2003;41:2164–2171.
16. Nakagami H, Liao JK. Statins and myocardial hypertrophy. *Coron Artery Dis* 2004;15: 247–250.
17. Maack C, Kartes T, Kilter H, Schafers HJ, Nickenig G, Bohm M, Laufs U. Oxygen free radical release in human failing myocardium is associated with increased activity of rac1-GTPase and represents a target for statin treatment. *Circulation* 2003;108:1567–1574.
18. Ferrari R, Guardigli G, Mele D, Percoco GF, Ceconi C, Curello S. Oxidative stress during myocardial ischaemia and heart failure. *Curr Pharm Des* 2004;10:1699–1711.
19. Ijsselmuiden AJ, Musters RJ, de RG, van HL, rse-Baas F, van SM, Leyte A, Tangelder GJ, Laarman GJ, Paulus WJ. Circulating white blood cells and platelets amplify oxidative stress in heart failure. *Nat Clin Pract Cardiovasc Med* 2008;5:811–820.
20. Khatri JJ, Johnson C, Magid R, Lessner SM, Laude KM, Dikalov SI, Harrison DG, Sung HJ, Rong Y, Galis ZS. Vascular oxidant stress enhances progression and angiogenesis of experimental atheroma. *Circulation* 2004;109:520–525.
21. Witztum JL, Steinberg D. Role of oxidized low density lipoprotein in atherogenesis. *J Clin Invest* 1991;88:1785–1792.
22. Malinski T. Understanding nitric oxide physiology in the heart: a nanomedical approach. *Am J Cardiol* 2005;96:13i–24i.
23. Hare JM, Stamler JS. NO/redox disequilibrium in the failing heart and cardiovascular system. *J Clin Invest* 2005;115:509–517.
24. Martinez-Ruiz A, Lamas S. S-nitrosylation: a potential new paradigm in signal transduction. *Cardiovasc Res* 2004;62:43–52.
25. Kannan K, Jain SK. Oxidative stress and apoptosis. *Pathophysiology* 2000;7:153–163.
26. Keller JN, Kindy MS, Holtsberg FW, St Clair DK, Yen HC, Germeyer A, Steiner SM, Bruce-Keller AJ, Hutchins JB, Mattson MP. Mitochondrial manganese superoxide dismutase prevents neural apoptosis and reduces ischemic brain injury: suppression of peroxynitrite production, lipid peroxidation, and mitochondrial dysfunction. *J Neurosci* 1998;18:687–697.

27. Shiomi T, Tsutsui H, Matsusaka H, Murakami K, Hayashidani S, Ikeuchi M, Wen J, Kubota T, Utsumi H, Takeshita A. Overexpression of glutathione peroxidase prevents left ventricular remodeling and failure after myocardial infarction in mice. *Circulation* 2004;109:544–549.
28. Lud CJ, Harrington B, Ordonez S. Bcl-2 overexpression attenuates dopamine-induced apoptosis in an immortalized neural cell line by suppressing the production of reactive oxygen species. *Synapse* 2000;35:228–233.
29. Fu Y, Gu Y, Zheng Z, Wasteneys G, Yang Z. Arabidopsis interdigitating cell growth requires two antagonistic pathways with opposing action on cell morphogenesis. *Cell* 2005;120:687–700.
30. Kokoszka JE, Coskun P, Esposito LA, Wallace DC. Increased mitochondrial oxidative stress in the Sod2 (+/–) mouse results in the age-related decline of mitochondrial function culminating in increased apoptosis. *Proc Natl Acad Sci U S A* 2001;98:2278–2283.
31. Goossens V, Stange G, Moens K, Pipeleers D, Grooten J. Regulation of tumor necrosis factor-induced, mitochondria- and reactive oxygen species-dependent cell death by the electron flux through the electron transport chain complex I. *Antioxid Redox Signal* 1999;1:285–295.
32. von HR, Li PF, Dietz R. Signaling pathways in reactive oxygen species-induced cardiomyocyte apoptosis. *Circulation* 1999;99:2934–2941.
33. Long X, Goldenthal MJ, Wu GM, Marin-Garcia J. Mitochondrial Ca2+ flux and respiratory enzyme activity decline are early events in cardiomyocyte response to H2O2. *J Mol Cell Cardiol* 2004;37:63–70.
34. Hickson-Bick DL, Sparagna GC, Buja LM, McMillin JB. Palmitate-induced apoptosis in neonatal cardiomyocytes is not dependent on the generation of ROS. *Am J Physiol Heart Circ Physiol* 2002;282:H656–H664.
35. Cesselli D, Jakoniuk I, Barlucchi L, Beltrami AP, Hintze TH, Nadal-Ginard B, Kajstura J, Leri A, Anversa P. Oxidative stress-mediated cardiac cell death is a major determinant of ventricular dysfunction and failure in dog dilated cardiomyopathy. *Circ Res* 2001;89:279–286.
36. Migliaccio E, Giorgio M, Mele S, Pelicci G, Reboldi P, Pandolfi PP, Lanfrancone L, Pelicci PG. The p66shc adaptor protein controls oxidative stress response and life span in mammals. *Nature* 1999;402:309–313.
37. Cha YM, Dzeja PP, Shen WK, Jahangir A, Hart CY, Terzic A, Redfield MM. Failing atrial myocardium: energetic deficits accompany structural remodeling and electrical instability. *Am J Physiol Heart Circ Physiol* 2003;284:H1313–H1320.
38. Osorio JC, Stanley WC, Linke A, Castellari M, Diep QN, Panchal AR, Hintze TH, Lopaschuk GD, Recchia FA. Impaired myocardial fatty acid oxidation and reduced protein expression of retinoid X receptor-alpha in pacing-induced heart failure. *Circulation* 2002;106:606–612.
39. Ananthakrishnan R, Moe GW, Goldenthal MJ, Marin-Garcia J. Akt signaling pathway in pacing-induced heart failure. *Mol Cell Biochem* 2005;268:103–110.
40. Chandel NS, McClintock DS, Feliciano CE, Wood TM, Melendez JA, Rodriguez AM, Schumacker PT. Reactive oxygen species generated at mitochondrial complex III stabilize hypoxia-inducible factor-1alpha during hypoxia: a mechanism of O2 sensing. *J Biol Chem* 2000;275:25130–25138.
41. Marin-Garcia J, Goldenthal MJ, Damle S, Pi Y, Moe GW. Regional distribution of mitochondrial dysfunction and apoptotic remodeling in pacing-induced heart failure. *J Card Fail* 2009;15:700–708.
42. Elahi MM, Naseem KM, Matata BM. Nitric oxide in blood. The nitrosative-oxidative disequilibrium hypothesis on the pathogenesis of cardiovascular disease. *FEBS J* 2007;274:906–923.
43. Sharma R, Davidoff MN. Oxidative stress and endothelial dysfunction in heart failure. *Congest Heart Fail* 2002;8:165–172.
44. Landmesser U, Dikalov S, Price SR, McCann L, Fukai T, Holland SM, Mitch WE, Harrison DG. Oxidation of tetrahydrobiopterin leads to uncoupling of endothelial cell nitric oxide synthase in hypertension. *J Clin Invest* 2003;111:1201–1209.
45. Ziolo MT, Kohr MJ, Wang H. Nitric oxide signaling and the regulation of myocardial function. *J Mol Cell Cardiol* 2008;45:625–632.
46. Silberman GA, Fan TH, Liu H, Jiao Z, Xiao HD, Lovelock JD, Boulden BM, Widder J, Fredd S, Bernstein KE, Wolska BM, Dikalov S, Harrison DG, Dudley SC, Jr. Uncoupled cardiac nitric oxide synthase mediates diastolic dysfunction. *Circulation* 2010;121:519–528.

47. Katusic ZS, d'Uscio LV, Nath KA. Vascular protection by tetrahydrobiopterin: progress and therapeutic prospects. *Trends Pharmacol Sci* 2009;30:48–54.
48. Moens AL, Kass DA. Therapeutic potential of tetrahydrobiopterin for treating vascular and cardiac disease. *J Cardiovasc Pharmacol* 2007;50:238–246.
49. Chalupsky K, Cai H. Endothelial dihydrofolate reductase: critical for nitric oxide bioavailability and role in angiotensin II uncoupling of endothelial nitric oxide synthase. *Proc Natl Acad Sci U S A* 2005;102:9056–9061.
50. Vasquez-Vivar J. Tetrahydrobiopterin, superoxide, and vascular dysfunction. *Free Radic Biol Med* 2009;47:1108–1119.
51. Moens AL, Takimoto E, Tocchetti CG, Chakir K, Bedja D, Cormaci G, Ketner EA, Majmudar M, Gabrielson K, Halushka MK, Mitchell JB, Biswal S, Channon KM, Wolin MS, Alp NJ, Paolocci N, Champion HC, Kass DA. Reversal of cardiac hypertrophy and fibrosis from pressure overload by tetrahydrobiopterin: efficacy of recoupling nitric oxide synthase as a therapeutic strategy. *Circulation* 2008;117:2626–2636.
52. Moe G, Konig A, Liu P, Jugdutt BI. Selective type 1 angiotensin II receptor blockade attenuates oxidative stress and regulates angiotensin II receptors in the canine failing heart. *Mol Cell Biochem* 2008;317:97–104.
53. Chen Y, Saari JT, Kang YJ. Weak antioxidant defenses make the heart a target for damage in copper-deficient rats. *Free Radic Biol Med* 1994;17:529–536.
54. Radi R, Turrens JF, Chang LY, Bush KM, Crapo JD, Freeman BA. Detection of catalase in rat heart mitochondria. *J Biol Chem* 1991;266:22028–22034.
55. Antunes F, Han D, Cadenas E. Relative contributions of heart mitochondria glutathione peroxidase and catalase to H(2)O(2) detoxification in in vivo conditions. *Free Radic Biol Med* 2002;33:1260–1267.
56. Phung CD, Ezieme JA, Turrens JF. Hydrogen peroxide metabolism in skeletal muscle mitochondria. *Arch Biochem Biophys* 1994;315:479–482.
57. Judge S, Judge A, Grune T, Leeuwenburgh C. Short-term CR decreases cardiac mitochondrial oxidant production but increases carbonyl content. *Am J Physiol Regul Integr Comp Physiol* 2004;286:R254–R259.
58. Radi R, Bush KM, Freeman BA. The role of cytochrome c and mitochondrial catalase in hydroperoxide-induced heart mitochondrial lipid peroxidation. *Arch Biochem Biophys* 1993;300:409–415.
59. Zhou Z, Kang YJ. Cellular and subcellular localization of catalase in the heart of transgenic mice. *J Histochem Cytochem* 2000;48:585–594.
60. Fernandez-Checa JC, Garcia-Ruiz C, Colell A, Morales A, Mari M, Miranda M, Ardite E. Oxidative stress: role of mitochondria and protection by glutathione. *Biofactors* 1998;8:7–11.
61. Vaage J, Antonelli M, Bufi M, Irtun O, DeBlasi RA, Corbucci GG, Gasparetto A, Semb AG. Exogenous reactive oxygen species deplete the isolated rat heart of antioxidants. *Free Radic Biol Med* 1997;22:85–92.
62. Hasinoff BB, Schnabl KL, Marusak RA, Patel D, Huebner E. Dexrazoxane (ICRF-187) protects cardiac myocytes against doxorubicin by preventing damage to mitochondria. *Cardiovasc Toxicol* 2003;3:89–99.
63. Kang YJ. The antioxidant function of metallothionein in the heart. *Proc Soc Exp Biol Med* 1999;222:263–273.
64. Ali MM, Frei E, Straub J, Breuer A, Wiessler M. Induction of metallothionein by zinc protects from daunorubicin toxicity in rats. *Toxicology* 2002;179:85–93.
65. Okuda M, Lee HC, Kumar C, Chance B. Comparison of the effect of a mitochondrial uncoupler, 2,4-dinitrophenol and adrenaline on oxygen radical production in the isolated perfused rat liver. *Acta Physiol Scand* 1992;145:159–168.
66. Casteilla L, Rigoulet M, Penicaud L. Mitochondrial ROS metabolism: modulation by uncoupling protein. *IUBMB Life* 2001;52:181–188.
67. Papa S, Skulachev VP. Reactive oxygen species, mitochondria, apoptosis and aging. *Mol Cell Biochem* 1997;174:305–319.

68. Vidal-Puig AJ, Grujic D, Zhang CY, Hagen T, Boss O, Ido Y, Szczepanik A, Wade J, Mootha V, Cortright R, Muoio DM, Lowell BB. Energy metabolism in uncoupling protein 3 gene knockout mice. *J Biol Chem* 2000;275:16258–16266.
69. Hoerter J, Gonzalez-Barroso MD, Couplan E, Mateo P, Gelly C, Cassard-Doulcier AM, Diolez P, Bouillaud F. Mitochondrial uncoupling protein 1 expressed in the heart of transgenic mice protects against ischemic-reperfusion damage. *Circulation* 2004;110:528–533.
70. Teshima Y, Akao M, Jones SP, Marban E. Uncoupling protein-2 overexpression inhibits mitochondrial death pathway in cardiomyocytes. *Circ Res* 2003;93:192–200.
71. Ungvari Z, Gupte SA, Recchia FA, Batkai S, Pacher P. Role of oxidative-nitrosative stress and downstream pathways in various forms of cardiomyopathy and heart failure. *Curr Vasc Pharmacol* 2005;3:221–229.
72. Pacher P, Liaudet L, Mabley JG, Cziraki A, Hasko G, Szabo C. Beneficial effects of a novel ultrapotent poly(ADP-ribose) polymerase inhibitor in murine models of heart failure. *Int J Mol Med* 2006;17:369–375.
73. Cooper JM, Schapira AH. Friedreich's Ataxia: disease mechanisms, antioxidant and Coenzyme Q10 therapy. *Biofactors* 2003;18:163–171.
74. Santos DL, Moreno AJ, Leino RL, Froberg MK, Wallace KB. Carvedilol protects against doxorubicin-induced mitochondrial cardiomyopathy. *Toxicol Appl Pharmacol* 2002;185:218–227.
75. Lerman-Sagie T, Rustin P, Lev D, Yanoov M, Leshinsky-Silver E, Sagie A, Ben-Gal T, Munnich A. Dramatic improvement in mitochondrial cardiomyopathy following treatment with idebenone. *J Inherit Metab Dis* 2001;24:28–34.
76. Sayed-Ahmed MM, Salman TM, Gaballah HE, bou El-Naga SA, Nicolai R, Calvani M. Propionyl-L-carnitine as protector against adriamycin-induced cardiomyopathy. *Pharmacol Res* 2001;43:513–520.
77. Shite J, Qin F, Mao W, Kawai H, Stevens SY, Liang C. Antioxidant vitamins attenuate oxidative stress and cardiac dysfunction in tachycardia-induced cardiomyopathy. *J Am Coll Cardiol* 2001;38:1734–1740.
78. Lipshultz SE, Rifai N, Dalton VM, Levy DE, Silverman LB, Lipsitz SR, Colan SD, Asselin BL, Barr RD, Clavell LA, Hurwitz CA, Moghrabi A, Samson Y, Schorin MA, Gelber RD, Sallan SE. The effect of dexrazoxane on myocardial injury in doxorubicin-treated children with acute lymphoblastic leukemia. *N Engl J Med* 2004;351:145–153.
79. Geromel V, Darin N, Chretien D, Benit P, DeLonlay P, Rotig A, Munnich A, Rustin P. Coenzyme Q(10) and idebenone in the therapy of respiratory chain diseases: rationale and comparative benefits. *Mol Genet Metab* 2002;77:21–30.
80. Hausse AO, Aggoun Y, Bonnet D, Sidi D, Munnich A, Rotig A, Rustin P. Idebenone and reduced cardiac hypertrophy in Friedreich's ataxia. *Heart* 2002;87:346–349.
81. Rustin P, Munnich A, Rotig A. Quinone analogs prevent enzymes targeted in Friedreich ataxia from iron-induced injury in vitro. *Biofactors* 1999;9:247–251.
82. Shoffner JM, Wallace DC. Oxidative phosphorylation diseases and mitochondrial DNA mutations: diagnosis and treatment. *Annu Rev Nutr* 1994;14:535–568.
83. Ogasahara S, Yorifuji S, Nishikawa Y, Takahashi M, Wada K, Hazama T, Nakamura Y, Hashimoto S, Kono N, Tarui S. Improvement of abnormal pyruvate metabolism and cardiac conduction defect with coenzyme Q10 in Kearns-Sayre syndrome. *Neurology* 1985;35:372–377.
84. Mortensen SA, Vadhanavikit S, Baandrup U, Folkers K. Long-term coenzyme Q10 therapy: a major advance in the management of resistant myocardial failure. *Drugs Exp Clin Res* 1985;11:581–593.
85. Hargreaves IP. Ubiquinone: cholesterol's reclusive cousin. *Ann Clin Biochem* 2003;40: 207–218.
86. Castro P, Vukasovic JL, Chiong M, Díaz-ya az-Araya G, Alcaino H, Copaja M, Valenzuela R, Greig D, Perez O, Corbalan R, Lavandero S. Effects of carvedilol on oxidative stress and chronotropic response to exercise in patients with chronic heart failure. *Eur J Heart Fail* 2005;7:1033–1039.

87. Chin BS, Gibbs CR, Blann AD, Lip GY. Neither carvedilol nor bisoprolol in maximally tolerated doses has any specific advantage in lowering chronic heart failure oxidant stress: implications for beta-blocker selection. *Clin Sci (Lond)* 2003;105:507–512.
88. Nakamura K, Kusano K, Nakamura Y, Kakishita M, Ohta K, Nagase S, Yamamoto M, Miyaji K, Saito H, Morita H, Emori T, Matsubara H, Toyokuni S, Ohe T. Carvedilol decreases elevated oxidative stress in human failing myocardium. *Circulation* 2002;105:2867–2871.
89. Kono Y, Nakamura K, Kimura H, Nishii N, Watanabe A, Banba K, Miura A, Nagase S, Sakuragi S, Kusano KF, Matsubara H, Ohe T. Elevated levels of oxidative DNA damage in serum and myocardium of patients with heart failure. *Circ J* 2006;70:1001–1005.
90. Chin BS, Langford NJ, Nuttall SL, Gibbs CR, Blann AD, Lip GY. Anti-oxidative properties of beta-blockers and angiotensin-converting enzyme inhibitors in congestive heart failure. *Eur J Heart Fail* 2003;5:171–174.
91. Bauersachs J, Widder JD. Endothelial dysfunction in heart failure. *Pharmacol Rep* 2008;60:119–126.
92. Bauersachs J, Schafer A. Endothelial dysfunction in heart failure: mechanisms and therapeutic approaches. *Curr Vasc Pharmacol* 2004;2:115–124.

Nutrient Selection in the Management of Canine Atopic Dermatitis

John Kuck

Abstract Atopy, or atopic dermatitis is a common allergic skin condition affecting dogs in North America. As a result of type-I hypersensitivity reaction, individuals afflicted with atopy experience intense pruritus, bacterial and fungal skin infection, and lifelong therapy to decrease the severity of the disease. Oxidative stress resulting from and being caused by allergic inflammation is thought to perpetuate skin irritation and magnify the severity of disease manifestation. Traditional anti-inflammatory and immune-suppressant therapies are effective in most cases at controlling signs of disease but often carry significant costs and side effects for the patient. Recent interest in alternative or complementary therapies for atopy has resulted in proliferation of dermatology-specific antioxidant preparations for veterinary patients. Compounds of particular interest include proanthocyanidins, omega three fatty acids, alpha lipoic acid, and spirulina. As clinical experience with the use of such treatment alternatives grows and as this experience gathers support by scientific research, antioxidants may become commonplace adjunct therapy in the treatment of many allergic skin conditions.

Keywords Atopic dermatitis • Oxidative stress • Antioxidant • Proanthocyanidin • Omega three fatty acids • Alpha lipoic acid • Spirulina • TBARS • Total antioxidant power

Introduction

Allergic dermatitis remains the most common presenting complaint attended to by veterinary practitioners in North America with regard to diseases of the skin. Atopy, or sensitivity to inhaled or percutaneously absorbed environmental allergens

J. Kuck (✉)
Aspen Animal Hospital, Nutrient Selection in the Management of Canine Atopic Dermatitis, Aspen, CO, USA
e-mail: john_kuck@yahoo.com

L. Mandelker and P. Vajdovich (eds.), *Studies on Veterinary Medicine*, Oxidative Stress in Applied Basic Research and Clinical Practice 5,
DOI 10.1007/978-1-61779-071-3_10,

is second only to flea bite hypersensitivity as the most common form of allergic dermatitis. Atopy displays a strong familial and breed predilection and appears to be polygenic in origin. Treatment of atopic dermatitis has evolved with our understanding of the disease and traditionally has been based upon immune modulation and anti-inflammatory therapies. Increased interest among veterinarians and pet owners in alternative treatment modalities has developed rapidly over the last decade and has resulted in a proliferation of natural-based products for treatment of skin diseases, namely antioxidants. The perceived ambiguity of just how natural products work on a biochemical and physiological level has made product recommendation difficult for many veterinarians attempting to meet the demand for rational alternative and complementary care. A review of the medical literature does prove rewarding for an investigator looking for scientific data, but the search can be time consuming. The purpose of this chapter is to summarize some of the more promising nutrient therapies for atopic dermatitis as well as review the disease itself, comparing and contrasting the mechanisms of traditional and complementary approaches. Although not an exhaustive summary, this chapter may aid the general practitioner in selecting adjunct approaches to the treatment of atopic dermatitis. For most patients alternative therapies will never supplant traditional care, but they can complement chronic disease management, improve client relations, build trust, and reduce the side effects and cost associated with standard therapies.

Hypersensitivity reactions occur as a result of very complex sequences of events that are beyond the scope of this chapter. Outlined here are the major cellular and biochemical interactions simplified for the purpose of review and highlighting where antioxidant and nutrient therapies can be of benefit both in a theoretical and practical sense. In healthy individuals, immune surveillance of benign environmental antigens results in the formation of either IgG or IgM, effectively binding and facilitating the removal of foreign material without a systemic response. In atopic individuals the immune system forms IgE antibodies against some antigens, initiating the cascade of events that cumulatively result in allergy, otherwise known as atopic dermatitis. IgE formation in healthy individuals is generally reserved for response to parasitic infestation. This response, referred to as a type-I hypersensitivity reaction, can be divided into two distinct phases: the immediate and delayed phase responses.

The immediate phase is seen just minutes following allergen exposure and is initiated as previously sensitized IgE-activated mast cells degranulate in response to intimate contact with the allergen. Mast cells release their granules of histamine as well as prostaglandin D_2 into the capillary beds of the skin causing capillaries to dilate, endothelium to contract, and tissue edema to form as a result of vascular leak. The late phase response in type-I hypersensitivity reactions begins 2–4 h postallergen exposure and involves the recruitment of inflammatory leukocytes, primarily eosinophils, helper T lymphocytes (T_H2s, CD4 T cells), and neutrophils, into affected areas of the skin. Tissue macrophages elaborate tumor necrosis factor alpha which in turn increases the expression of leukocyte adhesion molecules, allowing escape of leukocytes into the interstitium through previously dilated capillaries [1]. This response is perpetuated through continued allergen exposure and complicating sequelae such as pyoderma and secondary yeast proliferation.

Oxidative stress is a term assigned to a state in which an individual's ability to reduce oxidants in situ has been diminished by disease or the production of oxidants has overwhelmed the normal defense mechanisms, disrupting homeostasis. Free radicals are defined as independent molecules with one or more unpaired electrons. As with most inflammatory diseases, the production of several species of radical oxygen and nitrogen are produced in above normal and locally toxic levels in patients suffering with atopy. It has been demonstrated that significant disruptions to homeostasis occur in human pediatric patients suffering with atopic dermatitis [21]. Both markers of oxidative stress (8-hydroxy-2′-deoxyguanosine, nitrite/nitrate) and antioxidant status (selenium) experienced significant derangements when compared to healthy controls [2].

Cellular respiration is the main source of oxygen radical formation but many enzymatic pathways are capable of producing various radical species. Superoxide radicals, hydroxyl radicals, hydrogen peroxide, hypochlorous acid, and nitric oxide all result from and have the potential to incite inflammation during a sensitivity reaction [3]. A quantitative assessment of the magnitude of oxidative stress as well as the antioxidant response of patients may eventually be within the reach of practitioners, but for now it remains the realm of the researcher. Interest in measuring oxidative stress and quantifying its effects has led to recent advancements in methods of testing for oxidized by-products of DNA, protein, and lipids. These assays may assume a commercial form available through veterinary diagnostic laboratories in the future allowing the practitioner to identify individuals under oxidative stress and assist in selecting therapeutic intervention [4]. Until that time, subjective assessment of clinical response remains our most practical tool for guiding therapy.

Treatment for allergic dermatitis, whether due to flea bite sensitivity or atopic dermatitis, has been directed toward intervening at crucial steps along the chain of events leading to disease manifestation. Glucocorticoids, long the mainstay therapy for allergic conditions, affect leukocyte kinetics, phagocytic defenses, cell-mediated immunity, humeral immunity, and production of inflammatory mediators [5]. Cyclosporine binds to specific intracellular receptors in T-lymphocytes. This action inhibits synthesis of cytokines interleukin 2 (IL-2), interleukin 3 (IL-3), and tumor necrosis factor (TNF) resulting in a suppression of the activated T-lymphocytes [22]. Subsequent generations of specific T-cell inhibitors, namely tacrolimus and pimecrolimus, act similarly and carry several advantages over cyclosporine; specifically a much smaller molecular size allowing penetration of the epidermis when given topically [6].

Topical therapies have also been deployed with varying amounts of success. These include antiseborrheic agents, topical glucocorticoids, and antimicrobial preparations. Antihistamines have limited use against allergic dermatitis in veterinary patients and this limitation is most likely due to our inability to recognize the immediate phase response during which antihistamine therapy is most likely to be effective. A positive response to antihistamine therapy in atopic canine patients is thought to occur in approximately 10–15% of cases.

Control of secondary conditions is also often necessary and these most often include systemic and topical antibiotics, antifungals, ceruminolytics, and antiseborrheals.

Hyposensitization therapy has been a useful tool for decreasing the magnitude of allergic response in humans with environmental sensitivities. Repeated administration of antigen injected subcutaneously has been demonstrated to decrease IgE levels and increase IgG titers in allergic individuals. Specific T-cell tolerance of antigen is thought to be induced by changing the predominant phenotype of antigen-specific T-cells from T_H2 to T_H1 [1]. T_H1 cells evoke cell-mediated immunity and phagocyte-dependent inflammation whereas T_H2 mediated reactions evoke strong antibody responses and eosinophilic accumulation, with the latter carrying a much greater implication for allergic diseases. Similar results have been achieved in some veterinary patients but overall the response to hyposensitization is considered to be sporadic. The expense and inconvenience of allergen testing and follow-up injections is cost prohibitive for many clients. The low response rates are thought to be due to incomplete identification of allergens in individuals sensitive to numerous environmental components.

Beginning with the French Paradox in which researchers observed the antiatherosclerotic properties of oligomeric proanthocyanidins (OPCs), these compounds have received close attention for their free radical scavenging and anti-inflammatory properties. Found in most brightly colored fruits and vegetables, OPCs are available in a pharmaceutical grade preparation refined from grape seeds. Subsequent research has demonstrated that this antioxidant action also has direct effects on inflammation. Decreased levels of interleukin-6, tumor necrosis alpha, and prostaglandin E_2 have been correlated directly to proanthocyanidins in experimental models of inflammation [7, 8, 19]. They have also been demonstrated to inhibit tissue macrophage activation, which is a main source of tumor necrosis factor alpha [7]. One study also suggested OPCs can stabilize mast cells inhibiting degranulation [9, 24]. This would suggest that OPCs possess properties that disrupt skin inflammation at several steps leading to allergic dermatitis.

The case for deployment of OPCs to combat inflammation is compelling enough that one company, Animal Health Options of Golden, Colorado, who also is a sponsor of this project, has based an entire product line around the compound. In its original form, Proanthozone became the first veterinary-specific antioxidant product to be placed under the scrutiny of university-level research. The paper, which appeared in *JAVMA*, revealed that in cats challenged with acetaminophen, Proanthozone decreased the magnitude of oxidative stress and resulting Heinz body anemia at labeled dosages [10]. The next-generation OPC product, Proanthozone Derm contains alpha lipoic acid, omega three fatty acids, and spirulina in addition to proanthocyanidin in the hopes of bringing additive and synergistic effects to a safe natural product designed specifically to suppress allergic inflammation.

A pilot study evaluating the effectiveness of Proanthozone alone as a treatment for atopic dermatitis yielded mixed results. Of the seven patients completing the study, one showed major improvement as assessed by the owner. Five of the study subjects showed moderate improvement as assessed by the owner. Only one owner indicated no benefit from Proanthozone therapy alone (Beco L., 2006, Sponsored by Animal Health Options, Golden, CO. 2006). All test subjects were subjected to appropriate drug withdrawal periods for glucocorticoids, cyclosporine, antihistamines, and immunotherapy prior to enrollment. Although owner assessment of patients completing the study was positive overall, a percentage of the original study population withdrew from the study due to a lack of response or worsening of the condition. Regardless, the results warrant further investigation as withdrawal from the study may have represented compliance or expectation discrepancies among the population of owners. As most nutritional therapies have a wide margin of safety, responses may also be dose-dependent and this represents another topic worthy of further investigation.

Perhaps the most thoroughly researched nutritional supplement for treatment of allergic skin conditions, omega three fatty acid's effects on skin health and inflammation are linked to the metabolism of inflammatory leukotrienes as well as maintenance of the skin as a barrier to allergen and infection. Omega threes are found in fish oils and are available in a wide variety of veterinary formulations including many commercial dermatologic diets. Often taken as a daily nutritional supplement for benefits in preventing certain cancers and the promotion of healthy hair and skin, in high doses omega threes can have a more direct effect on allergic inflammation. Eicosapentanoic acid (EPA) and Dihomogammalinoleic acid (DHGLA) compete with arachidonic acid as a substrate for the proinflammatory lipoxygenase and cyclo-oxygenase enzyme pathways. EPA and DHGLA metabolites are less inflammatory than those of arachidonic acid resulting in a net anti-inflammatory effect [11]. Omega three fatty acids are also thought to affect IL-2 activation of monocytes [11, 12]. The daily dose for omega three fatty acids is generally well accepted and considered to be around 22 mg/kg/day [13].

Alpha lipoic acid (lipoate) was heralded as a supernutrient when it was found that in its reduced form, dihydrolipoate, it acts as a powerful antioxidant when administered orally. Crucial to antioxidant recycling in vivo, alpha lipoic acid increases levels of vitamin C, glutathione, and, as a result, vitamin E [14]. This net effect naturally elevates the body's own defense against chronic inflammatory damage. Alpha lipoic acid also has been shown to suppress the proliferation of human keratinocytes, a finding potentially helpful in suppressing some of the secondary changes accompanying chronic dermatitis [15]. Interleukin-2 is another inflammatory cytokine involved in the late phase of type-I hypersensitivity that perpetuates the immune response. Alpha lipoic acid is also known to suppress the release of interleukin-2 and interleukin-4 by human peripheral lymphocytes [16, 20]. Dosage recommendations for alpha lipoic acid in canines range from 1.0 to 5.0 mg/kg/day [3].

Spirulina represent a class of cyanobacteria found in tropical and subtropical lakes worldwide. Thought to have provided a nutritious food source for Mesoa-merican peoples, NASA and the European Space Agency now are investigating the possibility of spirulina cultivation during long-term space flights. Therapeutically, spirulina has been potentially useful as an antihistamine in treating hay fever in humans [17, 23]. Much of the data available on the therapeutic effects of spirulina have been gleaned from experimentally induced allergic rhinitis in rats. In these test subjects spirulina demonstrated an ability to increase the phagocytic activity of macrophages, stimulate the production of antibodies and cytokines, increase accumulation of NK cells into tissue, and the activation and mobilization of T and B lymphocytes. Spirulina also reduced the inflammatory reaction of nasal mucosa in treatment groups when compared to negative controls. Mast cell degranulation was also reduced in the treatment group compared to that of the negative control group [17, 18, 25].

Data from clinical trials on Proanthozone Derm are not yet available at this writing but will be available from the manufacturer in the near future. In the proposed study, the investigators will be quantitatively assessing the total antioxidant power (TAP) and thiobarbituric acid reactive substances (TBARS) in control versus treatment groups that include both healthy and atopic individuals. The TBARS assay is the most widely accepted assay for measuring lipid peroxidation as an assessment of oxidative stress. TAP refers to the total radical scavenging capacity of exogenous antioxidant supplementation. The goals of this study are twofold: to determine whether Proanthozone Derm can quantitatively reduce oxidative stress in atopic individuals and if so, how this affects the clinical course of disease. The implications of the study are potentially very important and could significantly add to our understanding of the role oxidative stress plays in manifestation of atopic dermatitis.

Often a multimodal approach to treating allergic dermatitis will yield the best results, requiring both traditional and alternative therapies. Secondary or complicating factors such as secondary pyoderma, yeast overgrowth, otitis externa, or underlying endocrinopathy must be dealt with specifically with appropriate conventional therapies. As parasitic hypersensitivity can appear clinically similar to atopy, all patients should be screened thoroughly to rule out ectoparasitism. Case selection for complementary care is crucial to success because response times can be somewhat prolonged. Failure of alternative therapy is often due to compliance problems in which frustrated clients do not continue with therapy when immediate results are not achieved. This is particularly true in clients accustomed to responses achieved with the administration of steroids and so client expectations should be carefully assessed and considered. In these cases, combination therapy is best with an induction and maintenance phase implemented through concurrent steroid or other lymphocyte inhibitor, and antioxidant therapies followed by maintenance with an appropriate dermatology-specific antioxidant preparation. Treatment goals, for example, a reduction in steroid dose and resulting side effects versus complete disease control, should also be clear in the mind of the veterinarian and the client at the onset of therapy as these will also vary from case to case.

Clinical data on nutritional products has been hard to come by, mostly for reasons pertaining to the ready availability of naturally based components. The incentive for investment in expensive research data is small when readily available natural ingredients are the subject. Therefore, much of the research on natural compounds has been conducted through the benefit of federal research grants. Patents have been issued for novel combinations of natural ingredients, however, and for this reason more research from the private sector may be on the way. Product selection should be based upon ingredient rationale, quality controls, and the needs of the individual patient. Properly applied, antioxidant and complementary modalities can add greatly to the health of atopic patients, build solid client relationships and fulfill a need for alternatives to traditional therapies that carry their own set of drawbacks.

References

1. Abbas, A.K., Lichtman, A.H., Pillai, S. (2007) *Cellular and Molecular Immunology*, 6th Edition. Saunders Elsevier, Philadelphia
2. Omata, N. (2001) Increased oxidative stress in childhood atopic dermatitis. *Life Sci.* **69**, 223–8
3. Mandelker, L. (ed.) (2008) Oxidative stress: the role of mitochondria, free radicals, and antioxidants. *Veterinary Clinics of North America: Small Animal Practice.* **38**, 1
4. Armstrong, D. (ed.) (2002) *Free Radical and Antioxidant Protocols, Methods in Molecular Biology*, vol. 186. Humana Press, Totowa
5. Patricia D.W. (2003) Medical Management of Canine Pruritus. Western Veterinary Conference Proceedings
6. Keith A.H. (2005) Cyclosporine Therapy I: Immune Mediated Dermatoses. The North American Veterinary Conference. Proceedings, Orlando
7. Terra, X., Valls, J., Vitrac, X., Merrillon, JM., Arola, L., Ardevol, L., Blade, C., Fernendez-Larrea, J., Pujadas, G., Salvado, J., Blay, M. (2007) Grape seed procyanidins act as anti-inflammatory agents in endotoxin-stimulated RAW 264.7 macrophages by inhibiting NFkB signaling pathway. *J Agric Food Chem.* **55**, 4357–65
8. Li, W.G., Zhang, X.Y., Wu, Y.J., Tian, X. (2001) Antiinflammatory effect and mechanism of proanthocyanidin from grape seeds. *Acta Pharmacol Sin.* **12**, 1117–20
9. Allison, R.W., Lassen, E.D., Burkhard, M.J., Lappin, M.R. (2000) Effects of a bioflavanoid dietary supplement on acetaminophen induced oxidative injury to feline erythrocytes. *JAVMA.* **217** (8), 1157–61
10. Kondo, K. Uchida, R. Tokutake, S. Maitani, T. (2006) Polymeric grape-seed procyanidins, but not monomeric catechins and oligomeric procyanidins, impair degranulation and membrane ruffling in RBL-2 H3 cells. *Bioorganic & Medicinal Chemistry* **14**, 641–9
11. Mueller, R.S., Fieseler, K.V., Fettman, M.J., Zabel, S., Rosychuk, R.A.W., Ogilvie, G.K., Greenwalt, T.L. (2006) Effects of omega 3 fatty acids on canine atopic dermatitis. *J Small Anim Pract.* **45**, 293–7
12. Simopoulos, A.P. (2006) Evolutionary aspects of diet, the omega-6/omega-3 ratio and genetic variation: nutritional implications for chronic diseases. *Biomed Pharmacother.* **60**, 502–7
13. *Fatty Acids, Essential/Omega* | *Fish Oil/Vegetable Oil Dietary Supplements.* Plumbs Veterinary Drug Handbook Address (URL): http://www.vin.com/Members/Drug/VDH.plx?ID=174
14. Packer, L., Witt, E.H., Tritschler, H.J. (1995) Alpha-Lipoic acid as a biological antioxidant. *Free Rradic Biol Med.* **19**, 227–50
15. Fuchs, J., Milbradt, R. (1994) Antioxidant inhibition of skin inflammation induced by reactive oxidants: evaluation of the redox couple dihydrolipoate/lipoate. *Skin Pharmacol.* **7**, 278–84
16. Venkatraman, M.S., Chittiboyina, A., Meingassner, J., Ho, C.I., Varani, J., Ellis, C.N., Avery, M.A., Pershadsingh, H.A., Kurtz, T.W., Benson, S.C. (2004) Alpha-Lipoic acid-based PPARgamma agonists for treating inflammatory skin diseases. *Arch Dermatol Res.* **296**, 97–104
17. Chen, L.L., Zhang, S.F., Huang, D.N., Tan, J.Q., He, S.H. (2005) Experimental study of spirulina platensis in treating allergic rhinitis in rats. *Zhong Nan Da Xue Xue Bao Yi Xue Ban.* **30**, 96–8
18. Khan, Z., Bhadouria, P., Bisen, P.S. (2005) Nutritional and therapeutic potential of Spirulina. *Curr Pharm Biotechnol.* **5**, 373–9
19. Li, WG., Zhang, XY., Wu, YJ., Tian, X., (2001) Antiinflammatory effect and mechanism of proanthocyanidin from grape seeds. *Acta Pharmacol Sin.* **12**, 1117–20
20. Kim, GD., Kim, TH., Jang, AH., Ahn, HJ., Park, CS. (2010) Alpha lipoic acid suppresses the development of DNFB – induced atopic dermatitis like symptoms in NC/Nga mice. *Experimental Dermatology*. Dec. 17 [E published ahead of print]

21. Dryden, G.W. Jr., Deaciuc, I., Arteel, G., McClain, C.J. (2005) Clinical implications of oxidative stress and antioxidant therapy. *Curr Gastroenterol Rep.***7**, 308–16
22. Nesbitt G. (2002) Antipruritic Therapy. Proceedings, Tufts Animal Expo Tufts Dermatology and Allergy Service, Walpole, Massachusetts
23. Karkos, P.D., Leong, S.C., Karkos, C.D., Sivaji, N., Assimakopoulos D.A. (2008) Spirulina in clinical practice: evidence-based human applications. *Evidence-Based Complementary and Alternative Medicine.* Sept 14. Epub ahead of print
24. Remirez, D., Ledón, N., González, R. (2002) Role of histamine in the inhibitory effects of phycocyanin in experimental models of allergic inflammatory response. *Mediators Inflamm.* **11**, 81–5
25. Kim, H.M., Lee, E.H., Cho, H.H., Moon, Y.H. (1998) Inhibitory effect of mast cell-mediated immediate-type allergic reactions in rats by spirulina. *Biochem Pharmacol* **55**, 1071–6

Avian Antioxidants and Oxidative Stress: Highlights from Studies of Food, Physiology, and Feathers

Kevin J. McGraw

Abstract As generally fast-moving animals with high metabolic rates, birds have been the subject of many studies of oxidative balance. For a given lifespan and body size, birds tend to experience less oxidative stress than other comparable endothermic animals such as mammals, due at least in part to superior antioxidant machinery. Here, recent research is reviewed, with emphases on avian oxidative stress and antioxidants in the context of life-history evolution, paying special attention to oxidant balance as it relates to growth, reproduction, and aging. Dietary antioxidants such as carotenoids can enhance these life-history features in both free-ranging and domesticated species, but patterns are not as clear for tocopherol intake. Studies of endogenous antioxidants such as hormones (e.g., melatonin) and enzymes (e.g., superoxide dismutase) are much more limited, with fewer in wild birds, but for some systems (e.g., sperm, aorta) and domesticated species these molecules offer a strong defense from oxidative damage in the context of reproduction, nutritional state, and toxin exposure. Uric acid and glutathione have also emerged as focal avian antioxidants to track, alongside more general assays of "total antioxidant capacity". More studies are encouraged that comanipulate oxidative stress and antioxidant supplies/actions, to better understand the interactions and coevolution of these biological molecules and processes.

Keywords Avian • Carotenoid • Antioxidant capacity • Uric acid • Unsaturated fatty acids • Feathers • Flavonoids

K.J. McGraw (✉)
School of Life Sciences, Arizona State University, Tempe, AZ 85287-4501, USA
e-mail: kevin.mcgraw@asu.edu

L. Mandelker and P. Vajdovich (eds.), *Studies on Veterinary Medicine*, Oxidative Stress in Applied Basic Research and Clinical Practice 5,
DOI 10.1007/978-1-61779-071-3_11,

Introduction

To date protection from oxidative stress has been studied in many organisms, and since the 1950s and the inception of the free radical theory of aging [35] biologists have sought to find ideal model systems for investigating oxidative damage as it pertains to long-lived, endothermic humans. We have gained tremendous information for various basic and applied purposes from studies of domesticated mammals (see other chapters in this text), but given the high metabolic rates and unusual resistance to oxidative damage that many birds show [45, 74], it is valuable to consider their patterns of antioxidant accumulation and action as well as their sources of and coping mechanisms for oxidative stress, relative to the human condition and potential interventionist solutions to our aging pathologies. Detailed historical accounts of studies of avian antioxidants and oxidative damage are presented elsewhere, especially in domesticated birds [92], so the goal here is to synthesize much of the new information published mostly on free-living birds in the context of life-history trade-offs (e.g., growth, mating, aging) and evolution. For additional reading on the topic, also see [19, 26], and [68].

Dietary Antioxidants

Water- and fat-soluble antioxidant nutrients abound in the avian diet, from plant vitamins (e.g., vitamin C), tocopherols, polyphenols, and terpenoids (e.g., carotenoids) available to herbivores to similar forms sequestered in animal tissues for carnivores. Concentrations tend to be higher in plant foods, with the exception of vitamin A, which is comparatively enriched in animal tissues [56] as a metabolic derivative of plant carotenoids. Fundamentally, the full spectrum of dietary antioxidant types in birds is the same as that available to the broad range of herbivorous, carnivorous, and omnivorous mammal species, but because birds are less commonly considered to be herbivorous than mammals [65] one would expect their exceptional longevity not be strongly food-driven.

Some birds show clear dietary preferences for antioxidants. Given the superior color vision of diurnal birds such as passerines compared to mammals generally (except trichromatic primates), food preferences for colorful foods may confer antioxidant benefits on consumers, given that plant pigments such as carotenoids and anthocyanins can be antioxidants. A recent study on European blackcaps (*Sylvia atricapilla*) showed preferential consumption of artificial food rich in extracts of anthocyanin, a common pigment in their diet of small dark-colored berries [87]. When present in food, colorless fruit flavonoids also are preferred by blackcaps and enhance the responsiveness of their immune system [14]. Other birds are known to prefer colorful foods (e.g., red, blue, and black fruits in songbirds; [66, 70, 80]), although the benefits of such selections are not yet clear. Perhaps the oddest example of this is the selective ingestion of ungulate feces by Egyptian vultures (*Neophron percnopterus*), whose diet of decaying animal tissue is typically deficient in antioxidants and whose yellow facial coloration is dependent upon a diet rich in yellow carotenoid pigments from plants that are only partially digested by (and thus defecated by) grazers [71].

Based on its molecular properties, vitamin E is viewed as one of the most effective radical scavengers in animals. Concentrated most in leafy green vegetable foods, high intake/accumulation has clear beneficial effects on antioxidant capacity of serum in young broiler chickens (*Gallus domesticus*; [17]) and on growth rate in wild barn swallow (*Hirundo rustica*) nestlings [24]. Recent antioxidant studies of vitamin E circulation in adult wild birds, however, have been less conclusive. A comparative study of nearly 100 New World bird species has revealed that vitamin E levels poorly associate with other circulating micromolecular antioxidant metrics [15] and with life-history variation (e.g., clutch size, survival rate, incubation period, nestling period, basal metabolic rate, body mass, and tropical versus temperate habitat; [18]). Some have speculated on the valuable lipid-protecting role of vitamin E in birds, where flight and fatty acid metabolism are linked [15], for which little support has been found, but an additional possibility is that rapid, fine-scale dynamics of vitamin E action and recycling (by vitamin C, unique among dietary antioxidants; [31]) has been difficult to capture in these coarse comparative studies where species can be represented by a single blood sample. Short-term physiological studies coupling vitamins C and E are now warranted in wild birds to better understand vitamin E contributions to antioxidant protection.

Carotenoids have clearly been the dietary antioxidant class that has received the most attention in birds. This has largely stemmed from the fact that birds deposit carotenoid pigments into the integument to develop colorful red-to-yellow colors in feathers, bills, legs, and skin, which are used in visual communication (e.g., mate selection; [39]). In various species, only those individuals who accumulate large supplies of carotenoids from food are able to maintain proper health and advertise such nutritional and somatic gains to prospective rivals or mates by developing brilliant coloration [57, 62]; thus, tracking carotenoid intake and use has become vital for understanding these "honest advertisement" models of sexual selection.

Despite a link between signal honesty and antioxidant action in the mid-1990s [59], the antioxidant benefits of carotenoids in the context of signaling was essentially ignored empirically until a decade later (see further discussion by [38, 60]). In a dietary carotenoid supplementation experiment on captive zebra finches (*Taeniopygia guttata*), which display sexually selected red beak coloration, Alonso-Alvarez et al. [2] found that elevations in systemic carotenoid levels covaried positively with increases in antioxidant defense, measured as the susceptibility of red blood cells to free radical attack (also see evidence in [47] in captive greenfinches, *Carduelis chloris*). Since this time, there has been considerable debate as to the antioxidant potency of carotenoids in birds [37, 48, 49, 78]. A meta-analysis, admittedly from few studies and species at this stage of investigation, yielded an insignificant antioxidant role for carotenoids in birds [20], although the authors admit that more emphasis on antioxidant/oxidative stress balance is required to comprehensively evaluate what antioxidant metrics truly indicate quantitatively [22]. In fact, it is possible that other dietary antioxidants, such as vitamin E, have more potent effects on sexual signaling traits including carotenoid coloration than the carotenoids themselves (e.g., [37, 77], but see [54]).

Despite the outburst of studies in this area showing pervasive benefits of carotenoids for growth (i.e., embryonic stages when metabolism and levels of susceptible biomolecules, e.g., polyunsaturated fatty acids, are high), health, reproduction, and survival [12, 63, 64, 86]), we still know very little about carotenoid supplies and demands in free-ranging birds. In future work, special attention should be placed on carotenoid variation in the avian diet [40, 75], interactions with other antioxidants (e.g., vitamins), and the variety of ways that carotenoids could interact with radicals [61] and affect system performance (i.e., immunopermissive actions at the receptor, cellular, and genetic levels).

Aside from assessing levels of individual dietary antioxidant types, some researchers have developed a technique whereby a cumulative estimate of "antioxidant capacity", largely capturing micromolecular dietary forms circulating through avian plasma or serum, can be obtained, independent of key endogenously synthesized antioxidants such as uric acid (see more below). These residual values [16, 17] are obviously confounded by physiological use or storage of these compounds, but nonetheless given the importance of this "antioxidant capacity" metric (i.e., TEAC) in the human literature [67], it warrants consideration here. Residual TEAC scores (i.e., non-uric-acid antioxidant capacity) were very weakly associated with life-history variation in the aforementioned comparative study [16] compared to raw TEAC values and to uric acid concentrations. Hence, very different information can be gained by analyses of single molecule types versus quenching actions of the antioxidant milieu, and it will only be by assessing complex antioxidant (and oxidative stress) dynamics within free-ranging bird species that we will gain the necessary insights to frame antioxidant nutrition and physiology in evolutionary and in human contexts.

Nondietary Endogenous Antioxidants

Although dietary antioxidants essentially serve as the second line of (radical propagation) protection from oxidative damage, internally produced molecules can function in the initial, intermediate, and final stages of defense from free radical attack. Enzyme systems such as superoxide dismutase (SOD) work to prevent radical formation initially. The neurohormone melatonin, the tripeptide glutathione, and the nitrogen-rich waste product uric acid all can act as dietary forms in offsetting radical attack. Finally, various other enzyme systems (e.g., lipases, proteases, DNA-repair enzymes) can repair or delete radical-damaged tissue. Although much of our foundation for understanding these mechanisms in birds still lies in poultry research [92], a few investigations have been done of late at all levels of endogenous antioxidant activity in nongamebirds and in wild birds, which inform us on comparative patterns between birds and mammals and key life-history investments where an oxidative balance must be struck.

Endogenous Radical-Protecting Enzymes

One of the broader investigations of antioxidant enzyme systems among birds was done on seminal fluid and spermatozoa in five species from a phylogenetically ancient group comprised of gamebirds (chicken, turkey, guinea fowl) and waterfowl (duck and goose; [93]). Avian sperm have higher levels of unsaturated fatty acids (UFAs) than other tissues and thus are particularly exposed to lipid peroxidation, which may influence sperm survival and fertilization capacity. Surai et al. [93] measured SOD and glutathione peroxidase (GTP) activities in spermatozoa from these bird species and found some correspondence between free-radical-attack susceptibility and offsetting enzyme action; for example, levels of UFAs, SOD, and GTP were highest in duck spermatozoa, whereas all were lowest in turkey. This suggests important coevolutionary links between oxidative stress and antioxidant demands in a vital cell type, and one that may have to endure special longevity compared to organisms such as mammals, as avian sperm survive while stored in tubules within female reproductive tracts for a period of up to several weeks prior to fertilization.

Quail, given their ancestral muscle-demanding flight requirements, have been used as a model for understanding oxidative stress and muscle performance, by adding weights to bodies to induce hypertrophy and then removing them to induce atrophy; here, CAT, SOD, and GTP activity is most elevated under hypertrophic muscle conditions and then declines with increasing atrophy [85], suggesting that these enzymes protect tissue when it is most stressed. Quail have also served as experimental models for studying atherosclerosis, and cholesterol supplementation increases SOD levels in aorta but does not affect catalase or GTP in the aorta or myocardium [33]. They have also been used to show a genetic basis to oxidative stress risk in aortic endothelial cells, as cells from atherosclerosis-susceptible strains of quail exhibit lower GTP, SOD, and catalase activity [41]. Within species these patterns suggest high levels are protective in the short term for individuals; however, birds are touted to have lower levels of these enzymes than mammals [42], probably reflecting a longer-term coevolutionary response with oxidative stress susceptibility and inputs of other antioxidants.

Enzyme activity in individuals may be traded off against other energetically demanding activities, such as breeding. Wiersma et al. [97] nicely demonstrated an ecologically relevant determinant of SOD and GTP activity; they manipulated the number of nestlings raised (i.e., brood size) by captively breeding zebra finches and found that SOD and GTP activities decreased by approximately 25% in both sexes when having to feed more young.

Endogenous antioxidants also do not necessarily work independently of nutritional state, and the intake of specific antioxidant nutrients. Turkeys have been used as a model for understanding the effects of alcohol on congestive cardiomyopathy [27], and here alcohol intake increases SOD, GTP, and catalase (see more below in "Endogenous Cell/Tissue Repair Enzymes") enzyme expression and action, perhaps working to minimize lipid peroxidation and repair tissue in the heart. Intake of oxidants (from rancid fish), beyond enzyme-deactivating capabilities, was thought to lead to cardiac myopathy and death in four rehabilitating brown pelicans (*Pelecanus occidentalis*; [32]). Foods low in UFAs (e.g., canola oil) promote antioxidant enzyme activities in turkeys, whereas those with high UFA concentrations (e.g., tallow) decrease them [82]. Also, ingestion of other environmental compounds, such as polluting toxins, can affect endogenous antioxidant defenses; high SOD, CAT, and GTP levels were found in white stork (*Ciconia ciconia*) chicks from areas near copper manufacturing plants in Poland [52, 53].

Endogenous Antioxidant Molecules

Glutathione is synthesized in cells (especially from liver) from amino acid precursors and, among many other key biological functions, stands as an important free-radical scavenger. By serving as an electron donor, it is converted to its oxidized state (glutathione disulfide), whose concentrations in birds (as in other animals) are often assessed relative to its reduced form as a metric of oxidative stress. For example, experimental exposure to flame-retardant chemicals (PBDEs) and mercury toxicity increased oxidized glutathione in young American kestrels (*Falco sparverius*; [30]) and in mallard ducks (*Anas platyrhynchos*; [43]), respectively; in addition, supplementation with the antioxidant vitamin C decreased oxidized glutathione in heart of Japanese quail [99]. Although the majority of past studies have emphasized its use under laboratory/experimental conditions, Isaksson et al. [50] investigated glutathione status of small songbirds (great tits, *Parus major*) in an urban ecological setting. Adult birds living closer to the city center were paler in carotenoid coloration (see above) and had higher oxidized glutathione levels than those living more rurally, but the same pattern was not evident in nestlings [50]. Interestingly, glutathione also plays a role as a cysteine precursor in white-crowned sparrows that use high levels of this sulfur amino acid in feather (keratin) synthesis [69].

Melatonin is a monoamine hormone produced primarily in the pineal gland of the brain that regulates biorhythms, growth, and sexual activity in birds, in addition to serving as a potent endogenous antioxidant [79]. For example, in vitro melatonin administration effectively protected heterophils of ring doves (*Streptopelia risoria*) from induced lipid peroxidative damage [83]; it can also work to upregulate antioxidant enzymes in chickens (GTP; [81]) and ring doves (SOD; [84]). Melatonin also may slow aging in birds, as in vivo supplementation reduced superoxide anion levels in very old ring doves [25]. As with other antioxidants reviewed above, melatonin levels are also influenced by environmental conditions, such as high electromagnetic fields (in captive kestrels; [29]) or immobilization stress (in ring doves; [9]). Research on melatonin in the context of mating has also been conducted, with Bertrand et al. [10] finding that melatonin supplementation reddens the sexually attractive beak of adult male zebra finches. Comparative studies of melatonin and fitness in wild birds are now needed to confirm these patterns established in captive domesticated models.

Uric acid – the main nitrogenous waste product in birds – is the third main type of endogenous avian antioxidant, and based on a flurry of recent studies is among the best-studied antioxidant molecules in wild birds [91]. In fact, high levels of uric acid have been invoked to explain in part the exceptional longevity of birds [96]. In numerous, mostly passerine, bird species, total antioxidant activity of plasma sampled is dominated by the action of uric acid ([4, 16, 17, 47, 5]; see [90] for a discussion of poultry). Cohen et al. [16] found in their large study of avian life-history variation that, as with total antioxidant capacity, uric acid levels were associated among species with a "live-fast, die-young" strategy. Within species, uric acid varies by sex, weight, and heredity in several strains of domesticated turkeys [36] and according to levels of metabolic activity (e.g., flight) in pigeons [88] and to handling stress in wild-caught house sparrows (*Passer domesticus*) and gray catbirds (*Dumetella carolinensis*; [16]). Directions of relationships between stress and antioxidants were not equal for these two songbird species, however, indicating that uric acid concentrations reflect dynamic (and potentially species- or context-specific) antioxidant production, mobilization, and consumption processes. A cell-mediated immune challenge was found not to be a source of change for uric acid levels in red-legged partridges (*Alectoris rufa*), however [78].

Two recent insights on uric acid in birds may further our understanding of antioxidant physiology in this group. First, allantoin is an oxidative product of uric acid and the balance between these two molecules was found to be linked to exercise status of captive white-crowned sparrows (*Zonotrichia leucophrys gambelii*; [94]), suggesting that, as in humans, this ratio may be an even better indicator of oxidative stress in birds than uric acid. Moreover, in a study of a fruit-eating passerine species (yellow-vented bulbul, *Pycnonotus xanthopygos*), waste was found to contain proportionally more ammonia than uric acid, especially when water intake was high and protein intake was low [95]. This occurred by postrenal recovery of nitrogen and led to high levels of plasma uric acid, all of which is occurring in a group that lacks the enzyme to synthesize vitamin C. Taken together, these authors suggest that bulbuls show a specialized uric acid antioxidant recycling strategy that compensates for vitamin C deficiency [95].

In humans, the presence of apocarotenoids – endogenously formed intermediates in the conversion process of β-carotene to retinal – can serve as an endogenous molecular indicator of the antioxidant protection afforded to individuals [100]; however, despite the presence of apocarotenoids in chickens [89], this approach has not yet been applied to birds.

Endogenous Cell/Tissue Repair Enzymes

This form of defense is the least studied of the antioxidant mechanisms in birds, and what little has been done has focused on environmental toxins and glutathione-S-transferase (GST). It is unclear why less attention has been paid to these enzymes, as spectrophotometric assays appear straightforward [34] but require terminal sampling and tissue harvesting. The prediction is that GST and other repair enzymes will increase in activity after exposure to and damage from oxidative stress. GST levels in erythrocytes increase in chickens exposed to arsenic and endosulfan (an organochlorine insecticide; [1]) and in the liver of quail that have been treated with cyanobacterial toxins [76]. In a very nice urban ecological study of tissue antioxidants in great tits, Isaksson et al. [51] sampled lungs and liver of birds from urban and rural sites and only lung CAT (not GST) tended to increase in urban animals. Much more life-history information is needed on the causes and consequences of repair enzyme activity in birds.

Oxidative Stress

Birds have emerged as a model taxon for understanding how oxidative damage shapes somatic maintenance in animals. They have notably high lifetime energy expenditures, largely a function of endothermy and flight (e.g., oxygen consumption), but perhaps unexpectedly have long lifespans and slower rates of aging compared to similarly sized endotherms (such as mammals, including bats; [45, 46]). Researchers now suspect that strong antioxidant mechanisms (above) combine with reduced oxidant-generation processes ("free radical production hypothesis of aging" [6–8]) to contribute to the high maximum lifespan potential of birds.

Methods for assessing ROS damage to the avian body appear consistent with other taxa, especially including measurements of lipid peroxidation and DNA damage in various cells and tissues (e.g., [42, 51, 73, 92]). Short-term in vitro assessments of free radical intermediate formation, using electron spin resonance, have also been used [55], but mitochondrial free radical production has been of particular interest to avian biologists, because birds appear to have lower rates of ROS production in the mitochondria of tissues such as heart, brain, and lung [6]. In fact, it is suggested that the mechanism behind low rates of ROS generation in birds like pigeons is that the ROS generator sites (e.g., iron–sulfur centers) maintain a low degree of reduction [6]. At the molecular level, DNA repair, low rates of ROS production near DNA, capacities for brain neurogeneration, and low rates of accumulation of advanced glycosylation end-products are also invoked to explain extreme avian longevity [6, 46, 58]; the delayed maturity and slow annual fecundity of some birds (e.g., in seabirds) also interplay at the behavioral/evolutionary level [46].

Some of the more ecologically relevant assessments and manipulations of oxidative stress have been performed in wild birds. For example, in free-ranging, long-lived alpine swifts (*Apus melba*; [11]), red-blood-cell resistance to oxidative stress (using the KRL test) was positively related to annual survival in adult males and to clutch size in females. Moreover, females with low resistance to oxidative stress laid eggs with a similarly poor resistance, which were then less likely to hatch. A different study of nestling birds (wild Eurasian kestrels, *Falco tinnunculus*) showed that, early in life, oxidative stress is closely associated with the number of competing siblings in the nest [23]. Induction of avian oxidative stress, for experimental purposes, has also taken behaviorally naturalistic forms, in addition to traditional biochemical routes (e.g., oxygen, hydrogen peroxide, paraquat, gamma-radiation; [48, 73]). These include increasing the number of offspring in a nest, which was shown to elevate parental workload as well as oxidative stress in domesticated zebra finches [97]. Alonso-Alvarez et al. [3, 4] also showed in zebra finches that testosterone elevation as well as more frequent breeding weakens red blood cell resistance to oxidative stress. A meta-analysis of avian studies has also produced one of the best comparative pieces of evidence that immune system activation affects oxidative stress [21].

As suggested above, in addition to the body and eggs, sperm are argued to be especially susceptible to oxidative stress, given their high concentration of polyunsaturated fatty acids and limited DNA repair ability [93]. Oxidative stress is thought to be a major cause of infertility in male chickens [98], and a few recent studies of birds have considered reduction of oxidative stress in bird semen [13]. Supplemental vitamin E can limit oxidative stress in the semen of chickens [28], although the same was not true of vitamin C in turkey seminal fluid [72]; instead ascorbic acid appeared to delay the formation of multinucleated giant cells that are indicative of infection. Clearly, more work is needed on sperm and oxidative stress in free-ranging birds.

References

1. Aggarwal, M., Naraharisetti, S.B., Sarkar, S.N., Rao, G.S., Degen, G.H., and Malik, J.K. 2009. Effects of subchronic coexposure to arsenic and endosulfan on the erythrocytes of broiler chickens: a biochemical study. Archives of Environmental Contamination and Toxicology 56, 139–148.
2. Alonso-Alvarez, C., Bertrand, S., Devevey, G., Gaillard, M., Prost, J., Faivre, B., and Sorci, G. 2004. An experimental test of the dose-dependent effect of carotenoids and immune activation on sexual signals and antioxidant activity. American Naturalist 164, 651–659.
3. Alonso-Alvarez, C., Bertrand, S., Devevey, G., Prost, J., Faivre, B., Chastel, O., and Sorci, G. 2006. An experimental manipulation of life-history trajectories and resistance to oxidative stress. Evolution 60, 1913–1924.
4. Alonso-Alvarez, C., Bertrand, S., Faivre, B., Chastel, O., and Sorci, G. 2007. Testosterone and oxidative stress: the oxidation handicap hypothesis. Proceedings of the Royal Society of London B 274, 819–825.
5. Alonso-Alvarez, C., Perez-Rodriguez, L., Garcia, J.T., and Vinuela, J. 2009. Testosteronemediated trade-offs in the old age: a new approach to the immunocompetence handicap and carotenoid-based sexual signalling. Proceedings of the Royal Society B 276, 2093–2101.

6. Barja, G. 1998. Mitochondrial free radical production and aging in mammals and birds. Annals of the New York Academy of Sciences 854, 224–238.
7. Barja, G. 2002. Rate of generation of oxidative stress-related damage and animal longevity. Free Radical Biology and Medicine 33, 1167–1172.
8. Barja, G., Cadenas, S., Rojas, C., Lopez-Torres, M., and Perez-Campo, R. 1994. A decrease of free radical production near critical targets as a cause of maximum longevity in animals. Comparative Biochemistry and Physiology: Biochemistry and Molecular Biology 108, 501–512.
9. Barriga, C., Marchena, J.M., Lea, R.W., Harvey, S., and Rodriguez, A.B. 2002. Effect of stress and dexamethasone treatment on circadian rhythms of melatonin and corticosterone in ring dove (*Streptopelia risoria*). Molecular and Cellular Biochemistry 232, 27–31.
10. Bertrand, S., Faivre, B., and Sorci, G. 2006. Do carotenoid-based sexual traits signal the availability of non-pigmentary antioxidants? Journal of Experimental Biology 209, 4414–4419.
11. Bize, P., Devevey, G., Monaghan, P., Doligez, B., and Christe, P. 2008. Fecundity and survival in relation to resistance to oxidative stress in a free-living bird. Ecology 89, 2584–2593.
12. Blount, J.D., Houston, D.C., Surai, P.F., and Møller, A.P. 2004. Egg-laying capacity is limited by carotenoid pigment availability in wild gulls *Larus fuscus*. Biology Letters 271, S79–S81.
13. Blount, J.D., Møller, A.P., and Houston, D.C. 2001. Antioxidants, showy males and sperm quality. Ecology Letters 4, 393–396.
14. Catoni, C., Schaefer, H.M., and Peters, A. 2008. Fruit for health: the effect of flavonoids on humoral immune response and food selection in a frugivorous bird. Functional Ecology 22, 649–654.
15. Cohen, A.A. and McGraw, K.J. 2009. No simple measures for antioxidant status in birds: complexity in inter- and intraspecific correlations among circulating antioxidant types. Functional Ecology 23, 310–320.
16. Cohen, A.A., Hau, M., and Wikelski, M. 2008. Stress, metabolism, and antioxidants in two wild passerine bird species. Physiological and Biochemical Zoology 81, 463–472.
17. Cohen, A., Klasing, K., and Ricklefs, R. 2007. Measuring circulating antioxidants in birds. Comparative Biochemistry and Physiology Part B 147, 110–121.
18. Cohen, A.A., McGraw, K.J., Wiersma, P., Williams, J.B., Robinson, W.D., Robinson, T.R., Brawn, J.D., and Ricklefs, R.E. 2008. Interspecific associations between circulating antioxidant levels and life-history variation in birds. American Naturalist 172, 178–193.
19. Costantini, D. 2008. Oxidative stress in ecology and evolution: lessons from avian studies. Ecology Letters 11, 1238–1251.
20. Costantini, D. and Møller, A.P. 2008. Carotenoids are minor antioxidants in birds. Functional Ecology 22, 367–370.
21. Costantini, D. and Møller, A.P. 2009. Does immune response cause oxidative stress in birds? A meta-analysis. Comparative Biochemistry and Physiology A 153, 339–344.
22. Costantini, D. and Verhulst, S. 2009. Does high antioxidant capacity indicate low oxidative stress? Functional Ecology 23, 506–509.
23. Costantini, D., Casagrande, S., De Filippis, S., Brambilla, G., Fanfani, A., Tagliavini, J., and Dell'Omo, G. 2006. Correlates of oxidative stress in wild kestrel nestlings (*Falco tinnunculus*). Journal of Comparative Physiology B 176, 329–337.
24. de Ayala, R.M., Martinelli, R., and Saino, N. 2007. Vitamin E supplementation enhances growth and condition of nestling barn swallows (*Hirundo rustica*). Behavioral Ecology and Sociobiology 60, 619–630.
25. del Pillar Terron, M., Paredes, S.D., Barriga, C., Ortega, E., Reiter, R.J., and Rodriguez, A.B. 2005. Oral administration of melatonin to old ring doves (*Streptopelia risoria*) increases plasma levels of melatonin and heterophil phagocytic activity. Journals of Gerontology Series A 60, 44–50.
26. Dowling, D.K. and Simmons, L.W. 2009. Reactive oxygen species as universal constraints in life-history evolution. Proceedings of the Royal Society of London Series B 276, 1737–1745.

27. Edes, I., Piros, G., Forster, T., and Csanady, M. 1987. Alcohol-induced congestive cardiomyopathy in adult turkeys: effects on myocardial antioxidant defence systems. Basic Research in Cardiology 82, 551–556.
28. Eid, Y., Ebeid, T., and Younis, H. 2006. Vitamin E supplementation reduces dexamethasome-induced oxidative stress in chicken semen. British Poultry Science 47, 350–356.
29. Fernie, K.J. and Bird, D.M. 2001. Evidence of oxidative stress in American kestrels exposed to electromagnetic fields. Environmental Research 86, 198–207.
30. Fernie, K.J., Shutt, J.L., Mayne, G., Hoffman, D., Letcher, R.J., Drouillard, K.G., and Ritchie, I.J. 2005. Exposure to polybrominated diphenyl ethers (PBDEs): changes in thyroid, vitamin A, glutathione homeostasis, and oxidative stress in American kestrels (*Falco sparverius*). Toxicological Sciences 88, 375–383.
31. Fukuzawa, K., Ikebata, W., and Sohmi, K. 1993. Location, antioxidant and recycling dynamics of alpha-tocopherol in liposome membranes. Journal of Nutritional Science and Vitaminology 39, S9–S22.
32. Giri, D.K., Miller, D.L., Thompson, L.J., Mailler, L., Styer, E., and Baldwin, C. 2007. Superoxide dismutase expression and oxidative damage in a case of myopathy in brown pelicans (*Pelecanus occidentalis*). Journal of Veterinary Diagnostic Investigation 19, 301–304.
33. Godin, D.V., Nichols, C.R., Hoekstra, K.A., Garnett, M.E., and Cheng, K.M. 2003. Alterations in aortic antioxidant components in an experimental model of atherosclerosis: a time-course study. Molecular and Cellular Biochemistry 252, 193–203.
34. Habig, W.H., Pabst, M.J., and Jakoby, W.B. 1974. Glutathione-S-transferases, the first enzymatic step in mercapturic acid formation. Journal of Biological Chemistry 249, 7130–7139.
35. Harman, D. 1956. Aging: a theory based on free radical and radiation chemistry. Journal of Gerontology 11, 298–300.
36. Hartman, S., Taleb, S.A., Geng, T., Gyenai, K., Guan, X., and Smith, E. 2006. Comparison of plasma uric acid levels in five varieties of the domestic turkey, *Meleagris gallopavo*. Poultry Science 85, 1791–1794.
37. Hartley, R.C. and Kennedy, M.W. 2004. Are carotenoids a red herring in sexual display? Trends in Ecology and Evolution 19, 353–354.
38. Hill, G.E. 1999. Is there an immunological cost to carotenoid-based ornamental coloration? American Naturalist 154, 589–595.
39. Hill, G.E. 2006. Female mate choice for ornamental coloration. In Bird Coloration, Volume II, Function and Evolution (Hill, G.E. and McGraw, K.J., eds). Harvard University Press, Cambridge, MA, pp. 137–200.
40. Hill, G.E., Inouye, C.Y., and Montgomerie, R. 2002. Dietary carotenoids predict plumage coloration in wild house finches. Proceedings of the Royal Society of London Series B 269, 1119–1124.
41. Hoekstra, K.A. 2008. Inflammatory genes and antioxidant response of aortic endothelial cells to minimally modified LDL in atherosclerosis-susceptible and –resistant Japanese quail. FASEB Journal 22, 924.3.
42. Hoffman, D.J. 2002. Role of selenium toxicity and oxidative stress in aquatic birds. Aquatic Toxicology 57, 11–26.
43. Hoffman, D.J. and Heinz, G.H. 1998. Effects of mercury and selenium on glutathione metabolism and oxidative stress in mallard ducks. Environmental Toxicology and Chemistry 17, 161–166.
44. Holmes, D.J. and Martin, K. 2009. A bird's-eye view of aging: what's in it for ornithologists? Auk 126, 1–23.
45. Holmes, D.J. and Ottinger, M.A. 2003. Birds as long-lived animal models for the study of aging. Experimental Gerontology 38, 1365–1375.
46. Holmes, D.J., Fluckiger, R., and Austad, S.N. 2001. Comparative biology of aging in birds: an update. Experimental Gerontology 36, 869–883.
47. Horak, P., Saks, L., Zilmer, M., Karu, U., and Zilmer, K. 2007. Do dietary antioxidants alleviate the cost of immune activation? An experiment with greenfinches. American Naturalist 170, 625–635.

48. Isaksson, C. and Andersson, S. 2008. Oxidative stress does not influence carotenoid mobilization and plumage pigmentation. Proceedings of the Royal Society of London Series B 275, 309–314.
49. Isaksson, C., McLaughlin, P., Monaghan, P., and Andersson, S. 2007. Carotenoid pigmentation does not reflect total non-enzymatic antioxidant activity in plasma of adult and nestling great tits, *Parus major*. Functional Ecology 21, 1123–1129.
50. Isaksson, C., Ornborg, J., Stephensen, E., and Andersson, S. 2005. Plasma glutathione and carotenoid coloration as potential biomarkers of environmental stress in great tits. EcoHealth 2, 138–146.
51. Isaksson, C., Sturve, J., Almroth, B.C., and Andersson, S. 2009. The impact of urban environment on oxidative damage (TBARS) and antioxidant systems in lungs and liver of great tits, *Parus major*. Environmental Research 109, 46–50.
52. Kaminski, P., Kurhalyuk, N., Jerzak, L., Kasprzak, M., Tkachenko, H., Klawe, J.J., Szady-Grad, M., Koim, B., and Wisniewska, E. 2008. Ecophysiological determinations of antioxidant enzymes and lipoperoxidation in the blood of white stork *Ciconia ciconia* from Poland. Environmental Research 109, 29–39.
53. Kaminski, P., Kurhalyuk, N., Kasprzak, M., Jerzak, L., Tkachenko, H., Szady-Grad, M., Klawe, J.J., and Koim, B. 2009. The impact of element-element interactions on antioxidant enzymatic activity in the blood of white stork (*Ciconia ciconia*) chicks. Archives of Environmental Contamination and Toxicology 56, 325–337.
54. Karu, U., Saks, L., and Horak, P. 2008. Carotenoid-based plumage coloration is not affected by vitamin E supplementation in male greenfinches. Ecological Research 23, 931–935.
55. Klandorf, H., Rathore, D.S., Iqbal, M., Shi, X., and Van Dyke, K. 2001. Accelerated tissue aging and increased oxidative stress in broiler chickens fed allopurinol. Comparative Biochemistry and Physiology C 129, 93–104.
56. Klasing, K.C. 1998. Comparative Avian Nutrition. CAB International, Wallingford, Oxon, UK.
57. Kodric-Brown, A. 1989. Dietary carotenoids and male mating success in the guppy: an environmental component to female choice. Behavioral Ecology and Sociobiology 25, 393–401.
58. Ku, H.H. and Sohal, R.S. 1993. Comparison of mitochondrial pro-oxidant generation and anti-oxidant defenses between rat and pigeon: possible basis of variation in longevity and metabolic potential. Mechanisms of Ageing and Development 72, 67–76.
59. Lozano, G.A. 1994. Carotenoids, parasites, and sexual selection. Oikos 70, 309–311.
60. Lozano, G.A. 2001. Carotenoids, immunity, and sexual selection: comparing apples and oranges? American Naturalist 158, 200–203.
61. Martinez, A. and Barbosa, A. 2008. Antiradical power of carotenoids and vitamin E: testing the hydrogen atom transfer mechanism. Journal of Physical Chemistry B 112, 16945–16951.
62. McGraw, K.J. 2006. Mechanics of carotenoid-based coloration. In Bird Coloration, Volume I, Mechanisms and Measurements (Hill, G.E. and McGraw, K.J. eds). Harvard University Press, Cambridge, MA, pp. 177–242.
63. McGraw, K.J. and Ardia, D.R. 2003. Carotenoids, immunocompetence, and the information content of sexual colors: an experimental test. American Naturalist 162, 704–712.
64. McGraw, K.J., Adkins-Regan, E., and Parker, R.S. 2005. Maternally derived carotenoid pigments affect offspring survival, sex ratio, and sexual attractiveness in a colorful songbird. Naturwissenschaften 92, 375–380.
65. McNab, B.K. 1988. Complications inherent in scaling the basal rate of metabolism in mammals. Quarterly Review of Biology 63, 25–54.
66. McPherson, J.M. 1988. Preferences of cedar waxwings in the laboratory for fruit species, colour and size: a comparison with field observations. Animal Behaviour 36, 961–969.
67. Miller, N.J., Rice-Evans, C., Davies, M.J., Gopinathan, V., and Milner, A. 1993. A novel method for measuring antioxidant capacity and its application to monitoring the antioxidant status in premature neonates. Clinical Science 84, 407–412.

68. Monaghan, P., Metcalfe, N.B., and Torres, R. 2009. Oxidative stress as a mediator of life history trade-offs: mechanisms, measurements and interpretation. Ecology Letters 12, 75–92.
69. Murphy, M.E. and King, J.R. 1984. Dietary sulfur amino acid availability and molt dynamics in white-crowned sparrows. Auk 101, 164–167.
70. Murray, K.G., Winnett-Murray, K., Cromie, E.A., Minor, M., and Meyers, E. 1993. The influence of seed packaging and fruit color on feeding preferences of American robins. Plant Ecology 107–108, 217–226.
71. Negro, J.J., Grande, J.M., Tella, J.L., Garrido, J., Hornero, D., Donazar, J.A., Sanchez-Zapata, J.A., Benitez, J.R., and Barcell, M. 2002. Coprophagy: an unusual source of essential carotenoids. Nature 416, 807–808.
72. Neuman, S.L., Orban, J.I., Lin, T.L., Latour, M.A., and Hester, P.Y. 2002. The effect of dietary ascorbic acid on semen traits and testis histology of male turkey breeders. Poultry Science 81, 265–268.
73. Ogburn, C.E., Austad, S.N., Holmes, D.J., Kiklevich, J.V., Gollahon, K., Rabinovitch, P.S., and Martin, G.M. 1998. Cultured renal epithelial cells from birds and mice: enhanced resistance of avian cells to oxidative stress and DNA damage. Journal of Gerontology: Biological Sciences 53A, B287–B292.
74. Ogburn, C.E., Carlberg, K., Ottinger, M.A., Holmes, D.J., Martin, G.M., and Austad, S.N. 2001. Exceptional cellular resistance to oxidative damage in long-lived birds requires active gene expression. Journal of Gerontology Series A 56, B468–B474.
75. Olson, V.A. and Owens, I.P.F. 2005. Interspecific variation in the use of carotenoid-based coloration in birds: diet, life history and phylogeny. Journal of Evolutionary Biology 18, 1534–1546.
76. Paskova, V., Adamovsky, O., Pikula, J., Skocovska, B., Band'ouchova, H., Horakova, J., Babica, P., Marsalek, B., and Hilscherova, K. 2008. Detoxification and oxidative stress responses along with microcystins accumulation in Japanese quail exposed to cyanobacterial biomass. Science of the Total Environment 398, 34–47.
77. Perez, C., Lores, M., and Velando, A. 2008. Availability of nonpigmentary antioxidant affects red coloration in gulls. Behavioral Ecology 19, 967–973.
78. Perez-Rodriguez, L., Mougeot, F., Alonso-Alvarez, C., Blas, J., Vinuela, J., and Bortolotti, G.R. 2008. Cell-mediated immune activation rapidly decreases plasma carotenoids but does not affect oxidative stress in red-legged partridges (*Alectoria rufa*). Journal of Experimental Biology 211, 2155–2161.
79. Poeggeler, B., Saarela, S., Reiter, R.J. et al. 1994. Melatonin – a highly potent endogenous radical scavenger and electron donor: new aspects of the oxidation chemistry of this indole accessed in vitro. Annals of the New York Academy of Sciences 738, 419–420.
80. Puckey, H.L., Lill, A., and O'Dowd, D.J. 1996. Fruit color choices of captive Silvereyes (*Zosterops lateralis*). Condor 98, 780–790.
81. Reiter, R.J. 1991. Pineal melatonin: cell biology of its synthesis and of its physiological interactions. Endocrine Reviews 12, 151–180.
82. Renerre, M., Poncet, K., Mercier, Y., Gatellier, P., and Metro, B. 1999. Influence of dietary fat and vitamin E on antioxidant status of muscles of turkeys. Journal of Agricultural and Food Chemistry 47, 237–244.
83. Rodriguez, A.B., Nogales, G., Marchena, J.M., Ortega, E., and Barriga, C. 1999. Suppression of both basal and antigen-induced lipid peroxidation in ring dove heterophils by melatonin. Biochemical Pharmacology 58, 1301–1306.
84. Rodriguez, A.B., Nogales, G., Ortega, E., and Barriga, C. 1998. Melatonin controls superoxide anion level: modulation of superoxide dismutase activity in ring dove heterophils. Journal of Pineal Research 24, 9–14.
85. Ryan, M.J. and Always, S.E. 2008. Loading and unloading induces an age-dependent regulation of antioxidant enzymes and redox status in quail patagialis. Medicine and Science in Sports 40, S108.

86. Saino, N., Ferrari, R., Romano, M., Martinelli, R., and Møller, A.P. 2003. Experimental manipulation of egg carotenoids affects immunity of barn swallow nestlings. Proceedings of the Royal Society of London Series B 270, 2485–2489.
87. Schaefer, H.M., McGraw, K., and Catoni, C. 2008. Birds use fruit colour as honest signal of dietary antioxidant rewards. Functional Ecology 22, 303–310.
88. Schwilch, R., Jenni, L., and Jenni-Eiermann, S. 1996. Metabolic responses of homing pigeons to flight and subsequent recovery. Journal of Comparative Physiology B 166, 77–87.
89. Sharma, R.V., Mathur, S.N., and Ganguly, J. 1976. Studies on the relative biopotencies and intestinal absorption of different apo-beta-carotenoids in rats and chickens. Biochemistry Journal 158, 377–383.
90. Simoyi, M.F., van Dyke, K., and Klandorf, H. 2002. Manipulation of plasma uric acid in broiler chicks and its effect on leukocyte oxidative activity. American Journal of Physiology 282, R791–R796.
91. Stinefelt, B., Leonard, S.S., Blemings, K.P., Shi, X., and Klandorf, H. 2005. Free radical scavenging, DNA protection, and inhibition of lipid peroxidation mediated by uric acid. Annals of Clinical and Laboratory Science 35, 37–45.
92. Surai, P.F. 2002. Natural Antioxidants in Avian Nutrition and Reproduction. Nottingham University Press, Nottingham, UK.
93. Surai, P.F., Blesbois, E., Grasseau, I., Chalah, T., Brillard, J.-P., Wishart, G.J., Cerolini, S., and Sparks, N.H.C. 1998. Fatty acid composition, glutathione peroxidase and superoxide dismutase activity and total antioxidant activity of avian semen. Comparative Biochemistry and Physiology Part B 120, 527–533.
94. Tsahar, E., Arad, Z., Izhaki, I., and Guglielmo, C.G. 2006. The relationship between uric acid and its oxidative product allantoin: a potential indicator for the evaluation of oxidative stress in birds. Journal of Comparative Physiology B 176, 653–661.
95. Tsahar, E., Martinez del Rio, C., Izhaki, I., and Arad, Z. 2005. Can birds be ammonotelic? Nitrogen balance and excretion in two frugivores. Journal of Experimental Biology 208, 1025–1034.
96. Vleck, C.M., Haussmann, M.F., and Vleck, D. 2007. Avian senescence: underlying mechanisms. Journal of Ornithology 148, 611–624.
97. Wiersma, P., Selman, C., Speakman, J.R., and Verhulst, S. 2004. Birds sacrifice oxidative protection for reproduction. Proceedings of the Royal Society of London Series B Supplement 271, S360–S363.
98. Wishart, G.J. 1984. Effects of lipid peroxide formation in fowl semen on sperm motility, ATP content and fertilizing ability. Journal of Reproduction and Fertility 71, 113–118.
99. Yenisey, C., Seyrek, K., Serter, M., Kargin-Kiral, F., and Bardakcioglu, H.E. 2006. Effects of dietary vitamin C supplementation on glutathione, malondialdehyde and nitric oxide concentrations in brain and heart of laying Japanese quails exposed to heat stress (34.8°C). Revue de Medecine Veterinaire 157, 149–152.
100. Yeum, K.-J., dos Anjos Ferreira, A.L., Smith, D., Krinsky, N.I., and Russell, R.M. 2000. The effect of α-tocopherol on the oxidative cleavage of β-carotene. Free Radical Biology and Medicine 29, 105–114.

Free Radicals and Antioxidants in Avian Diseases

Miklós Mézes and Krisztián Balogh

Abstract Oxygen and nitrogen free radicals have importance in the pathogenesis of different infectious and metabolic poultry diseases. The present review discusses the free radical-initiated oxidative damage of macromolecules (such as lipids, proteins, and DNA) as part of the unspecific immune response caused by some viral (Marek's disease, Newcastle diseases, or infectious bursal disease) or bacterial diseases (*Salmonella*, *Staphylococcus*, *Clostridium*, or *E. coli*), parasitic infections (coccidiosis), and some nutritional or metabolic diseases (ascites syndrome, nutritional encephalomalatia, knockdown syndrome of turkeys, or tibial dyschodroplasia). The effects of dietary antioxidants (vitamins E and C, and selenium) on free radical formation and their effect on humoral immune response against invading viruses and bacteria are also discussed.

Keywords Poultry • Marek's disease • Newcastle disease • Pulmonary hypertension (ascites) syndrome • Nutritional encephalomalacia • Knockdown syndrome • Tibial dyschondroplasia

Introduction

Excessive levels of reactive oxygen species (e.g., superoxide anion, hydrogen peroxide, and hypochlorous acid) and reactive nitrogen species (nitric oxide, nitroxyl radical, and nitrogen dioxide) can be stimulated by stressful conditions such as acute [62] or chronic [33, 63, 75, 77] hypo or hyperthermia, and clinical diseases [20] such as inflammatory processes [107]. Oxygen and nitrogen free radicals have importance in antimicrobial and antitumour defence, and they also contribute to the pathogenesis of diseases [80]. The cellular defence mechanisms against invading micro-organisms result in free radical formation [51]. The reactive species thus

M. Mézes (✉)
Department of Nutrition, Szent István University, H-2103 Gödöllő, Hungary
e-mail: Mezes.Miklos@mkk.szie.hu

L. Mandelker and P. Vajdovich (eds.), *Studies on Veterinary Medicine*, Oxidative Stress in Applied Basic Research and Clinical Practice 5,
DOI 10.1007/978-1-61779-071-3_12, © Springer Science+Business Media, LLC 2011

produced attack invading micro-organisms by nitration, oxidation, and chlorination reactions [91]. However, the level of reactive oxygen free radical production during the so-called respiratory burst mechanism in phagocytes exceeds that required for cell activation [12]. Additionally, inducible nitric oxide synthase (iNOS), which becomes activated as a consequence of the recognition of invading bacteria or viruses by the immune system, activates the cyclo-oxygenase (COX-2) enzyme through a binding and S-nitrosilation mechanism and results in proinflammatory processes [48].

There is some evidence that birds have a more potent antioxidant system than mammals against the detrimental effects of oxygen free radicals [49, 96]. Therefore, it was hypothesised that the changes occurring in the antioxidant systems of chickens upon stress induced by chronic corticosterone exposure [61] differ from those seen in mammals. In addition, the long life-spans and slow ageing rates of birds are paradoxical in view of their high metabolic rate and high body temperature, which are predicted to be responsible for accelerated ageing according to the free radical theory of ageing [37]. However, some research results support that birds have special adaptations for preventing age-related tissue damage caused by reactive oxygen species [39].

Antioxidants and Immunity

Numerous data indicating that several natural antioxidants (vitamin E, selenium, and carotenoids) are among the major immunostimulating agents and their requirements for such action usually higher than those for the bird's growth and development [99]. For instance, vitamin E improves the phagocytic function of macrophages [52], and vitamin E and selenium significantly improve antibody production [7] or affect lymphocyte function [58].

Free Radicals and Antioxidants in Avian Diseases

Most of the infectious or nutritional diseases generate oxygen free radical formation and consequently impair the antioxidant defence system in avian species. Some of the most important viral, bacterial, parasitic, and nutritional diseases and their free radical background are discussed in this chapter.

Marek's Disease

Marek's disease (MD) is a herpes virus-induced, naturally occurring lymphoproliferative disease of chickens [93]. Free radicals, such as nitric oxide (NO) produced by the enzyme NO synthase (NOS) are potent antiviral agents in addition to having immune regulating functions. NO inhibits the replication of Marek's disease virus (MDV) in vitro and in vivo [110]. iNOS gene expression during MDV infection is mediated during the cytolysis phase of infection, which finding implicates that NO is a critical factor involved in increasing the virulence of MDV. Jarosinski et al. [43] reported that chickens resistant to the development of MD had a greater potential to produce NO than MD-susceptible chickens, as was shown in vitro by measuring NO levels in chick embryo fibroblast cultures obtained from these chickens after treatment with lipopolysaccharide (LPS) and in vivo inoculation with MDV strains ranging from mild to very virulent. The results showed that the more virulent strains induced the highest level of NO in the blood plasma, suggesting a role of NO in the pathogenesis of MD induced by more virulent strains.

In a recent study [47], experimental MDV infection in chickens caused significantly increased DNA damage and elevated malondialdehyde and NO metabolite levels in the MD group as compared to the healthy controls. A positive correlation was also found between MDA levels and DNA damage in MDV-infected chickens. Of the antioxidants of small molecular weight, the concentration of reduced glutathione was lower in the MD group as compared to the control.

Newcastle Disease

Newcastle disease (ND) is a paramyxovirus-induced disease with varying pathogenicity to avian species from mild or unapparent infection to sudden death, and causing respiratory, digestive, ocular, and enteric problems [95]. After infection with Newcastle disease virus (NDV), chicken heterophils had a decreased ability to phagocytise bacteria in vitro but showed increased NO production, suggesting that NDV can stimulate heterophils to produce and/or utilise nitrogen intermediates but not oxygen intermediates [57], and that these free radicals activate viral replication [80]. Free radical formation during NDV infection supports that dietary butylated hydroxytoluene having antioxidant property prevents mortality of chickens exposed to NDV [10], possibly through the same mechanism as found in vitro in murine isolated macrophages where NDV-mediated NOS and NF-κB activation were blocked by another antioxidant, butylated hydroxyanisole [105].

Although the effects of butylated hydroxytoluene and butylated hydroxyanisole on virus infection were suspected to be direct, it cannot be excluded that protection occurs through other – unknown – effects on the metabolism of host organism [80]. It is well known that leukocytes utilise nitrogen free radicals to control microbial pathogens. Upon activation, activated NOS catalyses the production of NO radicals, which leads to the formation of reactive nitrogen intermediates. That system is activated and modulated by microbial products and a series of temporally expressed cytokines. The effector molecules, generated in the early innate immune response, are not specific to the invading pathogen and may also cause damage to the host. It is the critical balance of these processes in the initial stages of infection that determines the outcome of infectious disease [46]. However, NDV infection not only activates the above-mentioned process but causes DNA fragmentation in the mononuclear cells, shortening of sarcomeres in the heart muscle containing debris from degenerated mitochondria, and the condensation of chromatin in cardiomyocytes, which are typical signs of apoptosis [55, 56] that has been proposed to be a free-radical-mediated process [32].

Infectious Bursal Disease

Infectious bursal disease (IBD) is a highly contagious disease of young chickens caused by the IBD virus. IBD has free radical mediated pathogenesis as in the case of other virus-induced diseases. That finding was proven by those findings when antioxidant treatment, such as high doses of vitamin E and selenium decreased the morbidity and mortality of IBD [78].

Salmonella Infection

The Gram-negative bacterium *Salmonella enteridica* invades and colonises chicken intestines [101]. Its LPS elicit both humoral and cellular immune responses, inducing an acute phase response [40] and activating an oxidative burst mechanism in chicken heterophil granulocytes through a toll-like receptor mediated signal mechanism [25]. The oxidative burst mechanism results in reactive oxygen radical formation [89] which may cause free radical mediated damages.

Staphylococcus Infection

The Gram-positive enterotoxigenic bacterium *Staphylococcus aureus* causes arthritis, dermatitis, and septicaemia in poultry [98]. Its toxic metabolite, lipoteichoic acid (LTA), activates the acute phase response [40] and the oxidative burst mechanism in chicken heterophil granulocytes [25]. The oxidative burst mechanism, as was mentioned above, possibly causes free radical mediated damage in different tissues, besides exerting local effects on invading bacteria and granulocytes.

Clostridium perfringens Infection

The free radical mediated effects of *Clostridium perfringens* are supported by findings indicating that the production of reactive oxygen species dramatically increased in circulating polymorphonuclear (PMN) leukocytes as by the chemiluminescence method after challenge with *C. perfringens* in broiler chickens. It was also found that 14- to 21-day-old chickens orally challenged with *C. perfringens* had higher leucocyte chemiluminescence responses and more severe intestinal lesions than unchallenged birds [79].

Necrotic Enteritis

Necrotic enteritis is primarily associated with the rapid proliferation of α- and β-toxin produced by *Clostridium perfringens* strains type A or C; however, the presence of this anaerobic bacterium does not lead directly to disease [16]. At a critical population size of *Clostridium perfringens* in the intestine of poultry, its toxin production commences and the epithelial tissues of the gastrointestinal tract develop a necrotised appearance [27]. The α-toxin has phospholipase activity [104], which causes epithelial cell membrane damage [19]. The membrane damage may be caused by the release of inositol triphosphate and diacylglycerol, which activate intracellular protein kinase C. The latter activates endogenous phospholipases and the arachidonic acid pathway, resulting in inflammatory response and eventually cell death. The importance of the arachidonic acid cascade mechanism is supported by findings that n-3 fatty acids, which are known to inhibit eicosanoid, mainly prostaglandin E_2 and interleukin 1β, synthesis [65], decrease the occurrence and pathogenesis of necrotic enteritis [4]. According to the above-mentioned facts, the hydrolysis of membrane phospholipids is a key event in the pathogenesis of necrotic enteritis.

Nutritional stress, such as feeds contaminated with mycotoxins well known to be free radical generators [99], challenges the antioxidant status of poultry, rendering them more susceptible to pathogenic bacteria such as *Clostridium perfringens.* Although enterocytes and colonocytes are exposed to oxygen free radicals, their susceptibility may vary, and distal ileal enterocytes and colonocytes are more prone to inflammation and oxidative stress associated with these radicals [6]. Brush border membrane and colonic apical membrane are susceptible to superoxide, which can cause peroxidation and degradation of membrane lipids, specifically certain phospholipids [81].

Escherichia coli Infection

Avian pathogenic *Escherichia coli* is a Gram-negative bacterium causing airsacculitis, polyserositis, septicaemia, and other mainly extraintestinal diseases in chickens, turkeys, and other avian species. *E. coli* can be found in the intestinal microflora of healthy birds, and most of the diseases associated with *E. coli* infection are secondary to environmental and host-related predisposing factors. Experimental infection studies have shown that the air-exchange regions of the lung and the air sacs are important sites of entry of *E. coli* into the bloodstream of birds during the initial stages of infection and that resistance to phagocytosis may be an important mechanism in the development of the disease [17]. In an in vitro model system it was found that *E. coli* inhibits macrophages through a classical, antiapoptotic NF-κB activation, which delays the phagocytosis-induced cell death of macrophages upon ingestion of *E. coli*. This finding was supported by the partial inhibition of NF-κB activation which did not influence the rate of uptake of bacteria but was followed by the increased production of oxygen radicals and enhanced intracellular killing in macrophages [34]. However, the *E. coli* genome contains a specific oxy-R sequence that improves the tolerance of the bacteria against oxidative damage, in that case the oxidative killing mechanism [111]. Intravenous administration of *E. coli*-derived LPS significantly increased the concentration of malondialdehyde and 3-nitro-tyrosine in plasma of poultry [106]. There is experimental evidence that the ingestion of *E. coli* by PMN leukocytes causes apoptosis of the latter cells, which is related to the respiratory burst mechanism and the formation of intracellular reactive oxygen intermediates [108].

Coccidiosis

Coccidiosis is an avian disease caused by protozoal parasites of the *Coccidia* subclass, *Eimeriidae* family, giving rise to intestinal damage. A potential oxidative destruction of mucosal tissue was suggested, supported by the increased activity of NADPH oxidase and increased production of O_2^- during *Eimeria* infection in the intestinal mucosa [1]. It has been demonstrated that the sporozoites of *Eimeria* spp. are very sensitive to superoxide ions [76]. The ability of the mucosal leukocytes to produce hydrogen peroxide as the precursor of superoxide formation also depends on the cellular immune status of chicken; for example, naive (specific pathogen free) birds produce less hydrogen peroxide than their hyperimmune counterparts [82]. Changes in the concentration of NO, another free radical, indicated that the virulence of *Eimeria* strains was not directly related to NO production [1]. With these findings it can be explained why the supplementation of L-arginine, which is the substrate of iNOS [74], did not reverse the pathological signs of *Eimeria* infection [2].

It was found [53] that the infection of broilers with coccidia intensifies the processing of free radicals in the blood and the liver homogenate. This was evident from the increased levels of reduced glutathione and lipid peroxidation and also from the activities of superoxide dismutase, glutathione peroxidase, and glutathione reductase. Changes in the enzyme activities were more pronounced in the blood than in the liver of the infected birds. Changes in the content of some low-molecular-weight antioxidants, such as vitamins A, C, and E, occur in the host in coccidiosis [23]. In the case of *Eimeria* infection and mucosal damage, α-tocopherol absorption is also impaired, which can only partially be restored by increasing the vitamin E content of the diet. However, in contrast to previous findings that α-tocopherol enhances the immunocompetence of chickens and thereby enhances their immunity to several diseases including coccidiosis [22], supplementation of the feed with up to 200 mg kg^{-1} α-tocopheryl-acetate had no consistent beneficial effect on the pathological changes caused by mild or severe infection with *Eimeria* [3]. Experimentally induced *Eimeria tenella* infection caused impairment of the lipid peroxide and antioxidant status of chickens, namely increased concentration of malondialdehyde and catalase activity, and decreased superoxide dismutase activity in *Eimeria*-infected birds, compared to healthy chickens [31].

Pulmonary Hypertension (Ascites) Syndrome

Pulmonary hypertension syndrome (PHS) of broilers, also called ascites, is an increase in intravascular hydrostatic pressure occurring secondary to right ventricular failure (RVF). In response to increased pressure, the transudate leaks out of blood vessels and accumulates in the abdominal cavity, thus producing ascites. However, the mechanism of cell injury underlying the pathogenesis of the syndrome is not clearly understood. The development of cardiac failure in an apparently healthy young broiler chicken can be explained by a series of factors such as genetic selection for a rapid growth rate, high feed efficiency, and a large pectoral muscle mass, all requiring high oxygen levels [14, 66]. The modern chicken has a small lung volume to body weight ratio, causing an inability of the respiratory system to respond to the broiler's elevated oxygen needs, which leads to hypoxia and respiratory acidosis [14]. An increase in free radicals occurs as a result of the pulmonary hypertension-induced RVF. RVF suggests that the increase in free radicals caused the heart disease [44].

Pulmonary hypertension (PH) is caused by increased blood flow or increased resistance to blood flow in the lung [45]. During hypoxia, various mechanisms increase the production of free radicals including lipid peroxide, hydrogen peroxide, and superoxide. This is supported by the fact that thiobarbituric acid reactive substances (TBARS), as indicators of cellular lipid peroxidation, increased by more than 100% in the heart and liver of broilers affected by ascites syndrome as compared to control birds [18]. However, it has to be mentioned that TBARS is not an absolutely specific indicator of lipid peroxidation because thiobarbituric acid may react with aldehydes other than malondialdehyde, one of the end-products of lipid peroxidation [42].

Tissue damage secondary to hypoxia attracts white blood cells, which in turn release more free radicals, causing further damage [8, 35]. Enkvetchakul et al. [21] observed inflammatory cell infiltration in various tissues of chickens with PHS. An alternative hypothesis concerning the cause of mortality in PHS is heart muscle damage because of the abnormal mitochondrial hydrogen peroxide accumulation [68]. This hypothesis supports findings according to which changes in the permeability of the sarcolemmal membrane, due to lipid peroxidation, led to the accumulation of calcium in the mitochondria [94]. In ascitic broilers, there was ultrastructural evidence of a calcium overload in the mitochondria of their cardiomyocytes [67]. Lipid peroxidation may therefore develop during ascites because the antioxidant status of the birds is compromised [21]. These changes may contribute to the influx of calcium and changes in the permeability of the mitochondrial membranes and, consequently, to the deposition of hydrogen peroxide in the mitochondrial matrix of cardiomyocytes. Basal hydrogen peroxide production, as an indicator of electron leak, was higher in breast and heart muscle mitochondria of birds suffering from PHS than in controls, and the differences in electron leak in the mitochondria of muscles were magnified by the inhibition of electron transport at Complex I and III (cyt b_{562}). Complex I activity was lower in the heart muscle mitochondria of birds with ascites syndrome as compared to controls but there was no difference in Complex II activity. Thus, mitochondria isolated from the heart muscle of birds with ascites syndrome exhibited site-specific defects in electron transport within Complex I and III that could contribute to a lower respiratory chain coupling. These findings prove the inefficient cellular use of oxygen that may contribute to the development of ascites syndrome in broilers [102].

Acidosis will also affect cellular membrane integrity and reduce free radical elimination, hence exacerbating the negative effect of free radicals. Higher plasma lipid peroxide values have been reported in PHS broilers [8], and Enkvetchakul et al. [21] demonstrated impaired antioxidant status as well as lower pulmonary and hepatic tocopherol and glutathione levels in PHS broilers. Generation of hydroxyl radicals ($OH^{\bullet}$) was found in broiler chickens with ascites induced experimentally using exposure to low temperature or administration of triiodothyronine (T_3). It was suggested that reactive oxygen species, such as $OH^{\bullet}$ ions, may be involved in the pathogenesis of ascites syndrome in broiler chickens [5].

The free radical theory concerning the pathogenesis of ascites syndrome is supported by findings that dietary antioxidants are useful in the prevention of PHS but only at high doses. For instance, the feeding of a diet containing 87 mg kg^{-1} vitamin E was ineffective [9] but with high vitamin E (250 mg kg^{-1}) and selenium (0.3 mg kg^{-1}, in the form of seleno-methionine) content decreased the incidence of PHS syndrome in chickens (10% vs. 0.9%) after a short-term exposure to heat stress [88]. In other studies [54, 109], dietary vitamin C at 500 mg kg^{-1} feed reduced PHS mortality induced by cool environmental temperatures and feeding of a thyroid hormone.

Nutritional Encephalomalacia

Nutritional encephalomalacia is a disease of young poultry, reaching its highest incidence during the third week of life [11]. The condition is obviously caused by peroxidative dysfunction of the cerebellum in absolute or relative deficiency of vitamin E, and is readily induced by diets supplying linoleic acid [11, 90]. However, in the presence of adequate vitamin E supplementation linoleic acid, even when added as rancid fat, did not provoke lipid peroxidation in the cerebellum and nutritional encephalomalacia [28]. Susceptibility of the brain to that nutritional deficiency may be caused by the fact that among the chicken tissues the brain showed the highest susceptibility to iron-induced oxidation [100].

The chicken brain contains high amounts of polyunsaturated, mainly n-6, fatty acids [64]. Recently it has been found [87] that fatty acids have an ability to scavenge superoxide anions in an unsaturation-dependent manner. Polyunsaturated fatty acids of the n-3 series resulted in a lower formation of reactive oxygen species as compared with saturates, monounsaturates, or polyunsaturates of the n-6 series. The concentration of α-tocopherol and the activities of antioxidant enzymes (superoxide dismutase and glutathione peroxidase) are very low in the brain [29, 72], and α-tocopherol content of the liver, acting as the main storage site during embryonic development, decreases rapidly during the first week posthatching [71]. Avian vitamin E deficiency has several other clinical manifestations including exudative diathesis and degenerative myopathy of the pectoral muscles, heart, or gizzard, and all of these pathological changes are caused by free radical mediated membrane disturbances [50].

Knockdown Syndrome of Turkeys

Turkey knockdown was defined as any condition identified in a turkey flock that has affected the neuromuscular or skeletal systems to a degree that a turkey is unable to walk or stand properly. Knockdown syndrome may be associated with numerous feed, management, or disease factors alone or in combination. Dosage of monensin was proposed as one of the main causes [13], but feed restriction, heat stress, copper, mycotoxins, and sodium chloride in feed have all been suggested as contributing factors.

The diagnosis of turkey knockdown syndrome is based on the clinical signs and histopathological lesions associated with knockdown. Affected birds were found to be recumbent, demonstrated paresis, and were unable to vocalise. Post-mortem examination revealed few significant lesions although pallor of the adductor muscles and petechiation in the adductor and gastrocnemius muscles were noted. Additional results indicated that changes in vitamin E status and in vitamin E content of the blood plasma correlated with the occurrence of knockdown [24, 70]. Of the other parameters of the biological antioxidant defence system, Cu/Zn-superoxide dismutase activity in the blood plasma also tended to be lower in turkcys suffering from knockdown syndrome as compared to the controls [70]. The same changes were found in monensin poisoning of chicken with impairment of the antioxidant status and elevated lipid peroxidation [92].

Tibial Dyschondroplasia

Tibial dyschondroplasia (TD) is a widespread abnormality found in rapidly growing meat-type poultry [59]. TD is a metabolic cartilage disease of young poultry in which endochondral bone formation is disrupted leading to the retention of a noncalcified, avascular plug of cartilage in the tibial growth plate. It is characterised by the persistence of cartilage below the growth plate. The TD lesion arises from a failure of growth plate chondrocyte differentiation, which results in an accumulation of prehypertrophic cells [85]. The cause of TD is believed to be a genotype and nutrition interaction that can be influenced by other factors such as mycotoxins, for example, fusarochromanone [38], or the presence of oxidised fat in the diet, which will cause the degradation of fat-soluble vitamins, including vitamins D, A, and E [60]. Oxidised fat intake as a possible oxidative stress effect also impairs the presence of reduced glutathione which requires both for keeping the sulfhydryl groups in the cells, and also acts as an important part of the antioxidant defence mechanism [69]. The effect of reduced glutathione in the pathogenesis of TD was proved by Rath et al. [84] using thiram, a glutathione inhibitor, for inducing this metabolic cartilage disease in poultry.

It is also important to mention that the osteolytic action of osteoclasts may be enhanced in the presence of free radicals [30]. Experimental studies have shown that addition of the active vitamin D_3 metabolite 1,25-dihydroxycholecalciferol [86] or its primary metabolite, 25-hydroxycholecalciferol [85], reduced the incidence and severity of TD in broilers. In addition, ascorbic acid would also be a potential preventive factor against TD because it participates in the biosynthesis of stable cross-linked collagen, stimulates the hydroxylation of 25-hydroxycholecalciferol to the active D_3 metabolite [26], and also acts as an antioxidant [36]. However, the availability of copper is reduced by high doses of ascorbic acid [15]. Copper is required for connective tissue development, and in the case of copper deficiency fibrils will not be properly cross-linked, and the resulting structure may be weakened or may fail [83].

Antioxidants as Prevention Against Avian Diseases

Based on the above-mentioned results it can be concluded that antioxidant (vitamins E and C and selenium) supplementation may improve the efficiency of vaccination, namely antibody production, against viral or bacterial infections [97], and vitamin E improve humoral immunity and phagocytic activity [103]. In addition to the above-mentioned antioxidants, vitamin D_3 can also be used as an effective prevention against some nutritional diseases [41, 73].

Summary

The purpose of the present review was to summarise the oxygen and nitrogen free radical background of some important infectious and metabolic poultry diseases. It is well known that most viral and bacterial diseases cause free radical formation and lipid peroxidation in different tissues as part of the immune response mechanism.

However, that mechanism becomes uncontrolled by the endogenous antioxidant defence mechanisms and may cause free radical-mediated damages.

Dietary antioxidants, such as vitamins E and C and selenium, have a dual effect on the above-mentioned processes. One is a direct antioxidant effect controlling the formation and/or scavenging the free radicals and the other is mainly to improve the humoral immune response of the organism against invading viruses and bacteria, and partly improve the antibody formation after vaccination against them.

References

1. Allen PC. Production of free radical species during *Eimeria maxima* infections in chickens. Poult Sci 1997; 76(6):814–821.
2. Allen PC. Effects of daily oral doses of L-arginine on coccidiosis infections in chickens. Poult Sci 1999; 78:1506–1509.
3. Allen PC, Fetterer RH. Interaction of dietary vitamin E with *Eimeria maxima* infections in chickens. Poult Sci 2002; 81(1):41–48.
4. Allen PC, Lydon J, Danforth HD. Effects of components of *Artemisia annua* on coccidia infections in chickens. Poult Sci 1997; 76(8):1156–1163.
5. Arab HA, Jamshidi R, Rassouli A, Shams G, Hassanzadeh MH. Generation of hydroxyl radicals during ascites experimentally induced in broilers. Br Poult Sci 2006; 47(2):216–222.
6. Baskar L, Balasubramanian KA. Effect of oxidant exposure on isolated rat colonocytes. Scand J Gastroenterol 1995; 30(3):332–336.
7. Bassiouni AA, Zaki MM, Hady MM. Effect of vitamin E and selenium on the immune response of chickens against living Newcastle disease vaccine. Vet Med J Giza 1990; 38(2):145–155.
8. Bottje W, Enkvetchakul B, Wideman R. Antioxidants, hypoxia and lipid peroxidation involvement in pulmonary hypertension syndrome (ascites). Novus Nutr Update 1995; 5(2):1–11.
9. Bottje WG, Erf GF, Bersi TK, Wang S, Barnes D, Beers KW. Effect of dietary DL-α-tocopherol on tissue α- and γ-tocopherol and pulmonary hypertension syndrome (ascites) in broilers. Poult Sci 1997; 76:1506–1512.
10. Brugh M. Butylated hydroxytoluene protects chickens exposed to Newcastle disease virus. Science 1977; 197(4310):1291–1292.
11. Budowski P, Leighfield MJ, Crawford MA. Nutritional encephalomalacia in the chick: an exposure of the vulnerable period for cerebellar development and the possible need for both ω6- and ω3-fatty acids. Br J Nutr 1987; 58:511–520.
12. Burdon RH. Superoxide and hydrogen peroxide in relation to mammalian cell proliferation. Free Radic Biol Med 1995; 18:775–794.
13. Cardona CJ, Galey FD, Bickford AA, Charlton BR, Cooper GL. Skeletal myopathy produced with experimental dosing of turkeys with monensin. Avian Dis 1993; 37(1):107–117.
14. Chabot R. Le syndrome de l'ascite chez les poulets à chair. Mémoire de Maîtrise, Faculté de Médecine vétérinaire de l'Université de Montréal à St-Hyacinthe; 1992, 50 pp.
15. Carlton WW, Henderson W. Studies in chickens fed a copper-deficient diet supplemented with ascorbic acid, reserpine and diethylstilbestrol. J Nutr 1965; 85(1):67–72.
16. Devitt RM, Brooker JD, Amovic T, Sparks NHC. Necrotic enteritis; a continuing challenge for the poultry industry. World Poult Sci J 2006; 62(2):221–247.
17. Dho-Moulin M, Fairbrother JM. Avian pathogenic *Escherichia coli* (APEC). Vet Res 1999; 30(2–3):299–316.
18. Díaz-Cruz A, Nava C, Villanueva R, Serret M, Guinzberg R, Piña E. Hepatic and cardiac oxidative stress and other metabolic changes in broilers with the ascites syndrome. Poult Sci 1996; 75(7):900–903.

19. Donelli G, Fiorentini C, Matarrese P, Falzano L, Cardines R, Mastrantinio P, Payne DW, Titball RW. Evidence for cytoskeletal changes secondary to plasma membrane functional alterations in the *in vitro* cell response to *Clostridium perfringens* epsilon toxin. Comp Immunol Microbiol Infect Dis 2003; 26(1):145–156.
20. Dröge W. Free radicals in the physiological control of cell function. Physiol Rev 2002; 82:47–95.
21. Enkvetchakul B, Bottje W, Moore R, Huff W. Compromised antioxidant status associated with ascites in broilers. Poult Sci 1993; 72:2272–2280.
22. Erf GF, Bottje WG, Bersi TK, Headrick MD, Fritts CA. Effects of dietary vitamin E on the immune system in broilers: altered proportions of CD4 T cells in the thymus and spleen. Poult Sci 1998; 77:529–537.
23. Evans P, Halliwell B. Micronutrients: oxidant/antioxidant status. J Nutr 2001; 121:324–338.
24. Evans, RD, Edson, RK, Watkins, KL, Robertson, JL, Meldrum JB, Novilla, MN (2000): Turkey knockdown in successive flocks. Avian Dis 44:730–736.
25. Farnell MB, He H, Kogut MH. Differential activation of signal transduction pathways mediating oxidative burst by chicken heterophils in response to stimulation with lipopolysaccharide and lipoteichoic acid. Inflammation 2003; 27(4):225–231.
26. Farquharson C, Berry JL, Maker EB, Seawright E, Whitehead CC. Mechanism of ascorbic acid regulation of growth plate chondrocyte differentiation: a possible role in the prevention of tibial dyschondroplasia. Proceedings of the WPSA UK Branch Spring Meeting, Scarborough; 1997, pp. 33–34.
27. Frame DD, Bickford AA. An outlook of coccidiosis and necrotic enteritis in 16-week-old cage-reared layer replacement pullets. Avian Dis 1986; 30:601–602.
28. Fuhrmann H, Schultheis S, Drommer W, Kaup FJ, Sallmann HP. Tissue lipid peroxidation in nutritional encephalomalacia of broiler chickens [In German]. Zentralbl Veterinärmed 1996; 43A(1):9–21.
29. Gaál T, Mézes M, Noble RC, Dixon J, Speake BK. Development of antioxidant capacity in tissues of the chick embryo. Comp Biochem Physiol 1995; 112B:711–716.
30. Garret I, Boyce R, Oreffo L, Bonewald J, Poser J, Mundy G. Oxygen-derived free radicals stimulate osteoclastic bone resorption in rodent bone *in vitro* and *in vivo*. J Clin Invest 1990; 85:632–639.
31. Georgieva NV, Koinarski V, Gadjeva V. Antioxidant status during the course of *Eimeria tenella* infection in broiler chickens. Vet J 2006; 172:488–492.
32. Gouazé V, Mirault ME, Carpentier S, Salvayre R, Levade T, Andrieu-Abadie N. Glutathione peroxidase-1 overexpression prevents ceramide production and partially inhibits apoptosis in doxorubicin-treated human breast carcinoma cells. Mol Pharmacol 2001; 60:488–496.
33. Gradinski Vrbanac B, Emanovic D, Milinkovic Tur S, Stojevic Z, Dobranic V, Zupancic Z, Susic V, Greguric J. Influence of hypothermia on chicken erythrocyte lipid peroxidation in vivo. Vet Med 1999; 44(5):129–132.
34. Groesdonk HV, Schlottmann S, Richter F, Georgieff M, Senftleben U. *Escherichia coli* prevents phagocytosis-induced death of macrophages via classical NF-κB signaling, a link to T-cell activation. Infect Immun 2006; 74(10):5989–6000.
35. Halliwell B, Gutteridge JMC. Role of free radicals and catalytic metal ion in human disease: an overview. Meth Enzymol 1990; 186:1–85.
36. Halliwell B, Gutteridge JMC. Free radicals in biology and medicine, 4th edition. Oxford: Oxford University Press; 2008, pp. 160–166.
37. Harman D, Heidrick ML, Eddy DE. Free radical theory of aging: effect of free-radical-reaction inhibitors on the immune response. J Am Geriatr Soc 1977; 25:400–407.
38. Haynes JS, Walser MM. Ultrastructure of *Fusarium*-induced tibial dyschondroplasia in chickens: a sequential study. Vet Pathol 1986; 23(3):499–505.
39. Holmes DJ, Flückiger R, Austad SN. Comparative biology of aging in birds: an update. Exp Gerontol 2001; 36:869–883.

40. Holt PS, Gast RK. Comparison of the effects of infection with *Salmonella enteritidis*, in combination with an induced molt, on serum levels of the acute phase protein, α1 acid glycoprotein, in hens. Poult Sci 2002; 81:1295–1300.
41. Hörak P, Zilmer M, Saks L, Ots I, Karu U, Zilmer K. Antioxidant protection, carotenoids and the costs of immune challenge in greenfinches. J Exp Biol 2006; 209:4329–4338.
42. Janero DR. Malondialdehyde and thiobarbituric acid-reactivity as diagnostic indices of lipid peroxidation and peroxidative tissue injury. Free Radic Biol Med 1990; 9(6):515–540.
43. Jarosinski KW, Yunis R, O'Connell PH, Markowski-Grimsrud CJ, Schat KA. Influence of genetic resistance of the chicken and virulence of Marek's disease virus (MDV) on nitric oxide responses after MDV infection. Avian Dis 2002; 46(3):636–649.
44. Julian RJ. Physiological, management and environmental triggers of the ascites syndrome: a review. Avian Pathol 2000; 29:519–527.
45. Julian RJ, Wilson JB. Right ventricular failure as a cause of ascites in broiler and roaster chickens. Proceedings of the IVth International Symposium of Veterinary Laboratory Diagnosticians, Amsterdam; 1986, pp. 608–611.
46. Karupiah G, Hunt NH, King NJ, Chaudhri G. NADPH oxidase, Nramp1 and nitric oxide synthase 2 in the host antimicrobial response. Rev Immunogenet 2000; 2(3):387–415.
47. Keles H, Fidan AF, Cigerci IH, Kucukkurt I, Karadas E, Dundar Y. Increased DNA damage and oxidative stress in chickens with Natural Marek's Disease. Vet Immunol Immunopathol 2009; doi:10.1016/j.vetimm.2009.07.003
48. Kim SF, Huri DA, Snyder SH. Inducible nitric oxide synthase binds, S-nitrosylates, and activates cyclooxygenase-2. Science 2005; 310:1966–1970.
49. Klandorf H, Probert IL, Iqbal M. In the defence against hyperglycaemia: an avian strategy. World Poult Sci J 1999; 55(3):251–268.
50. Klasing KC, Austic RE. Nutritional diseases. In: Saif YM, Barnes HJ, Glisson HR, et al. (editors): Diseases of poultry, 11th edition. Ames (IA): Iowa State University Press; 2003, pp. 1027–1054.
51. Kocyigit A, Keles H, Selek S, Guzel S, Celik H, Erel O. Increased DNA damage and oxidative stress in patients with cutaneous leishmaniasis. Mutat Res 2005; 585(1):71–78.
52. Konjucfa VK, Bottje WG, Bersi TK, Erf GF. Influence of dietary vitamin E on phagocytic functions of macrophages in broilers. Poult Sci 2004; 83(9):1530–1534.
53. Kostadinovic L, Popovic M, Pavkov S, Dimitrijevic S, Stevanovic D. Effect of coccidiosis on the enzymatic and non-enzymatic antioxidative systems in broilers. Acta Vet (Beograd) 1998; 48(5–6):323–330.
54. Ladmakhi MH, Buys N, Dewil E, Rahimi G, Decuypere E. The prophylactic effect of vitamin C supplementation on broiler ascites incidence and plasma thyroid hormone concentration. Avian Pathol 1997; 26(1):33–44.
55. Lam KM. Newcastle disease virus-induced apoptosis in the peripheral blood mononuclear cells of chickens. J Comp Pathol 1996; 114(1):63–71.
56. Lam KM. Ultrastructural changes in the cardiac muscle of chickens infected with the GB strain of Newcastle disease virus. J Comp Pathol 1996; 114(1):73–79.
57. Lam KM, Kabbur MB, Eiserich JP. Newcastle disease virus-induced functional impairments and biochemical changes in chicken heterophils. Vet Immunol Immunopathol 1996; 53(3–4):313–327.
58. Larsen HJ. Relations between selenium and immunity. Norw J Agric Sci 1993; 11(Suppl):105–119.
59. Leach RM, Lilburn MS. Current knowledge on the etiology of tibial dyschondroplasia in the avian species. Poult Sci 1992; 74(1):57–65.
60. Leeson S, Diaz G, Summers J. Poultry metabolic disorders and mycotoxins. Guelph: University Books; 1995, pp. 124–175.
61. Lin H, Decuypere E, Buyse J. Oxidative stress induced by corticosterone administration in broiler chickens (*Gallus gallus domesticus*). 1. Chronic exposure. Comp Biochem Physiol 2004; 139B:737–744.

62. Lin H, Decuypere E, Buyse J. Acute heat stress induces oxidative stress in broiler chickens. Comp Biochem Physiol 2006; 144A(1):11–17.
63. Maini S, Rastogi SK, Korde JP, Madan AK, Shukla SK. Evaluation of oxidative stress and its amelioration through certain antioxidants in broilers during summer. J Poult Sci 2007; 44(3):339–347.
64. Maldjian A, Cristofori L, Noble RC, Speake BK. The fatty acid composition of brain phospholipids from chicken and duck embryos. Comp Biochem Physiol 1996; 115B(2):153–158.
65. Mantzioris E, Cleland LG, Gibson RA, Neumann MA, Demasi M, James MJ. Biochemical effects of a diet containing foods enriched with n-3 fatty acids. Am J Clin Nutr 2000; 72(1):42–48.
66. Maxwell MH, Robertson GW. UK survey of broiler ascites and sudden death syndromes in 1993. Br Poult Sci 1998; 39(2):203–215.
67. Maxwell MH, Robertson GW, Mitchell MA. Ultrastructural demonstration of mitochondrial calcium overload in myocardial cells from broiler chickens with ascites and induced hypoxia. Res Vet Sci 1993; 54(2):267–277.
68. Maxwell MH, Robertson GW, Farquharson C. Evidence of ultracytochemical mitochondria-derived hydrogen peroxide activity in myocardial cells from broiler chickens with an ascites syndrome. Res Vet Sci 1996; 61(1):7–12.
69. Meister A, Anderson ME. Glutathione. Annu Rev Biochem 1983; 52:711–760.
70. Meldrum JB, Evans RD, Robertson JL, Watkins KL, Novilla MN. Alterations in levels of various host antioxidant factors in turkey knockdown syndrome. Avian Dis 2000; 44(4):891–895.
71. Mézes M. Investigations of vitamin E transport of different age groups of domestic fowl. Bull Univ Agric Sci Gödöllő 1988; 1:59–64.
72. Mézes M. Changes in lipid peroxidation and activities of some antioxidant enzymes of chicken brain homogenates during early postnatal development. In: Matkovics B, Boda D, Kalász H (editors): Oxygen free radicals and the tissue injury. Budapest: Akadémiai Kiadó; 1988, pp. 211–215.
73. Mézes M, Surai P, Sályi G, Speake BK, Gaál T, Maldjian A. Nutritional metabolic diseases of poultry and disorders of the biological antioxidant defence system. Acta Vet Hung 1997; 45(3):349–360.
74. Moncada S, Higgs A. The L-arginine-nitric oxide pathway. N Engl J Med 1993; 329:2002–2012.
75. Mujahid A, Pumford NR, Bottje W, Nakagawa K, Miyazawa T, Akiba Y, Toyomizu M. Mitochondrial oxidative damage in chicken skeletal muscle induced by acute heat stress. J Poult Sci 2007; 44(4):439–445.
76. Ovington KS, Smith NC. Cytokines, free radicals and resistance to *Eimeria*. Parasitol Today 1992; 8:422–426.
77. Pamok S, Aengwanich W, Komutrin T. Adaptation to oxidative stress and impact of chronic oxidative stress on immunity in heat-stressed broilers. J Thermal Biol 2009; doi:10.1016/j.jtherbio.2009.06.003
78. Panda SK, Rao AT. Effect of vitamin E – selenium combination on chickens infected with infectious bursal disease virus. Vet Rec 1994; 134(2):242–243.
79. Papp Z, Dahiya JP, Warren T, Widyar`atne G, Drew MD, Smits JE. Whole blood chemiluminescence response in broiler chickens on different experimental diets and challenged with *Clostridium perfringens*. Br Poult Sci 2009; 50(1):57–65.
80. Peterhans E. Oxidants and antioxidants in viral diseases: disease mechanisms and metabolic regulation. J Nutr 1997; 127(Suppl 5):962S–965S.
81. Prabhu R, Balasubramanian KA. Effect of oxidants on small intestinal brush border membranes and colonic apical membranes – a comparative study. Comp Biochem Physiol 2003; 134C:329–339.
82. Prowse SJ, Michalski WP, Fahey KJ. Enhanced H_2O_2 release from immune chicken leucocytes following infection with *Eimeria tenella*. Immunol Cell Biol 1992; 70(Pt. 1): 41–48.

83. Rath NC, Huff GR, Huff W, Balog JM. Factors regulating bone maturity and strength in poultry. Poult Sci 2000; 79:1024–1032.
84. Rath NC, Richards MP, Huff WE, Huff GR, Balog JM. Changes in the tibial growth plates of chickens with thiram-induced dyschondroplasia. J Comp Path 2005; 133(1):41–52.
85. Rennie JS, Whitehead CC. Effectiveness of dietary 25- and 1-hydroxy-cholecalciferol in combating tibial dyschondroplasia in broiler chickens. Br Poult Sci 1996; 37:413–421.
86. Rennie JS, Whitehead CC, Thorp BH. The effect of dietary 1,25-dihydroxy-cholecalciferol in preventing tibial dyschondroplasia in broilers fed diets imbalanced in calcium and phosphorus. Br J Nutr 1993; 69:809–816.
87. Richard D, Kefi K, Barbe U, Bausero P, Visioli F. Polyunsaturated fatty acids as antioxidants. Pharmacol Res 2008; 57:451–455.
88. Roch G, Boulianne M, De Roth L. Dietary antioxidants reduce ascites in broilers. World Poultry 2000; 16(11):18–22.
89. Rosen GM, Pou S, Ramos CL, Cohen MS, Britigan BE. Free radicals and phagocytic cells. FASEB J 1995; 9(2):200–209.
90. Sallmann HP, Fuhrmann H, Molnár S, Stegmanns T. Endogenous lipid peroxidation in broiler chickens under dietary loads. Fat Sci Technol 1991; 93:457–462.
91. Saran M, Beck-Speier I, Fellerhoff B, Bauer G. Phagocytic killing of microorganisms by radical processes: consequences of the reaction of hydroxyl radicals with chloride yielding chlorine atoms. Free Radic Biol Med 1999; 26:482–490.
92. Sályi G, Mézes M, Bánhidi G. Changes in the lipid peroxide status of broiler chickens in acute monensin poisoning. Acta Vet Hung 1990; 38(2):263–270.
93. Schat KA, Chen CLH, Calnek BW, Char D. Transformation of T-lymphocyte subsets by Marek's disease herpesvirus. J Virol 1991; 65:1408–1413.
94. Shen AC, Jennings RB. Kinetics of calcium accumulation in acute myocardial ischemic injury. Am J Pathol 1972; 67:441–452.
95. Shivaprasad HL. Pathology of birds – an overview. Proc. C.L. Davis Foundation Conference on Gross Morbid Anatomy of Animals, Washington (DC); AFIP 2002; 18–19.
96. Simoyi MF, Dyke KV, Klandorf H. Manipulation of plasma uric acid in broiler chicks and its effect on leukocyte oxidative activity. Am J Physiol 2002; 282R:791–796.
97. Singh H, Sodhi S, Kaur R. Effects of dietary supplements of selenium, vitamin-E or combinations of the two on antibody responses of broilers, Br Poult Sci 2006; 47:714–719.
98. Skeeles JK. *Staphylococcus*. In: Calnek, BW (editor): Diseases of poultry, 10th edition. Ames (IA): Iowa State University Press; 1997, pp. 247–253.
99. Surai PF. Natural antioxidants in avian nutrition and reproduction. Nottingham: Nottingham University Press; 2002, pp. 456–472.
100. Surai PF, Speake BK, Noble RC, Sparks NHC. Tissue-specific antioxidant profiles and susceptibility to lipid peroxidation of the newly hatched chick. Biol Trace Elem Res 1999; 68(1):63–78.
101. Suzuki S. Pathogenicity of *Salmonella enteritidis* in poultry. Int J Food Microbiol 1994; 21(1):89–105.
102. Tang Z, Iqbal M, Cawthon D, Bottje WG. Heart and breast muscle mitochondrial dysfunction in pulmonary hypertension syndrome in broilers (*Gallus domesticus*). Comp Biochem Physiol 2002; 132A:527–540.
103. Tengerdy RP, Brown JC. Effect of vitamin E and A on humoral immunity and phagocytosis in E. coli infected chicken. Poult Sci 1977; 56(3):957–963.
104. Titball RW, Rood JI. *Clostridium perfringens*: wound infections. Mol Med Microbiol 2002; 1–3:1875–1903.
105. Umansky V, Shatrov VA, Lehmann V, Schirrmacher V. Induction of NO synthesis in macrophages by Newcastle disease virus is associated with activation of nuclear factor-κB. Internat Immunol 1996; 8(4):491–498.
106. Ünlü A, Türközkan N, Cimen B, Karabicak U, Yaman H. The effect of *Escherichia coli*-derived lipopolysaccharides on plasma levels of malondialdehyde and 3-nitrotyrosine. Clin Chem Lab Med 2001; 39(6):491–493.

107. Vajdovich P. Free radicals and antioxidants in inflammatory processes and ischemia-reperfusion injury. Vet Clin North Am Small Anim Pract 2008; 38(1):31–123.
108. Watson RW, Redmond HP, Wang JH, Condron C, Bouchier-Hayes D. Neutrophils undergo apoptosis following ingestion of *Escherichia coli.* J Immunol 1996; 156(10):3986–3992.
109. Xiang RP, Sun WD, Wang JY, Wang XL. Effect of vitamin C on pulmonary hypertension and muscularisation of pulmonary arterioles in broilers. Br Poult Sci 2002; 43:705–712.
110. Xing Z, Schat KA. Inhibitory effects of nitric oxide and gamma interferon on *in vitro* and *in vivo* replication of Marek's disease virus. J Virol 2000; 74(8):3605–3612.
111. Yoon SJ, Park JE, Yang J-H, Park J-W. *OxyR* regulon controls lipid peroxidation-mediated oxidative stress in *Escherichia coli.* J Biochem Mol Biol 2002; 35(2):297–301.

Oxidative Stress in Ruminants

Pietro Celi

Abstract This chapter examines the role that oxidative stress plays in ruminant medicine. We examine how redox homeostasis in involved in some physiological functions and we discuss the implications of the impairment of oxidative status on ruminant health and production. The study of oxidative stress is a relatively young field of research in ruminant medicine. The understanding of the role of oxidants and antioxidants in physiological and pathological conditions is rapidly increasing. Oxidative stress is an active field of research in ruminant medicine and has been implicated in numerous disease processes including sepsis, mastitis, acidosis, ketosis, enteritis, pneumonia, respiratory, and joint diseases. Compared to human medicine, only a limited number of conditions have been investigated in regard to the effects of oxidative stress in ruminants. Studies in cattle have been sporadic and mainly concerned with mastitis, pneumonia, and retained placenta. More recently, studies have been focused on metabolic diseases that affect dairy cows during the peripartum period. Numerous and rapidly evolving methodologies for evaluating oxidative stress are available to researchers and clinicians, each with their own distinct advantages and disadvantages. Differences in models and methodologies make it difficult to make meaningful comparisons, even for studies that seem quite similar superficially. With this in mind, it is the goal of this chapter to summarise the present knowledge of oxidative stress in ruminant medicine and to examine the basis of and evidence for the role of oxidative stress in ruminant health and production, highlighting the need for continued research on oxidative stress in ruminant medicine. Clarity of understanding of the pathophysiology of oxidative stress in ruminants will allow the design of specific antioxidant therapies. Future research should focus on the establishment of a reference panel of biomarker of oxidative stress to be used in ruminant medicine. The development of an oxidative stress index as an approach in ruminant and veterinary medicine is also discussed.

P. Celi (✉)
University of Sydney, C04 MC Franklin Lab,
PMB 3, 425 Werombi Rd, Camden, NSW 2570, Australia
e-mail: pietro.celi@sydney.edu.au

L. Mandelker and P. Vajdovich (eds.), *Studies on Veterinary Medicine*,
Oxidative Stress in Applied Basic Research and Clinical Practice 5,
DOI 10.1007/978-1-61779-071-3_13,

Keywords Ruminants • Oxidative stress • Health • Disease • Production • Reproduction

Oxidative Stress in Ruminant Physiology

Measuring Oxidative Stress in Ruminants

Oxidative stress in veterinary medicine and particularly in ruminant health is a relatively young field of research. Numerous and rapidly evolving methodologies for evaluating oxidative stress, each with its own distinct advantages and disadvantages, are available to clinicians and scientists. However, differences in models and methodologies make it difficult to make meaningful comparisons, even for studies that seem quite similar. Methods for quantifying oxidative stress mostly include direct or indirect measures of oxidants and antioxidants. Given their often high reactivity, quantification of oxidants and antioxidants often requires specialised equipment and considerable experience. As oxidative stress is indicative of an imbalance between oxidants and antioxidants, methods for quantifying oxidative stress mostly include direct or indirect measures of oxidants and antioxidants ([174, 184]). In these sections, some principles and commonly used measures of oxidative stress and damage are briefly outlined.

Pro-oxidants

The most abundant free radicals in biological systems are the oxygen-centred free radicals and their metabolites, usually referred as "reactive oxygen metabolites" (ROMs; [184]). ROMs are formed continuously as normal by-products of cellular metabolism and, in low concentrations, they are essential for several physiological processes, including protein phosphorylation, transcription factor activation, cell differentiation, apoptosis, oocyte maturation, steroidogenesis, cell immunity, and cellular defence against micro-organisms ([1, 77, 184]). However, when produced in excess, ROMs can damage cell functionality as they can harm cellular lipids, proteins, and DNA ([184, 260]).

The plasma level of ROMs is considered an indicator of free radical production [184]. ROMs are a collective term that includes not only oxygen-centred free radicals such as superoxide anion and hydroxyl radical, but also some nonradical derivates of oxygen, such as hydrogen peroxide, and hypochlorous acid [214]. A ROMs kit has been developed to assess oxidant levels in plasma and other biological fluids. The ROMs test has been validated using electron spin resonance [3] and is now considered the "gold standard" for measuring total oxidative status. Electron spin resonance is not suitable for routine analysis as the method is complex and requires specific technical assistance not available in most laboratories. The utility of the ROMs assay in monitoring oxidative stress in goats [53, 57, 74], sheep [209], and dairy cows [27, 28, 208] has been reported.

The concentrations of individual oxidant components can be measured separately in the laboratory, but such measurements are time consuming, labour intensive, and costly. It has been shown that free radical analytical system (FRAS 4) technology offers a quick, simple, precise, and reliable method of assessing oxidative status in dairy cows [183] and in horses [56] which is particularly useful in the field where it is not always practical or possible to get samples to a laboratory immediately. The possibility of assessing oxidative stress directly in blood provides veterinarians with a simple and reliable method of measuring oxidative stress in clinical situations such as the monitoring of therapy and in the antioxidant supplementation of domestic animals. However, given the lack of reference values for ROMs in ruminants it is difficult to establish if and when these animals are experiencing oxidative stress. Therefore it is important to calculate the specific referral ranges because a correct biochemical evaluation of oxidative status is an essential premise to prevent and eventually to treat the effects of oxidative stress in ruminant medicine.

Advanced oxidation protein products (AOPP) are terminal products of proteins exposed to free radicals and arise from the reaction between plasma proteins and chlorinated oxidants mediated by a neutrophil enzyme myeloperoxidase [90, 201]. In humans, AOPP have been associated to several diseases including chronic renal failure [291], diabetes mellitus [140], diabetic nephropathy [241], coronary artery diseases [142], and obesity [15]. Chronic accumulation of AOPP has been demonstrated to promote inflammatory processes in diabetic kidney [241] and in chronic renal failure [291] indicating that AOPP might be a by-product of neutrophil activation during infections. Studies in ruminants have reported higher levels of AOPP in lambs [51] and dairy cows [52] supplemented with Yerba Mate (*Ilex Paraguanensis*).

More information about the role of protein oxidation in ruminants' health and production could be obtained by the comparison of AOPP with other indicators of protein oxidation, such as advanced glycation end-products (AGE). However, although a correlation between AGE and inflammatory parameters is usually not found or is only weak, the induction of proinflammatory activities caused by AOPP seems to be more intense [141, 292]. This suggests that oxidative stress is more closely linked to inflammation and acute phase reactions than the advanced glycation process and its end-products. AOPP could thus better describe acute inflammation, whereas AGE might serve more as a marker of chronic long-lasting damage [141]. These observations are highly relevant as a high level of AOPP could indicate the presence of an inflammatory process which could compromise the correct embryonic development in the dairy cow [182, 183].

Lipids, in particular those that are polyunsaturated, are prone to oxidation. Lipids are one of the most susceptible substrates to free radical damage and biomarkers of lipid peroxidation are considered the best indicators of oxidative stress [102]. Malondialdehyde (MDA) is one of the several low-molecular-weight end-products formed during the radical induced decomposition of polyunsaturated fatty acid [136]. MDA readily reacts with thiobarbituric acid producing a red pigment that can be easily measured by spectrophotometry in the form of thiobarbituric acid reactive substances (TBARS) [136].

It is worth noting that the MDA assays have been criticised for low specificity and artefact formation inasmuch as only a fraction of the MDA measured is generated in vivo. Furthermore, the TBARS assay, a common method used to measure MDA, is considered inaccurate, and returns results that differ according to the assay conditions used [119]. For example, studies in dairy cows have yielded contrasting results with some reports failing to show any significant changes in plasma MDA concentrations during the peripartum period [48, 49], whereas in other studies MDA or TBARS concentrations increased around calving [27, 28, 35].

This apparent discrepancy could also have been mainly due to the great individual variations observed in MDA concentrations measured in the studies by [48, 49]. Similarly, studies in transported cattle have failed to report a consistent change in MDA concentration. It seems that this discrepancy is partly due to the different methodologies employed to assess MDA. TBARS detect a wide range of lipid peroxidation products, and are rather unspecific for MDA [113]. High-pressure liquid chromatography would be expected to be highly specific and perhaps more accurate than the spectrophotometric procedures [173]. More recently an ELISA-based isoprostane (considered to be the most reliable markers of lipid oxidation; [187]), assay has become commercially available and might be able to shed more light on the role of lipid peroxidation during the peripartum period in ruminants.

Antioxidants

Endogenous antioxidants can be divided into three major groups [184]. The first group comprises enzymatic antioxidants including superoxide dismutase (SOD) and glutathione-peroxidase (GSH-Px), and represents the main form of intracellular antioxidant defence. Plasma GSH-Px activity contributes to the oxidative defence of animal tissues by catalysing the reduction of hydrogen and lipid peroxides [119] and is also considered an indicator of oxidative stress [268]. GSH-Px functions in cellular oxidation–reduction reactions to protect the cell membrane from oxidative damage caused by free radicals [92]. SOD catalyses the dismutation of superoxide to hydrogen peroxide (H_2O_2) and it is considered the first defence against pro-oxidants [119]. Studies in grazing sheep have shown that GSH-Px activity is influenced by soil and pasture characteristics [7] and by season [8].

In dairy goats SOD activity is decreased during the postpartum period probably as a consequence of lower peroxide generation as testified by the decrease in ROM concentrations [57]. Because SOD activity increases H_2O_2 production, protection from reactive oxygen would only be given by a simultaneous increase in catalase and GSH-Px activities and availability of glutathione [93, 156]. Studies in dairy goats have shown that blood GSH-Px activity is decreased during the postpartum period, suggesting that goats may have experienced some degree of oxidative stress and lipid peroxidation [53, 57]. Inasmuch as GSH-Px is directly targeted at removing H_2O_2 generated during the dismutation of free radicals [77], it would seem reasonable to see a parallel decrease in ROMs levels. ROMs levels indeed, decreased on week 4 postpartum, however, the concentrations in week 2 were significantly higher [57], which further indicated that the goats experienced oxidative stress during the early postpartum period. Even if blood GSH-Px activity were inhibited [53, 57], the organism could have been defended against oxidative stress by other alternative routes. For example, catalase is another antioxidant enzyme that can catabolise H_2O_2 [77].

The second group includes nonenzymatic protein antioxidants that are primarily found in plasma. They are mainly represented by sulfhydryl (SH) groups of albumin and are considered a significant element of the extracellular antioxidant defence system against oxidative stress as are protein SH groups [271]. The reducing properties of SH residue are known to be oxidised under oxidative stress and other physiological conditions [72, 73]. Total thiol groups of plasma represent the SH groups of albumin, L-cysteine, and homocysteine. Under physiological conditions, SH groups are the most chemically reactive sites and have strong reducing properties [189, 255].

Studies in dairy goats have shown that plasma albumin concentrations are significantly reduced during summer [74]. This finding is quite relevant considering that albumin is part of the antioxidant pool, it being a free radical scavenger [118]. Plasma albumin levels are also decreased during the peripartum period in dairy goats which further indicates that goats were exposed to oxidative stress during the peripartum period [53, 57]. Albumin is exclusively synthesised by the liver, and is the main source of plasma SH. The reduction in liver function that is usually observed in the early postpartum period might explain lower plasma SH and albumin levels. Studies in dairy cows have confirmed the antioxidant role played by albumin particularly near calving when animals usually do not receive any vitamin/mineral supplementation [48].

The third group is represented by the nonenzymatic low-molecular-weight antioxidants, and it is found mainly in plasma but also in other extracellular and intracellular fluids. The primary antioxidant capacity of serum is derived from nonenzymatic antioxidants such as glutathione, α-tocopherol, β-carotene, and uric acid [120]. In particular, GSH plays an important role in protecting cells against oxidative stress and toxic agents. It acts as substrate or cosubstrate in enzymatic reactions and it also reacts directly with free radicals and lipid peroxides [38].

For example, in a selenium deficiency situation, hepatic glutathione (GSH) synthesis is increased and this depletes cellular cysteine [40], so it may impair physiological processes (growth and wool production) where cysteine is required for protein synthesis. Indeed, GSH synthesis seems to compete with wool growth for cysteine [170]. Cysteine is vital for wool growth and is usually the first limiting amino acid for wool fibre synthesis [171]. Sheep selected for high wool production tend to have lower blood cysteine and GSH concentrations than those selected for low wool production [128, 290]. Low concentrations of GSH can be associated with impaired animal health as cysteine and GSH play a key role in the regulation of the immune response [79]. Therefore, considering that GSH is a reservoir of cysteine, selecting for both GSH concentrations and wool growth rate might result in improvements in both wool production and health status [170]. In the meantime, when selenium deficiency is diagnosed, its supplementation is recommended to improve the development of resistance and resilience of sheep to gastrointestinal parasites [54].

Because of the difficulty in measuring each antioxidant component separately and their interaction in the plasma, several methods have been developed to assess total antioxidant capacity (TAC). The measure of antioxidant capacity considers the cumulative action of all the antioxidants present in plasma and body fluids, thus providing an integrated parameter rather than the simple sum of measurable antioxidants [104]. Antioxidant capacity can be measured by means of several methods, such as trolox equivalent antioxidant capacity (TEAC; [185]), total radical-trapping antioxidant parameter (TRAP; [103]), oxygen radical absorbance capacity (ORAC; [44]), or the ferric reducing ability of plasma (FRAP) and biological antioxidant potential (BAP; [26]). The BAP test provides a global measurement of many antioxidants, including uric acid, ascorbic acid, proteins, α-tocopherol, and bilirubin [26]. In dairy cows BAP levels have been measured during early lactation [208] and mid-lactation [52], however, comparisons can hardly be done because those studies involved cows in different physiological phases. Also, the rearing system may determine important differences in terms of oxidative status of animals.

A pasture-based system could improve the oxidative status of animals due to the elevated antioxidant content of green grass [71, 100] and the physical exercise related with the grazing system has better antioxidant status [100]. Grazing may provide dairy cows with health benefits from certain vitamins and minerals. Vitamins A and E, and selenium are known antioxidants that play important roles in animal health and production. Although fresh forage is typically considered capable of supplying adequate levels of antioxidants for dairy cattle, the availability of these compounds for lactating grazing cows is diminished when pasture availability is not adequate to meet their energy requirements. In this situation the gap between energy required for milk production and energy intake is often met by supplementing cows with conserved forage such as silage. Silage is known for its poor antioxidant content [20] and thus this might expose cows to oxidative stress. A decrease in BAP levels was observed when dairy cows were fed corn silage and concentrate supplementation [52]. Concentrate supplementation with the diet has been proved to reduce pasture intake [21] and to increase the degradation of some antioxidants at the ruminal level [282]. Plasma BAP might have decreased also as a direct consequence of ROM increase. Indeed, changes in the components of antioxidant systems are often not the cause, but the consequence of the oxidative stress induced by higher free radical activity [275].

Plasma antioxidant status is the result of the interaction of many different compounds and systemic metabolic interactions [104]. As a single measure, TAC provides relevant information that may effectively describe the dynamic equilibrium between pro-oxidants and antioxidants in the plasma compartment [43, 104].

In the absence of reference values for ruminants, TAC can be used to evaluate the effect of treatments when the results are expressed as change with respect to the basal value [104]. Indeed, TAC is a useful tool for measuring stress in transported calves [211]. In conclusion the evaluation of TAC on a herd basis may be a useful indicator of animal welfare and may be more sensitive and reliable than the measure of a single parameter, which could reveal individual variations. Moreover, TAC could be used as a tool to evaluate the nutritional status of animals fed different diets or to evaluate the general nutritional status of animals throughout the year.

Oxidative Stress in Ruminant Health

Mastitis

Mastitis is the most devastating disease of dairy animals; despite improved management practices and dry cow therapy, mastitis remains a worldwide problem of major economic threat to dairy farmers. The inflammatory reaction accompanying mastitis is generally caused by bacterial infection. Severe inflammation damages the mammary secretory epithelium and cytotoxic radicals and proinflammatory cytokines are released by the phagocytic cells [159]. Reactive nitrogen intermediates are important radicals that play a complex role in the inflammatory process [110]. Excessive amounts of neutrophiles, macrophages, lymphocytes, eosinophiles, and various epithelial cells of mammary tissue in milk are considered a response of mammary tissue to micro-organisms in part of inflammation of the mammary gland [159, 249]. The level of some cytokines (TNF-α, IL-1b, IL-6, IL-8) and other molecules such as nitric oxide (NO) are reported to increase during infections [200, 216].

NO operates in a variety of tissues to regulate a diverse range of physiological processes, including the inflammatory response [67]. Macrophage and epithelial cells of the mammary gland produce a significant amount of NO; this inducible NO mediates inflammation during mastitis [32]. High NO activity has been observed in mastitic animals [16] and an increase in milk NO levels from cows with experimentally induced mastitis has been reported, and it has been proposed that the elevated NO concentration in milk resulted from the inflammatory response of the mammary gland [32]. An increase in NO concentrations in milk from mammary glands with subclinical mastitis also supports the relationship between elevation of NO levels and inflammation [16]. Somatic cell count (SCC) in milk is a well-known indicator reflecting mammary health and milk quality. A positive relationship between SCC and NO concentration has been reported [16, 32]. Changes in milk NO levels thus could be proposed as an additional diagnostic tool to detect inflammation during mastitis. Finally, it has been noted that the concentrations of nitrite and nitrate increase both in plasma and milk after intramammary infusion of *E. coli* bacteria or *E. coli* endotoxin [31, 162] and that this increase can be prevented by treatment with aminoguanidine, a specific inhibitor of iNOS [32].

NO production is increased during inflammatory diseases [176, 188] and it is activated by cytokines. Indeed, a positive correlation between intramammary and systemic TNF-α and NO production has been reported in an acute mastitis [31]. NO production is considered as a primary defence system [131, 203] as it has antimicrobial properties due to peroxynitrite, a reactive nitrogen metabolite, derived from oxidation of NO [25]; however, peroxynitrite can cause alterations in antioxidant balance in the organism when produced in excess [58]. During inflammatory diseases such as mastitis, there is an increase in lipid peroxidation which causes a decrease in levels of some antioxidant molecules leading consequently to oxidative stress [110, 162, 284]. During inflammatory diseases, high levels of NO react with superoxide anions leading to formation of peroxinitrite radical [25], and peroxinitrite radicals oxidise long-chain fatty acids in cell membranes leading to increase in lipid peroxidation and formation of free radicals [5, 279]. Therefore, increase in lipid hydroperoxide level following lipid peroxidation during experimentally induced acute mastitis in cattle clearly indicates that mastitis causes formation of free radicals.

Milk from mammary glands with subclinical mastitis has higher levels of total oxidant capacity and lower levels of TAC compared to milk from mammary glands without subclinical mastitis [16]. An increase in ROMs following experimentally induced acute mastitis with lipopolysaccharide injection has been reported, however, a decrease in ROMS production was noted after retinoid administration [115]. Therefore it seems that mastitis could induce oxidative stress leading to increase in formation of free radicals in milk.

Serum and milk concentrations of ascorbate levels have also been identified as oxidative stress biomarkers in bovine mastitis. Dairy cattle affected by acute and subclinical forms of mastitis showed a significant decrease in ascorbate concentration [158, 213]. This decrease in ascorbate concentration seems to be accompanied by an increase in the levels of lipid hydroperoxide in erythrocytes isolated from dairy cows with acute mastitis. The association of increased lipid peroxidation and lowered ascorbate levels further support that mastitis induced oxidative stress in the mammary gland. Ascorbic acid is the most important water-soluble antioxidant in mammals [229]. It is well known that cows can synthesise vitamin C and vitamin C is not a required nutrient for dairy cows, however, several studies are suggesting that vitamin C is related to mastitis. Cows with mastitis have lower concentrations of vitamin C in the plasma and milk [158, 284] and the severity of clinical signs is proportional with the magnitude of the decrease in concentrations [284]. Subcutaneously injected vitamin C may have therapeutic value for cows with mastitis [195, 213], however, in LPS induced mastitis ascorbate administration had marginal effect on clinical signs but recovery of milk production was improved by the treatment [58]. In view of the above-mentioned results, it would be interesting to attempt the combined treatment of vitamin C and aminoguanidine to prevent the consequences of udder-related oxidative stress and inhibit inflammation-induced iNOS in mastitis [174].

Antioxidant nutrition plays an important part in mastitis prevention because of the critical role of these micronutrients in mammary resistance to this disease

[36, 86, 232, 251, 283]. Micronutrients that are associated with antioxidant activity including vitamin A, vitamin E, β-carotene, selenium, zinc, and copper have been studied for their effect on mastitis [36, 86] and it has been reported that antioxidant supplementation could decrease the duration, incidence, and severity of clinical mastitis [251]. In a study of experimentally induced acute mastitis, an increase in antioxidant capacity by administering vitamin E and selenium was reported [192].

Vitamin E and selenium given prior to calving have also been proposed for prophylactic treatment of mastitis, because vitamin E is depleted as calving approaches and remains low during weeks 1 and 2 of lactation [252]. Precalving administration of vitamin E increased plasma α-tocopherol concentrations and neutrophils from vitamin E treated animals had a greater ability to kill bacteria [127]. Selenium, either alone or in combination with vitamin E increases bacterial killing by blood neutrophils [126]. GSH-Px is the selenoenzyme and it has been observed that blood concentration of selenium and GSH-Px activity negatively correlated with the prevalence of intramammary infection [87]. Low GSH-Px activity in mastitic cows with high SCCs and high level of prostaglandin formation has been recorded [127]. GSH-Px is the selenoenzyme and it has been observed that blood concentration of selenium and GSH-Px activity negatively correlated with the prevalence of intramammary infection [87]. These data further corroborate the involvement of oxidative stress in mastitis.

Nitrate Poisoning

Oxidative stress in dairy ruminants is also likely to be induced by feed composition and in particular its content of antioxidants or thiol-containing compounds. For example, forage grown in the presence of high levels of nitrogen fertiliser contains high concentrations of nitrates and nitrites, which can be converted into peroxinitrite, a reactive species known to modify proteins and stimulate lipid peroxidation [212]. The consumption of forage with elevated concentrations of nitrates and nitrites may have negative repercussions on ruminant health and welfare and may result in poor reproductive performances. Poor fertility, particularly embryo mortality and high rates of return to estrus, is commonly observed during spring grazing in dairy cows [269]. Spring grass is rich in rapidly degradable nitrogen, thus, cows turned out onto spring grass may experience a period of excess rapidly degradable nitrogen, which results in elevated rumen and plasma ammonia and plasma urea concentrations, which may be deleterious to fertility [163].

Feeding excess rapidly degradable nitrogen increases embryo mortality in sheep [180]. Ruminants may develop hemolytic anaemia (also known as "kale anaemia") when they have a high intake of brassica when they are introduced to the crop. Other manifestations include growth retardation in lambs. The principal hemolytic factor present in such crops has been attributed to S-methylcysteine sulphoxide (SMCO), its effects being mediated through formation of dimethyldisulphide (DMDS) in the rumen by micro-organisms exhibiting SMCO-lyase activity [248].

All brassicas contain SMCO to various degrees. Rumen microflora convert SMCO into a breakdown product that reacts with components of red blood cells. The effect of the red blood cell break-down is high levels of abnormal bodies in the red cells, called "Heinz bodies" and lowered blood haemoglobin concentration. The subsequent oxygen deficiency can cause sudden death or sickness. The problem is common where brassicas have not been in the ground long enough to mature before grazing. It has been reported that cattle grazing brassica present a reduction in erythrocyte GSH and copper levels [22] as well as increased levels of lipid peroxidation, increased erythrocyte osmotic fragility, and a decrease in lipid packing density [250]. It has also been reported that sheep with an inherited GSH deficiency, and thus with lower antioxidant capacity, become more anaemic when fed brassica and exhibit a higher Heinz body count compared with GSH-sufficient animals [265]. Such observations indicate that the hemolytic activity of DMDS may be mediated through the formation of oxidative free radical species within the red blood cells. Indeed, their membrane is rich in polyunsaturated fatty acids, which makes them a primary target for reactions involving free radicals. The observation that the susceptibility of red blood cells to free radical insult is influenced more by plasma chain-breaking antioxidants (vitamin E, ascorbate) than by iron-binding antioxidants (transferrin and lactoferrin; [186]) suggests that the former provide primary protection against hemolysis induced by peroxyl radicals. Indeed, supplementation with vitamin E can increase the resistance of red blood cells to oxidative stress in cows [275].

Transport Related Diseases

Farm livestock experience a variety of stressors that can modify normal behaviour and growth, leading to losses in performance. Transport is a critical phase in animal production and utilisation and often considered as one of the main causes of stress, raising considerable interest both in economic and animal welfare terms. Transportation of cattle is a routine management practice. However, mixing groups of unfamiliar animals and loading, unloading, and driving the cattle are associated with psychological stress, physical damage, and injury. The most commonly measured indicator of short-term stress is cortisol, but its levels are highly variable and comparisons between different studies are difficult [160, 161]. During transport, animals are exposed to a variety of potential stressors such as motion of the vehicle, noise, vibrations, centrifugal forces, rapidly changing light conditions, heat, cold, poor air quality, deck height, mixing of unfamiliar groups, poor road conditions, and the possible lack of water and feed [39, 123]. Moreover, heat stress has been recognised as one of the most common problems encountered during road transportation of livestock [246] and contributes to transportation-induced stress during summer months [23]. Animal health can be impaired by various pretransport and transport conditions and may cause injury, reduce performance, and can promote the development of diseases in animals [123].

Oxidative stress has been implicated in the pathophysiology of transport-related diseases [61, 211, 271, 285]. Stress is a main predisposing factor for respiratory diseases especially causing shipping fever syndrome in cattle; this respiratory syndrome is caused by viruses and bacteria [80]. It has been suggested that viruses can also affect the host cell redox status by increasing cellular pro-oxidants. Some viruses are known to activate monocytes and polymorphonuclear leukocytes to generate free radicals [234].

Stress of any origin, including transport, is capable of depleting the body's antioxidant resources [235]. It has been reported that vitamin E concentrations are reduced in transported steers, and the exposure to simulated dust storms after being transported further decreases its concentrations [60], also stress decreases α-tocopherol concentrations in cattle plasma, neutrophils, and red blood cells [235]. Vitamin E supplementation may be required after stress to restore α-tocopherol in tissues [198] and indeed it seems to increase average daily gain and decrease shipping fever mortality [99]. Concentration of ROMS is increased in white blood cells isolated from calves after shipping, likely as a result of enhanced respiratory burst [285] and TAC is decreased in transported calves [61]. A decrease in average daily gain and increased bovine respiratory disease in conjunction with decreased posttransport concentrations of serum vitamins A and E have also been reported [60]. These observations indicate that the maintenance of optimal antioxidant status is of crucial importance to decrease the susceptibility to transport-related disease and to sustain productive performances in cattle.

The stress reaction can influence various metabolic processes including increase of the body temperature, heart and respiration rate, and also the decrease of body weight. These metabolic processes accompany oxidative stress which appears as the consequence of excess ROM production. Information about the relationship between cortisol concentrations and the lipid peroxidation process in cattle subjected to the stress conditions induced by transportation is lacking. It seems that the transport-related increase in cortisol may directly influence the increase of free fatty acids, urea, β-hydroxybutyrate, and total plasma abumin [111, 161, 280].

Lipid peroxidation products such as MDA increase in calves after transportation and it has been associated with mortality and morbidity in shipped cattle [61]. It has been observed that plasma TBARS are increased after transportation [285]. Transport stress induces an increase in lipid peroxidation products not only in the peripheral circulation but also in white blood cells. The oxidative response of bovine neutrophils in vitro may mirror the destructive response observed during bovine respiratory disease and it is one of the crucial factors causing the damage of lung diseases [154, 286]. The decrease of lipid peroxidation intensity in leukocytes after transportation by antioxidants used in this study may protect immunological cells from the development of respiratory infection process [272]. The results of the study suggest that these antioxidants have a major impact on bovine white blood cells during animal transport and should be recommended as protective factors during transport [272]. Oxidative stress associated with cattle transport has also been evidenced by excessive accumulation of leukocyte lipid oxidation products [271]. It seems that transport stress induces lipid peroxidation processes in animal tissues; indeed, stress is directly related to lipid oxidation in muscle [179]. Considering the potential effects of lipid peroxidation on the adipose tissue [120] it would be important to establish whether serum MDA concentrations are related to carcass characteristics of cattle, especially in cattle that have been exposed to extensive environmental and biological stressors [61].

Processes involving weaning and transport serve as appropriate models of commonly encountered stressors in calves and their immature antioxidant defence system during the neonatal period could make them susceptible to oxidative stress [133]. It is likely that stressors such as marketing (through sale yards) and transportation to the feedlot precipitate oxidative stress in cattle, which reduces the antioxidant defence capacity and increases total body lipid peroxidation, resulting in the susceptibility of cattle to respiratory disease at the feedlot. Concentration of reactive oxygen species is increased in white blood cells isolated from calves after shipping, likely as a result of enhanced respiratory burst [287]. TAC decreases in transported calves, and it continues to decline for up to 4 weeks after transportation [61]. Dietary supplementation with vitamin E increases average daily gain [99] and decreases morbidity in received cattle [47, 217], highlighting the importance of antioxidants to ruminant health.

GSH-Px activity in post-weaned calves is highly associated with preweaning selenium availability [24]. In weaned and transported calves challenged with *Manhemia hemolytica*, signs of morbidity coincided with elevated plasma GSH-Px, however, no evidence of whole blood GSH-Px alteration in the face of *Manhemia* exposure was observed even if both plasma and whole blood GSH-Px activity were higher in calves that had been supplemented with selenium [259]. GSH-Px activity, therefore, is dependent on selenium status in addition to oxidative stress and adaptive responses.

Transportation stress significantly decreases serum TAC concentrations in calves which is a reflection of the reductant capability of the whole antioxidant defence system, with the assumption that all of the antioxidant mechanisms are synergistic [61]. A decrease in antioxidant concentrations could result in a decrease in the ability of calves to detoxify reactive metabolites or reactive oxygen species produced by cells during aerobic metabolism. It seems that oxidative stress modulates the intensity of inflammation in respiratory disease development, also causing damage to immunological cells and lung tissues [65, 69] and that it exerts a significant impact on the immune functions of cells due to increased phagocytic activity of neutrophils and macrophages [254].

Antioxidant administration may inhibit the inflammatory process by decreasing expression of cell adhesion molecules, and may reduce migration of neutrophils and inhibit proinflammatory cytokine production [84]. Administration of exogenous α-tocopherol incorporated in liposomes protected against lung edema in animal models of acute respiratory distress syndrome and against induced lung injury by lipopolysaccharide in a mouse model, which correlated with the antioxidant and oxidative status of cells [88, 221]. Vitamin E supplementation may enhance resistance to infectious agents or augment immune responsiveness. Vitamin E is known for its antioxidant function and can augment immunity through enhanced GSH-Px activity in leukocytes and stimulation of helper T-cells, chemotaxis, phagocytosis, and antibody production ([4, 184, 258]). Vitamin E supplementation to newly arrived calves has been associated with reduced serum acute-phase protein concentrations on day 7 for SAA and AGP and day 28 for Hap, SAA, and AGP [61]. It is likely that vitamin E supplementation decreased inflammatory mediators such as IL-6 and TNF-α, which stimulate the acute-phase response through reduction of oxidative stress. However, whether it is desirable to reduce the acute-phase response of stressed cattle must be questioned as haptoglobin provides antioxidant and antimicrobial activity and plays a role in stimulation of angiogenesis [75]. Therefore, during development of respiratory tract disease, the acute phase response is probably important in helping an animal resist respiratory tract pathogens.

Reproductive Disorders

In human medicine, there is growing evidence on the effects of oxidative stress in the female reproduction system [1]. ROMs can affect a variety of physiological functions in the reproductive tract, and excessive levels can result in precipitous pathologies affecting female reproduction. In ruminants and in the dairy cow in particular, oxidative stress has been associated with several pathological conditions, such as retained placenta, udder edema, and mastitis, which in turn may impair reproductive performance. Antioxidant status may be one determinant of reproductive function in dairy cattle. Administration of vitamin E or the combination of vitamin E and selenium has been reported to reduce the incidence of postpartum reproductive disorders such as retained foetal membranes, metritis, and cystic ovaries [9, 122] and to improve fertility [11, 19]. In other studies, however, there was no beneficial effect of administration of supplemental vitamin E alone or in combination with selenium on reproductive function [81, 152, 233].

ROM can be either beneficial or detrimental to reproductive events [1]. Low levels of ROM can be beneficial in promoting the binding of sperm to the *zona pellucida* [2, 68, 238]. In cattle, the superoxide anion and hydrogen peroxide are essential for sperm capacitation and the acrosome reaction, respectively, in vitro [202, 203]. However, high hydrogen peroxide concentrations reduced bull sperm motility in vitro [202] and may impair fertilisation and embryo development [206, 276]. The role of oxidative stress in the control of female reproduction has not been fully elucidated in ruminants, however, it seems that pro- and antioxidants can influence the reproductive axis at different levels.

ROMs are produced in the steroidogenic cells and mononuclear phagocytes in the *corpus luteum* and the increase in ROMs in the *corpus luteum* seems to be involved in luteolysis [153, 260]. For instance, a decline in expression of SOD, as well as catalase, occurs during apoptosis in regressed *corpus luteum* of the bovine ovary [225]. Although ROMs have been implicated as both positive and negative regulators of luteal cell steroidogenesis [45], it has been proposed that uncontrolled generation of ROMs, such as that elicited by $PGF2_{\alpha}$ [231], disrupts cellular homeostasis leading to functional and structural luteolysis. Data supporting a role for ROMs in the loss of progesterone synthesis are derived from numerous studies of the effects of oxidative stressors and antioxidant factors on luteal cell steroidogenesis in vitro [194, 274].

Furthermore, $PGF2_{\alpha}$-induced accumulation of ROMs occurs before the decrease in progesterone synthesis [230], suggesting a direct cause–effect relationship. In addition, ROMs have been reported to inhibit progesterone synthesis by the inhibition of cytochrome P450 scc, intracellular transport of cholesterol to mitochondria, and by impairing luteinising hormone receptors [260]. Considering that ROMs are generated in the *corpus luteum* and that SOD activity has been reported in the *corpus luteum*, it is reasonable to speculate that ROMs and SOD are involved in the regulation of the *corpus luteum* function. When an imbalance occurs between oxidants and antioxidants, as in oxidative stress, the *corpus luteum* activity can be impaired leading to premature luteolysis [260]. Indeed, ROMs can disrupt the integrity of luteal cells plasma membrane, which is often seen in the regression of the *corpus luteum*. The increase in ROMs during the regression phase of the *corpus luteum* could be ascribed to the decrease in SOD expression observed in the *corpus luteum* of pseudopregnant rats after $PGF2_{\alpha}$ administration [242]. Also $PGF2_{\alpha}$ may contribute to ROM increase because it induces lipid peroxidation [242] and stimulates superoxide radical production [262].

Also macrophages have been hypothesised to be involved, at least in part, in luteolysis. Macrophages produce ROMs and damage cells when they are activated. They increase in number in the regressing *corpus luteum* and produce cytokines that inhibit P4 production by luteal cells [260]. In addition, a decrease in blood flow occurs in the ovary during the *corpus luteum* regression and this may cause tissue damage via reactive oxygen species generation by the mechanism of ischemia-reperfusion injury [260].

Finally, ROMs are generated during normal embryo metabolism but an excess in ROM production that occurs during oxidative stress has been reported to impair embryo development leading often to embryo death [116]. Embryos have internal (SOD and GSH-Px) and external (transferrin, ascorbic acid present in the oviduct) mechanisms to protect themselves against ROM attack as well to dispose of ROM excess [116]. However, due to continuous and abundant ROM production during embryo development, OS may occur. ROMs can damage preimplantation embryos by increasing lipid peroxidation [116] and DNA fragmentation [261], by altering either enzymes or mitochondrial structures and functions, and leading often to embryonic death [1, 116].

Oxidative stress might also be considered one of the mechanisms that link inflammation with embryo mortality. For example, observations from cows experiencing mastitis suggest that activation of inflammatory or immune responses can lead to embryonic mortality [121]. These embryonic losses appear to follow activation of multiple pathways that disrupt the reproductive axis at several points including the hypothalamic–pituitary axis, ovary, oocyte, and the embryo [121]. Activation of inflammatory and immune responses leads to an increase in cytokine production, which in turn can increase secretion of other molecules detrimental for embryo survival and development, such as PGF2$_{\alpha}$ or NO [121]. Elimination of bacterial infections through phagocytosis involves activation of neutrophils which, once activated, undergo the respiratory burst releasing ROMs intended to target foreign pathogens, but the lack of specificity in these reactions may result in tissue damage to the host [174].

Pregnancy itself involves an increase in activated monocyte and macrophages, as well as markers of OS, consistent with a proinflammatory state [197]. A positive correlation between AOPP and C-reactive protein in pregnant women has been reported indicating that an association between inflammatory mechanisms and oxidative stress in pregnancy is a likely event [90]. Maternal and foetal metabolism during pregnancy is greater than at any other stage in the life cycle, because of increased mitochondrial activity in maternal tissues and the conceptus [18]. This is associated with an increase in the production of ROMs particularly in dairy cows during late gestation [27, 28, 48]. A reduction in feed intake reduces antioxidant intake and the endogenous synthesis of antioxidant proteins and peptides [296], whereas overfeeding results in an increase in ROM production as a consequence of increased oxidation of energy substrates [89]; in both scenarios the oxidative defence system is weakened and oxidative stress can result. In addition, a deficiency of antioxidant minerals (e.g., Se, Zn, Cu, or Fe) or vitamins (e.g., folic acid, vitamin B6, and vitamin B12) reduces the survival and growth of embryos and foetuses [13].

Oxidative stress during pregnancy has been reported in cases of intrauterine growth restriction (IUGR; [30]) and the situation is worsened in response to both underfeeding [48] and overfeeding [63]. As a consequence, a reduction in the bioavailability of tetra-hydrobiopterin (BH$_4$), an essential factor for endothelial NOS and a potent antioxidant, and NO in maternal and foetal tissues, develops [240]. This may contribute to insulin resistance in cows and sows during late gestation, because NO mediates the stimulatory effect of insulin on muscle glucose uptake and metabolism [138]. The recognition of oxidative stress in IUGR has led to the development of selective interventions. For example, dietary supplementation of selenium could enhance placental angiogenesis and foetal growth in underfed ewes [215]. This effect may result, in part, from an increase in the bioavailability of BH$_4$ and NO in the vascular system through an increase in the activity of selenium-dependent GSH-Px to remove hydrogen peroxide [240, 294]. Finally, increasing the biological availability of Zn, Cu, and Mn through attachment to short-chain peptides has been reported to improve reproductive performance of swine, partly by enhancing antioxidant functions [130]. Moreover, it is well known that immune cells generate a large amount of ROMs when stimulated [174, 201, 257]. At the moment, the reciprocal role of each actor in respect to the others is far from being elucidated. Several pathways might be activated but a central role of oxidation products seems likely.

In sheep early embryonic mortality is the main source of reproductive wastage [50]. This high embryonic mortality rate has enormous economic implications, increasing reproductive wastage and retarding genetic progress. The high rate of pregnancy failure is assumed to be a consequence of insufficient communication between the conceptus and the maternal environment. Biomarkers of oxidative stress seem to be involved in this network [55]. The oxidant status can influence early embryo development by modifying the key transcription factors and hence modifying gene expression [70]. Concentrations of ROM may also play a major role both in the implantation and fertilisation of eggs [239]. There is an increased interest in examining the role of oxidative stress in female reproduction because it may be a major link in the infertility puzzle. Oxidative stress modulates a range of physiological functions and its role in pathological processes affecting female reproduction. It is important to further elucidate the role of oxidative stress in infertility and embryonic losses in sheep and thus design strategies to overcome its adverse effects. Strategies to overcome oxidative stress are aimed at minimising the exposure of gametes to environments that generate free radicals. Spermatozoa are the exogenous source of ROMS and therefore they can play an active role in the regulation of the reproductive tract microenvironment.

Oxidative stress can also affect reproductive events through reactive nitrogen species such as NO [224]. An endogenous NO system exists in the fallopian tubes [223]. NO has a relaxing effect on smooth muscles and it has similar effects on tubular contractility. Deficiency of NO may lead to tubal motility dysfunction, resulting in retention of the ovum, delayed sperm transport, and infertility. Infertility associated with urogenital tract infections is associated with diminished sperm motility and viability. Increased NO levels in the fallopian tubes are cytotoxic to the invading microbes and also may be toxic to spermatozoa [222]. In addition, NO might participate in the regulation of uterine contraction [199]. Indeed, a decrease in antioxidants may contribute to impaired muscle tone and thus uterine contractibility. This can result in decreased transport of sperm to ova or retained placenta [184]. In a normal fertile woman, the contractions increase throughout the proliferative and periovulatory phases, and decrease in the secretory phase. From this point, NO is synergistic with progesterone and might relax uterine contractions in the secretory phase in a paracrine fashion. Studies on the effect of oxidative stress in the regulation of the reproductive physiology in sheep are lacking.

Therefore it seems that oxidative stress is involved in several events that occur in the maternal reproductive tract. The precise mechanisms by which oxidative stress regulates the physiology of reproduction are still not clear. Therefore we need to investigate the role that oxidative stress plays in the control of reproductive events in sheep. Clarity in the understanding of these processes will lead to the improvement of assisted reproductive technologies. Further characterisation of components involved in maternal–embryo communication will lead to the identification of these factors. This will allow strategies to be devised that will reduce early embryonic and economic losses.

Retained Placenta

Retention of the foetal membranes is a common, albeit poorly understood postpartum disorder that has a negative effect on reproductive efficiency and profitability of milk production in dairy cattle [164, 257]. The release of foetal membranes postpartum is a physiological process, involving loss of fetomaternal adherence, combined with contraction of uterine musculature [181]. Retention of the foetal membranes seems to be due to an abnormal physiological process for the release of foetal membranes or pathological factors that affect the loosening mechanism of the placentomes [181]. Disturbances of the lipid metabolism during late pregnancy are associated with a high frequency of retained placenta [107]. Retention of the foetal membranes can result in improper control of enzyme activities leading to disturbances in metabolic pathways of collagen or hyaluronic acid [83, 149, 150]. This, in turn, may directly influence the process of improper separation of cotyledonary membranes. These consequences may also include an imbalance between production and neutralisation of free radicals, which could appear during impaired hormonal metabolism. Retention of foetal membranes seems to be connected with an imbalance between production and neutralisation of pro-oxidants [146]. Impairment of the antioxidant defence systems seems to be responsible for retained foetal membranes [144, 151] as well as alterations in lipid and protein peroxidation products [145, 147]. The reduced incidence of retention of the foetal membranes after supplementation with Vitamin E and selenium [82, 139, 236] suggests a role of oxidative stress in the aetiology of retention of the foetal membranes.

Steroid and prostaglandin metabolic pathways can be a source of ROMs as well as being altered by influence of excess ROMs. Differences in steroid hormones as well as prostaglandin concentrations between animals unaffected and affected by retention of the foetal membranes have been reported [143, 146, 148]. There are reports describing the differences in $PGF2_{\alpha}$ and PGE_2 concentrations in placental tissue between released and retained cotyledonary membranes [166, 249]. Plasma progesterone is higher and the levels of oestrogen are lower in cows that retain the placenta than in animals that do not [59, 114]. The levels of 8-iso-$PGF2_{\alpha}$ are higher in retained than in nonretained placental tissues [146] and this elevation seems to be due to an increase in lipid peroxidation [145, 147]. Thus, the increase in concentrations of 8-iso-$PGF2_{\alpha}$ in bovine caruncle and cotyledon of the placenta may be the result of ROM imbalance and may cause tissue oxidative damage, accompanying the improper release of cotyledonary membranes.

Production Diseases

Several production diseases of the dairy cow are a manifestation of the cow's inability to cope with the metabolic demands of high production and they include conditions such as retained placenta, displacement of the abomasums, ketosis, acidosis, fatty liver, hypocalcaemia, hypomagnesaemia, and laminitis [193]. It seems that oxidative stress is involved in the pathogenesis of these diseases and that antioxidant therapy might be beneficial in reducing the incidence of these diseases during the transition period.

Acidosis is the most important nutritional problem of dairy cattle and is also a major challenge for feedlot cattle. Lameness is one of the most important health and welfare challenges for the dairy industry. Relatively cheap grains provide an excellent and economical energy source, but the use of these and increasing emphasis on high quality, lush pastures can place herds at increased risk of acidosis. This condition can appear in a variety of situations and with different clinical signs. Ruminal lactic acidosis is a clinical disorder of cattle that can result in rumenitis, metabolic acidosis, lameness, hepatic abscessation, pneumonia, and death. Of greater economic importance are losses that result from subclinical acidosis in dairy cattle, particularly those fed on pasture [37]. Acidosis can be involved in the aetiology of several diseases. The marked increase in ruminal acidity and osmolality can damage the ruminal and intestinal wall, decrease blood pH, and cause dehydration. Laminitis, polioencephalomalacia, and liver abscesses often accompany acidosis. Even after animals recover from a bout of acidosis, nutrient absorption may be retarded [205]. The above, together with a shift in rumen bacteria and possible defaunation [101] could induce oxidative stress.

Grazing may provide dairy cows with health benefits from certain vitamins and minerals. Although fresh forage is typically considered capable of supplying adequate levels of antioxidants for dairy cattle, the availability of these compounds for lactating grazing cows is diminished when pasture availability is not adequate to meet their energy requirements. In this situation the gap between energy required for milk production and energy intake is often met by supplementing cows with conserved forage such as silage. Silage is known for its poor content in antioxidants [20] and thus this might expose cows to oxidative stress. Considering that rumen microbes are predominantly anaerobes, with a less-developed antioxidant capacity than aerobe organisms, it is likely that the reduced antioxidant content of the silage might induce oxidative stress in the ruminal micro-organism, compromising their proteolytic activity and optimal growth. Because ROMs and antioxidants are involved in several physiological functions it might be beneficial to supplement cows with antioxidants [208].

A dramatic increase in energy requirements during the transition period makes dairy cows highly susceptible to negative energy balance. The metabolic adaptation to negative energy balance requires interactions of metabolic fuels and its failure may occur in various tissues such as the liver, adipose tissue, and others [124]. The failure of metabolic adaptation to negative energy balance is crucial in the development of oxidative stress as the antioxidant capacity is frequently not sufficient to neutralise the increase in free radicals [184]. Intensified processes of NEFA oxidation proceed in the liver and result in the increased production of ketone bodies and lipid accumulation in the liver (lipomobilisation syndrome). The production of reactive oxygen species is also increased and cows experience oxidative stress [27, 28]. The lipomobilisation syndrome causes hepatomegaly and development of fatty liver during lactation [124]. Cows with liver failure have raised hepatic lipoperoxidative processes and a low antioxidative status [191].

Paraoxonase is a high-density lipoprotein (HDL)-associated enzyme, which has antioxidative properties. A decrease in serum paraoxonase activity has been observed in cows with hepatomegaly [266] and it could be considered as a result of the metabolic imbalance taking place mainly in the liver. Lower PON activity could contribute to an overall reduction of antioxidant capacity in dairy cows throughout the transition period which could make them more susceptible to inflammatory disease and reproductive disorders in advanced lactation. A positive correlation between paraoxonase and milk production has been observed in dairy cows. Bionaz et al. [29] emphasise the importance of preventing inflammation during the transition period. High liver function during the first month of lactation should be promoted to obtain high performance in dairy cows. It seems that baseline plasma paraoxonase is a promising diagnostic index for identifying the early lactating cows whose usual liver functions may be impaired [29]. Cows diagnosed with low paraoxonase can be treated to promote their recovery; thus, their liver function may be improved so that they can cope with the increase in metabolic demands after parturition.

Elevated levels of NEFA in plasma, betaoxidation of fatty acids for energy production, and hyperketonemia are the major biochemical changes recorded in ketosis. The free fatty acids not only induce a state of oxidative stress, but also impair endogenous antioxidant defences by decreasing intracellular glutathione [35]. Also, elevated ketone body levels have been reported to influence cellular GSH and lipid peroxidation in vivo, and there was enhanced generation of superoxide radical by acetoacetate [135], suggesting that oxidative stress plays a vital role in clinical and subclinical ketosis in dairy cows inducing a state of oxidative stress [28]. With the increased blood level of NEFA in ketosis, it can be well presumed that oxidative stress may play, at least in part, a significant role in the progression of the disease process in a state of negative energy balance. Indeed, the supplementation with antioxidants improves the rate of recovery and metabolic health of cows affected by subclinical ketosis [226].

Abomasal displacement is one of the most important noninfection diseases in high-yielding dairy cows and occurs in approximately 3.5% of dairy cows each year. Abomasal displacement is a multifactorial disease with numerous nutritional risk factors. Cows exhibiting low feed intake during the peripartum period are at increased risk of displaced abomasum. Cows fed high concentrate-to-forage ratio diets in early lactation, diets that promote hypocalcemia, or diets with inadequate particle size are also at increased risk of having a displaced abomasum [107]. Cows with elevated blood levels of NEFA before calving are at increased risk of developing left displacement of the abomasum, suggesting that prepartum diets as well as early lactation diets are contributing to abomasal displacement [42]. Finally, oxidative stress seems to be involved in the pathogenesis of this disease [96]. It has been reported that the administration of ascorbic acid and α-tocopherol in dairy cows with left abomasal displacement results in a more rapid decline of fatty acid concentrations to normal levels than untreated animals. The resumption of ruminal activity was observed as early as 24 h after repositioning of the abomasums in the treated animals [95]. This observation indicates that enhancing the antioxidant status of the cows is a useful tool to treat abomasal displacement. Whether antioxidant supplementation prepartum can prevent this condition remains to be determined.

Milk fever and subclinical hypocalcaemia are the most important macromineral disorders that affect transition dairy cows. Milk fever occurs when calcium leaves the blood to support milk production faster than Ca can be put back into the blood from the diet, skeletal Ca stores, and renal conservation of calcium. The disease is characterised by an acute decline in blood Ca concentration to levels that no longer support nerve and muscle function. The result is a weak cow that is unable to contract smooth muscles (such as those responsible for abomasal contraction and closure of the teat sphincter) or skeletal muscle properly, which can result in a cow that is unable to stand [107]. An important component of the homeostatic control that regulates extracellular calcium is 1,25-dihydroxy vitamin D. Delayed or insufficient production of 1,25-dihydroxy vitamin D is thought to be a common cause of milk fever [109]. Because hydroxylation of cholecalciferol in the 1 and 25 positions is dependent on cytochrome P-450 enzymes [281], ROMs inactivation of these enzymes [258] may inhibit 1,25-dihydroxy vitamin D production and have implications for milk fever.

Parasitic Infections

Parasitic diseases seem to be a causative source of oxidative stress; indeed, several studies have reported on the presence of oxidative stress in humans and animals infected with parasites [237, 270] as well as the antioxidant defence mechanism that exists between parasites and the mammalian host [269]. In human allergic inflammatory diseases, such as helminthic infections, the associated influx of eosinophils has been implicated as a primary source of tissue damage [66], possibly via their potent ROM production [210, 297]. Ruminant gastrointestinal nematode infections are common and widespread, and their immunopathology closely resembles that of human gastrointestinal infection. However, research on intestinal parasites linking to oxidative stress in ruminants is still scarce.

The responsiveness to nematode infection varies considerably between ruminant species, with goats being markedly more susceptible and less capable of developing immune resistance than sheep [129, 132, 293]. Marked species differences in antioxidant status have been observed between sheep and goats and it seems that these differences were influenced by nematode infection [169]. Moreover, an intriguing relationship emerged between antioxidant status and differences in the relative susceptibility of sheep and goats to nematode infection. Compared with sheep, goats have higher plasma concentrations of albumin, SH groups, vitamins E and A, and TAC [169]. Thus it seems that the effect of nematode infection on total antioxidant status reflects increases in oxidative stress related to intestinal nematode infection. As goats have evolved as browsers, they are less likely than sheep (grazers) to be exposed to parasitic larvae [263]. Therefore goats might have been under less evolutionary pressure to develop natural resistance to gastrointestinal parasite infections. Therefore the higher antioxidant status in goats might allow them to counteract the potentially greater oxidative challenge in response to gastrointestinal parasite infection [169].

The nutritional status of the host can influence the rate of acquisition of immunity to parasitic infection [64] and a better understanding of the role of nutrition will be important if producers are to make better use of the host acquired immunity and reduce dependence on anthelmintics [273]. The role of selenium is of particular relevance especially in selenium-deficient areas. Sheep produced and run in these areas are more susceptible to diseases and infection due to immunodeficiency [220]. With escalating anthelmintic resistance, it is critical that more integrated approaches to parasite control are investigated.

Selenium is an essential element in the diet of animals and is important in host antioxidant defence and immune function [219, 253]. A deficiency can result in ill thriftiness, reproductive losses, reduced reproductive efficiency, wool yields, and reduced growth rate in young lambs throughout the growing period [218, 220, 287], and a severe selenium deficiency can lead to sudden death from white muscle disease [288]. Selenium is a component of the selenoenzyme GSH-Px, which acts as an antioxidant and helps prevent tissue damage caused by free radicals. GSH-Px has been used as an indicator of selenium status in animals due to the high correlation found between dietary selenium and the activity of this enzyme in plasma and red blood cells [8].

A deficiency in selenium has been reported to cause delayed adult worm expulsion and increased egg production of established female worms in mice inoculated with the gastrointestinal parasite *Heligmosomoides polygyrus*, suggesting an impaired intestinal response in the mice [253]. No differences in the overall immune response to *Haemonchus contortus* infection have been observed in sheep given an intraruminal selenium pellet [137]. Selenium supplementation on helminthes burdens of marginally selenium deficient suckling Angora goat kids did not influence the level of parasitism [91]. It seems then that selenium supplementation may not offer a useful additional means of controlling internal parasites of Angora goat kids. However, other studies have demonstrated that both selenium and vitamin E are required for specific IL-4-related changes in intestinal physiology that promote host protection against gastrointestinal nematode parasites in mice [17, 253].

In selenium-supplemented sheep lower faecal egg counts than the nonsupplemented counterparts have been observed, which suggests the selenium status of sheep may influence the rate of acquisition of resistance to parasitic infection [54]. The observed negative correlation between faecal egg counts and GSH-Px suggests that an increase in GSH-Px activity may reduce parasitic infection. However, as the correlation was not very strong it is likely that selenium status is not the only factor responsible for the development of resistance to gastrointestinal parasites in Merino sheep [54]. Indeed, there are a number of other factors that can also play a role in the development of resistance and resilience to gastrointestinal parasites such as age, climatic conditions, genetics, nutrition, and grazing management [41, 62, 273]. Further evaluation of the contribution of selenium to worm expulsion should contribute to an understanding of the role of oxidative stress in the development of resistance to gastrointestinal parasites.

Tropical theileriosis is a progressive lymphoproliferative disease of cattle caused by the protozoan parasite *Theileria annulata*. The parasite acts as a serious constraint to cattle production in endemic areas, causing lethal infections in exotic cattle and considerable mortality in indigenous and crossbred stocks [106, 190]. There is some evidence that oxidative stress and lipid peroxidation incorporate in pathogenesis of anaemia in theileriosis. An increase in oxidative stress and in lipid peroxidation in erythrocytes of cattle infected with *T. annulata* has been reported and it seems that this might be the cause of increased erythrocyte fragility due to membrane lysis [112]. The levels of methemoglobin, used as an index of erythrocytes oxidation, markedly increase at the onset of anaemia in experimental *Theileria sergenti* infection [246] and an inverse relationship has been observed between methemoglobin levels and PCV [243]. An increase in protein oxidation in the membrane of erythrocytes has also been reported in *T. sergenti*-infected cattle [296]. Finally, the levels of antioxidants in erythrocytes are decreased during the progression of anaemia in cattle infected with *T. sergenti* [245]. It seems that oxidative changes in erythrocytes are closely related to the pathogenesis of anaemia in theileriosis.

Lipid peroxidation in erythrocytes of affected cattle increases MDA production [14, 112, 196, 228]. Increased MDA concentration in erythrocytes of affected cattle may be an indication of elevated oxidative stress in theileriosis. High levels of MDA have been associated with erythrocyte infection rate with *T. annulata* and the severity of anaemia [14]. Similarly, in cattle with *T. sergenti* during the onset of anaemia, levels of MDA increased remarkably in proportion to the decrease of packed cell volume and increase of parasitemia [244]. Similar observations have been reported in calves infected with *T. sergenti* [117, 295]. These observations further support the hypothesis that oxidative damage to the erythrocytes plays a crucial role in the pathogenesis of anaemia in bovine theileriosis.

Leptospirosis

Leptospirosis, a zoonotic disease with worldwide distribution, is considered a toxin-mediated disease leading to lipid peroxidation as lypopolysaccaride of its membrane plays a role in the cytotoxicity [157, 167]. The induction of lipid peroxidation gives rise to an increase in MDA content [85]. This procedure activates cell-protective antioxidant defence mechanisms such as glutathione and uric acid [94]. An increase in NO levels in humans with severe leptospirosis has been reported [175]. In inflammatory conditions, NO production increases through stimulation of inducible NO synthase via activation of proinflammatory cytokines and causes NO mediated tissue injury by reacting with superoxide to generate peroxynitrite, a powerful cytotoxin [46]. It seems then, that NO may play a role in the pathogenesis of leptospirosis [175]. In cattle affected by leptospirosis an increase in MDA and a decrease in GSH pool have been observed [85] indicating that oxidative stress is involved in the pathogenesis of leptospirosis. The increase in uric acid and NO, and decreased albumin concentration in leptorpirosis cases [85] further support the role of oxidative stress in leptospirosis-affected animals.

Uric acid increases during acute oxidative stress and ischemia and its increased concentrations might be a compensatory mechanism that protects tissue against free radicals [6, 105]. A decrease in albumin may also be indicative of oxidative stress as albumin possesses a significant antioxidant activity [33] and the antioxidant effect of albumin against free radical induced blood haemolysis has also been reported [34]. The measurement of uric acid, albumin, and reduced GSH and MDA concentrations can therefore be used as indicators of oxidative stress in diseases such as leptospirosis

Oxidative Stress in Ruminant Metabolism and Production

Metabolic Changes Mediated by ROMs

Impairment of animal performance by ROMs may involve altered metabolism as much or more than actual cell damage. ROMs are involved in numerous signalling pathways [77] and some of these are involved in the control of anabolic and catabolic processes. For example, proteins that are sensitive to oxidative damage often react with ROMs through redox-active cysteine residues and because these cysteine residues undergo similar oxidative modification in response to oxidative changes in SH groups' redox status, disease-related changes in redox status may cause disruption of those physiological responses [78]. Both antioxidant defence and reactions catalysed by steroidogenic enzymes require reducing equivalents provided by NADPH [168, 209]. Excessive consumption of reducing equivalents by oxidative stress can lower $NADPH_2$ and increase NADP concentrations despite elevated activity of the monophosphate shunt, which generates the reduced form [155]. Consumption of reducing equivalents by ROM reactions can diminish the supply of NADPH available for important physiological processes.

Finally, the induction of the monophosphate shunt by increased ROM imbalance can divert glucose from other pathways. This possibility assumes greater importance when the requirement for glucose and the quantity available in the ruminant are considered. In dairy cows, pregnancy and lactation are physiological stages considered to induce metabolic stress [76]. Dairy cows can experience oxidative stress [28, 49, 208], which may be associated with metabolic diseases during the peripartum period [184]. A number of recent studies have reported variable levels of oxidative stress during the periparturient period in dairy cows [28, 48, 97] and it seems that ROMs and antioxidants may be involved in some relevant physiological functions such as milk yield [98]. The adoption of intensive methods of husbandry in dairy ruminants for higher milk yields is likely to increase the incidence of metabolic diseases. The study of the metabolic and nutritional factors that affect the oxidative status in ruminants is an interesting area of research and there is a growing body of evidence underpinning the pathophysiological consequences of oxidative stress in farm animals.

The animal feed industry is under increasing consumer pressure to reduce the use of antibiotics as feed additives. This is a natural consequence of the increasing demand for safe products for human consumption. The use of herbs as additives in livestock nutrition, as an alternative to other chemical compounds, is a new goal in livestock production [177]. The use of alternative feedstuffs naturally rich in antioxidants, in ruminant nutrition represents a novel management tool that is green, clean, ethical [178], and extremely easy to use. However, the effect of dietary antioxidant supplementation in ruminants has not yet been fully evaluated and thus, the role and the activity of natural antioxidants not commonly present in the diets of ruminants warrants further investigation.

Milk Production

In the postpartum period, cows are metabolically challenged as energy demands outstrip energy intake, and animals enter a state of negative energy balance [134]. This triggers mainly catabolic pathways which, at the cellular level, increase the production of ROMs [28]. There is evidence that oxidative stress is affected by body condition at calving [28] and by milk yield [49, 172] and diet [98]. Cows with a body condition score greater than 3 before calving mobilise more body reserves in the postpartum period and experience more oxidative stress than animals with lower scores [28]. Cows with high milk yields have higher concentrations of oxidative stress than lower-yielding animals [49, 172] and the feeding of high levels of starch to cows during early lactation increases oxidative stress, possibly due to changes in oxidative phosphorylation [98].

In cows fed two different diets designed to achieve restricted or high milk production, the concentration of antioxidants was low and the concentration of ROMs high in the first 2 weeks of lactation in both groups [208]. The concentration of antioxidants was lower during early lactation, in agreement with other reports [108, 165], and is presumably due, in part, to the utilisation of antioxidants in colostrum production [108]. Changes in free radical and antioxidant concentration appear to represent homeorhetic processes that normally occur in early lactation, although these were higher in the feed-restricted cows [208]. Given the lack of reference values for ROM and BAP in cows, and the fact that few studies have been carried out in this area, the causes of oxidative stress are difficult to identify. Therefore, medium- and high-energy balance and milk yield categories have been developed to further investigate the relationship between these variables and oxidative stress [208]. This approach reduces the effect of individual cow variation and also takes into account the effect of the feeding system.

For the energy balance category, the results indicated that an increased energy reserve mobilisation affected the level of oxidative stress but only in cows fed to achieve restricted milk production. Cows fed to achieve restricted milk production had higher levels of OS and of NEB relative to cows fed to achieve high milk production. Oxidative stress levels were related to lower BAP concentrations in cows fed to achieve restricted milk production, a finding that might suggest that only cows with very high body reserve mobilisation experience oxidative stress [208]. Indeed, cows in good body condition at calving and in high negative energy balance in early lactation had higher oxidative stress [28]. Similarly in humans, obesity is associated with high levels of pro-oxidants [280]. Therefore, the higher level of oxidative stress in cows in extreme negative energy balance might reflect a reduction in the precursors of endogenous anti-oxidants. Although milk production is associated with oxidative stress due to the increased cellular metabolism involved [49, 172], it has been found that oxidative stress was higher in cows fed to achieve restricted milk production or tended to be higher in cows fed to achieve high milk production when the yield was lower [208]. The high level of oxidative stress in low-producing cows was associated with high concentrations of ROM.

Dairy cows are exposed to various metabolic pressures of adaptation to high milk yield and environmental conditions and oxidative stress may develop. It has been proposed that the activation of antioxidant defences is a preparative mechanism against oxidative stress caused by physiological stress situations [125]. It seems, indeed, that the entire antioxidant system is under homeostatic control. However, when metabolic reactions are overloaded, free radicals accumulate and the disturbance of the redox balance may occur, resulting in oxidative stress.

Heat stress has also been suggested to affect oxidative status in dairy ruminants which can affect reproductive and productive performances. Evidence suggests that the effects of elevated temperatures on embryonic development involve changes in the metabolism of free radicals. Heat shock increases intracellular ROMs in cultured bovine embryos, which in turn delays or blocks embryo development [229]. Moreover, exposure of cultured mouse embryos to elevated temperatures causes a reduction in GSH [10]. The situation in vivo is still somewhat controversial. No evidence of a beneficial effect of supplemental vitamin E and selenium above estimated requirements on reproduction or milk yield of lactating dairy cows under cool or heat stress conditions or of vitamin E as an embryonic thermoprotectant or as a culture medium additive for enhancing development of preimplantation embryos in culture have been observed [207]. It is possible that supplementation of vitamin E and selenium are unlikely to enhance reproductive or lactational performance of lactating dairy cows receiving adequate dietary requirements of these nutrients.

Higher erythrocyte oxidant and antioxidant biomarkers in cows calving in summer than in cows calving in spring has been reported [27], which would indicate some effects of hot weather on the oxidative status of transition cows. Also higher plasma ROM concentrations have been reported in lactating dairy goats in summer compared to spring [74]. In another study heat stress did not increase lipid peroxidation or lipid-soluble antioxidant concentrations in blood of lactating cows [266]. However, the objective animals of these studies were of different physiological status, which might be responsible for the discrepancies of the results. Taken all together, these results may suggest that the effects of heat stress on oxidative status are probably related to the physiological phase of the animals, and also that these effects might be different at the local (embryonic development within the uterus) or systemic level (plasma oxidative stress biomarkers).

Current and Future Developments

The field of oxidative stress in ruminant medicine is still in the early stages of development. Oxidative stress has been associated with numerous conditions, however, there is a great deal to be discovered about its role in ruminant health and production. For example, if oxidative stress is a primary cause of pathological change or a consequence of disease processes still remains to be determined. Clarity of understanding of the pathophysiology of oxidative stress in ruminants will allow the design of specific antioxidant therapies.

Future research should focus on the establishment of a reference panel of oxidative stress biomarkers to be used in veterinary medicine. Future research should also address the issue of standardisation of techniques and methodologies to study oxidative stress. For example, tissue biopsy sampling techniques can be utilised only in particular cases in veterinary medicine, therefore a major challenge in veterinary oxidative stress research is to develop a set of blood biomarkers that can reliably reflect the tissue oxidative status in the individual animal. There are numerous techniques that can be used to measure oxidative damage [174], however, veterinary scientists and clinicians need to establish whether oxidative stress biomarkers measured in blood and/or serum provide a reasonable index of the general oxidative stress status. Considering that oxidative damage is likely to occur in only few tissues at the same time, high levels of serum oxidative stress biomarkers can be the consequence of (1) a generalised increase in oxidative stress in most of the tissues, resulting in an increase of the biomarker concentration in blood proportional to the extent of tissue damage; (2) the increase of oxidative stress just in a particular tissue which is the origin of the pathological dysfunction; and (3) an increase of oxidative damage produced specifically in the circulatory system [12]. Even if several animal studies seem to suggest that oxidative stress biomarkers change in the same direction in blood and tissues, this needs to be fully evaluated in ruminant species.

References

1. Agarwal, A., Gupta, S., Sharma, R., 2005. Role of oxidative stress in female reproduction. Reproductive Biology and Endocrinology 3, 28.
2. Aitken, R.J., Clarkson, J.S., Fishel, S., 1989. Generation of reactive oxygen species, lipid peroxidation, and human sperm function. Biology of Reproduction 41, 183–197.
3. Alberti, A., Bolognini, L., Macciantelli, D., Caratelli, M., 2000. The radical cation of N, N-diethyl-para-phenylendiamine: a possible indicator of oxidative stress in biological samples. Research on Chemical Intermediates 26, 253–267.
4. Allison, R.D., Laven, R.A., 2000. Effect of vitamin E supplementation on the health and fertility of dairy cows: a review. Veterinary Record 147, 703–708.
5. Al-Sa'doni, H., Ferro, A., 2000. S-nitrosothiols: a class of nitric oxide-donor drugs. Clinical Science 98, 507–520.
6. Ames, B.N., Cathcart, R., Schwiers, E., Hochstein, P., 1981. Uric acid provides an antioxidant defence in humans against oxidant- and radical-caused aging and cancer: a hypothesis. Proceedings of the National Academy of Sciences of the United States of America 78, 6858–6862.

7. Andrés, S., Jimenez, A., Mañé, M.C., Sánchez, J., Barrera, R., 1997. Relationships between some soil parameters and the blood glutathione peroxidase activity of grazing sheep. Veterinary Record 141, 267–268.
8. Andrés, S., Mañé, M.C., Sánchez, J., Barrera, R., Jimenez, A., 1999. Temporal variations in blood glutathione peroxidase (GSHPx) activity in sheep at pasture in a Mediterranean area. Veterinary Journal 157, 186–188.
9. Aréchiga, C.F., Ortíz, O., Hansen, P.J., 1994. Effect of prepartum injection of vitamin E and selenium on postpartum reproductive function of dairy cattle. Theriogenology 41, 1251–1258.
10. Aréchiga, C.F., Ealy, A.D., Hansen, P.J., 1995. Evidence that glutathione is involved in thermotolerance of preimplantation murine embryos. Biology of Reproduction 52, 1296–1301.
11. Aréchiga, C.F., Vázquez-Flores, S., Ortiz, O., Hernández-Cerón, J., Porras, A., McDowell, L.R., Hansen, P.J., 1998. Effect of injection of [beta]-carotene or vitamin E and selenium on fertility of lactating dairy cows. Theriogenology 50, 65–76.
12. Argüelles, S., García, S., Maldonado, M., Machado, A., Ayala, A., 2004. Do the serum oxidative stress biomarkers provide a reasonable index of the general oxidative stress status? Biochimica et Biophysica Acta – General Subjects 1674, 251–259.
13. Ashworth, C., Antipatis, C., 2001. Micronutrient programming of development throughout gestation. Reproduction 122, 527–535.
14. Asri Rezaei, S., Dalir-Naghadeh, B., 2006. Evaluation of antioxidant status and oxidative stress in cattle naturally infected with *Theileria annulata*. Veterinary Parasitology 142, 179–186.
15. Atabek, M., Keskin, M., Yazici, C., Kendirci, M., Hatipoglu, N., Koklu, E., Kurtoglu, S., 2006. Protein oxidation in obesity and insulin resistance. European Journal of Pediatrics 165, 753–756.
16. Atakisi, O., Oral, H., Atakisi, E., Merhan, O., Metin Pancarci, S., Ozcan, A., Marasli, S., Polat, B., Colak, A., Kaya, S., 2010. Subclinical mastitis causes alterations in nitric oxide, total oxidant and antioxidant capacity in cow milk. Research in Veterinary Science 89, 10–13.
17. Au, Y.J.K., Smith, A., Zhao, A., Madden, K.B., Elfrey, J., Sullivan, C., Levander, O., Urban, J.F., Shea-Donohue, T., 2005. Impact of vitamin E or selenium deficiency on nematode-induced alterations in murine intestinal function. Experimental Parasitology 109, 201–208.
18. Aurousseau, B., Gruffat, D., Durand, D., 2006. Gestation linked radical oxygen species fluxes and vitamins and trace mineral deficiencies in the ruminant. Reproduction Nutrition and Development 46, 601–620.
19. Baldi, A., Savoini, G., Pinotti, L., Monfardini, E., Cheli, F., Orto, V.D., 2000. Effects of vitamin E and different energy sources on vitamin E status, milk quality and reproduction in transition cows. Journal of Veterinary Medicine Series A 47, 599–608.
20. Ballet, N., Robert, J.C., Williams, P.E.V., 2000. Vitamins in forages. In: D.I. Givens, E. Owens., R.F.E. Axford and H.M. Omed (Eds.), Forage Evaluation in Ruminant Nutrition, CABI, Oxon, UK, pp. 399–431.
21. Bargo, F., Muller, L.D., Delahoy, J.E., Cassidy, T.W., 2002. Milk response to concentrate supplementation of high producing dairy cows grazing at two pasture allowances. Journal of Dairy Science 85, 1777–1792.
22. Barry, T.N., Reid, T.C., Millar, K.R., Sadler, W.A., 1981. Nutritional evaluation of kale (*Brassica oleracea*) diets: 2. Copper deficiency, thyroid function, and selenium status in young cattle and sheep fed kale for prolonged periods. The Journal of Agricultural Science 96, 269–282.
23. Beatty, D.T., Barnes, A., Taylor, E., Pethick, D., McCarthy, M., Maloney, S.K., 2006. Physiological responses of Bos taurus and Bos indicus cattle to prolonged, continuous heat and humidity. Journal of Animal Science 84, 972–985.
24. Beck, P.A., Wistuba, T.J., M.E., D., Gunter, S.A., 2005. Case study: effects of feeding supplemental organic or inorganic selenium to cow-calf pairs on selenium status and immune responses of weaned beef calves. Professional Animal Scientist 21, 114–120.

25. Beckman, J.S., Beckman, T.W., Chen, J., Marshall, P.A., Freeman, B.A., 1990. Apparent hydroxyl radical production by peroxynitrite: implications for endothelial injury from nitric oxide and superoxide. Proceedings of the National Academy of Sciences of the United States of America 87, 1620–1624.
26. Benzie, I.F.F., Strain, J.J., 1996. The ferric reducing ability of plasma (FRAP) as a measure of "antioxidant power": the FRAP assay. Analytical Biochemistry 239, 70–76.
27. Bernabucci, U., Ronchi, B., Lacetera, N., Nardone, A., 2002. Markers of oxidative status in plasma and erythrocytes of transition dairy cows during hot season. Journal of Dairy Science 85, 2173–2179.
28. Bernabucci, U., Ronchi, B., Lacetera, N., Nardone, A., 2005. Influence of body condition score on relationships between metabolic status and oxidative stress in periparturient dairy cows. Journal of Dairy Science 88, 2017–2026.
29. Bionaz, M., Trevisi, E., Calamari, L., Librandi, F., Ferrari, A., Bertoni, G., 2007. Plasma paraoxonase, health, inflammatory conditions, and liver function in transition dairy cows. Journal of Dairy Science 90, 1740–1750.
30. Biri, A., Bozkurt, N., Turp, A., Kavutcu, M., Himmetoglu, Z., Durak, I., 2007. Role of oxidative stress in intrauterine growth restriction. Gynecologic and Obstetric Investigation 64, 187.
31. Blum, J.W., Dosogne, H., Hoeben, D., Vangroenweghe, F., Hammon, H.M., Bruckmaier, R.M., Burvenich, C., 2000. Tumor necrosis factor-[alpha] and nitrite/nitrate responses during acute mastitis induced by *Escherichia coli* infection and endotoxin in dairy cows. Domestic Animal Endocrinology 19, 223–235.
32. Bouchard, L., Blais, S., Desrosiers, C., Zhao, X., Lacasse, P., 1999. Nitric oxide production during endotoxin-induced mastitis in the cow. Journal of Dairy Science 82, 2574–2581.
33. Bourdon, E., Blache, D., 2001. The importance of proteins in defense against oxidation. Antioxidants & Redox Signaling 3, 293–311.
34. Bourdon, E., Loreau, N., Blache, D., 1999. Glucose and free radicals impair the antioxidant properties of serum albumin. FASEB Journal 13, 233–244.
35. Bouwstra, R.J., Goselink, R.M.A., Dobbelaar, P., Nielen, M., Newbold, J.R., van Werven, T., 2008. The relationship between oxidative damage and vitamin E concentration in blood, milk, and liver tissue from vitamin E supplemented and nonsupplemented periparturient heifers. Journal of Dairy Science 91, 977–987.
36. Bowers, T.L., 1997. Nutrition and immunity part 2: the role of selected micronutrients and clinical significance. Veterinary Clinical Nutrition 4, 96–101.
37. Bramley, E., Lean, I.J., Fulkerson, W.J., Stevenson, M.A., Rabiee, A.R., Costa, N.D., 2008. The definition of acidosis in dairy herds predominantly fed on pasture and concentrates. Journal of Dairy Science 91, 308–321.
38. Briviba, K., Sies, H., 1994. Non-enzymatic antioxidants defence system. In: B. Frei (Ed.), Natural Antioxidants in Human Health and Disease, Academic Press, San Diego, CA, pp. 107–128.
39. Broom, D.M., 2000. Welfare assessment and welfare problem areas during handling and transport. In: T. Grandin (Ed.), Livestock Handling and Transport, CABI, Wallingford, UK, pp. 43–61.
40. Burk, R.F., 1983. Biological activity of selenium. Annual Review of Nutrition 3, 53–70.
41. Burke, J.M., Miller, J.E., Terrill, T.H., 2009. Impact of rotational grazing on management of gastrointestinal nematodes in weaned lambs. Veterinary Parasitology 163, 67–72.
42. Cameron, R.E.B., Dyk, P.B., Herdt, T.H., Kaneene, J.B., Miller, R., Bucholtz, H.F., Liesman, J.S., Vandehaar, M.J., Emery, R.S., 1998. Dry cow diet, management, and energy balance as risk factors for displaced abomasum in high producing dairy herds. Journal of Dairy Science 81, 132–139.
43. Cao, G., Prior, R.L., 1998. Comparison of different analytical methods for assessing total antioxidant capacity of human serum. Clinical Chemistry 44, 1309–1315.
44. Cao, G., Alessio, H.M., Cutler, R.G., 1993. Oxygen-radical absorbance capacity assay for antioxidants. Free Radical Biology and Medicine 14, 303–311.
45. Carlson, J.C., Wu, X.M., Sawada, M., 1993. Oxygen radicals and the control of ovarian *corpus luteum* function. Free Radical Biology and Medicine 14, 79–84.

46. Carrillo-Vico, A., Lardone, P.J., Naji, L., Fernández-Santos, J.M., Martín-Lacave, I., Guerrero, J.M., Calvo, J.R., 2005. Beneficial pleiotropic actions of melatonin in an experimental model of septic shock in mice: regulation of pro-/anti-inflammatory cytokine network, protection against oxidative damage and anti-apoptotic effects. Journal of Pineal Research 39, 400–408.
47. Carter, J.N., Gill, D.R., Krehbiel, C.R., Confer, A.W., Smith, R.A., Lalman, D.L., Claypool, P.L., McDowell, L.R., 2005. Vitamin E supplementation of newly arrived feedlot calves. Journal of Animal Science 83, 1924–1932.
48. Castillo, C., Hernandez, J., Bravo, A., Lopez-Alonso, M., Pereira, V., Benedito, J.L., 2005. Oxidative status during late pregnancy and early lactation in dairy cows. The Veterinary Journal 169, 286–292.
49. Castillo, C., Hernandez, J., Valverde, I., Pereira, V., Sotillo, J., Alonso, M.L., Benedito, J.L., 2006. Plasma malonaldehyde (MDA) and total antioxidant status (TAS) during lactation in dairy cows. Research in Veterinary Science 80, 133–139.
50. Celi, P., Bush, D.R., 2010. Pregnancy, lambing and survival. In: D. Cottle, (Ed.), International Sheep and Wool Handbook, Nottingham University Press, UK, pp. 223–257.
51. Celi, P., Raadsma, H.W., 2009. The effects of Yerba Mate (*Ilex paraguarensis*) supplementation on the productive performance of lambs. Ruminant Physiology. Digestion, metabolism, and effects of nutrition on reproduction and welfare. Proceeding of the XIth International Symposium in Ruminant Physiology, Clermont-Ferrand, France, pp. 804–805.
52. Celi, P., Raadsma, H.W., 2010. The effects of Yerba Mate (*Ilex paraguarensis*) supplementation on the productive performance of lactating dairy cows. Animal Production Science 50, 339–344.
53. Celi, P., Di Trana, A., Quaranta, A., 2008. Metabolic profile and oxidative status in goats during the peripartum period. Australian Journal of Experimental Agriculture 48, 1004–1008
54. Celi, P., Eppleston, J., Armstrong, A., Watt, B., 2010. Selenium supplementation increases resistance and resilience of Merino weaners to intestinal parasite infection in a selenium deficient area. Animal Production Science 50, 688–692.
55. Celi, P., Merlo, M., Da Dalt, L., Stefani, A., Barbato, O., Gabai, G., 2011. Plasma advanced oxidised protein products (AOPP) as a novel indicator of late embryonic mortality in dairy cows. Reproduction Fertility and Development, http://www.publish.csiro.au/view/journals/dsp_journals_pip_abstract.cfm?nid=44&ver=1&pip=RD10268.
56. Celi, P., Sullivan, M., Evans, D., 2010. The stability of the reactive oxygen metabolites (d-ROMs) and biological antioxidant potential (BAP) tests on stored horse blood. The Veterinary Journal 183, 217–218.
57. Celi, P., Trana, A.D., Claps, S., 2010. Effects of plane of nutrition on oxidative stress in goats during the peripartum period. The Veterinary Journal 184, 95–99.
58. Chaiyotwittayakun, A., Erskine, R.J., Bartlett, P.C., Herdt, T.H., Sears, P.M., Harmon, R.J., 2002. The effect of ascorbic acid and L-histidine therapy on acute mammary inflammation in dairy cattle. Journal of Dairy Science 85, 60–67.
59. Chew, B.P., Keller, H.F., Erb, R.E., Malven, P.V., 1977. Periparturient concentrations of prolactin, progesterone and the estrogens in blood plasma of cows retaining and not retaining fetal membranes. Journal of Animal Science 44, 1055–1060.
60. Chirase, N.K., Greene, L.W., Purdy, C.W., Loan, R.W., Briggs, R.E., McDowell, L.R., 2001. Effect of environmental stressors on ADG, serum retinol and alpha-tocopherol concentrations, and incidence of bovine respiratory disease of feeder steers. Journal of Animal Science 79, 188.
61. Chirase, N.K., Greene, L.W., Purdy, C.W., Loan, R.W., Auvermann, B.W., Parker, D.B., Walborg, E.F., Stevenson, D.E., Xu, Y., Klaunig, J.E., 2004. Effect of transport stress on respiratory disease, serum antioxidant status, and serum concentrations of lipid peroxidation biomarkers in beef cattle. American Journal of Veterinary Research 65, 860–864.
62. Colditz, I.G., Watson, D.L., Gray, G.D., Eady, S.J., 1996. Some relationships between age, immune responsiveness and resistance to parasites in ruminants. International Journal for Parasitology. Novel Approaches to the Control of Helminth Parasites of Livestock 26, 869–877.

63. Cole, D.J.A., 1990. Nutritional strategies to optimise reproduction in pigs. Journal of Reproduction and Fertility Suppl. 40, 67–82.
64. Coop, R.L., Holmes, P.H., 1996. Nutrition and parasite interaction. International Journal of Parasitology. Novel Approaches to the Control of Helminth Parasites of Livestock. 26, 951–962.
65. Corongiu, F.P., Poli, G., Dianzani, M.U., Cheeseman, K.H., Slater, T.F., 1986. Lipid peroxidation and molecular damage to polyunsaturated fatty acids in rat liver. Recognition of two classes of hydroperoxides formed under conditions in vivo. Chemico-Biological Interactions 59, 147–155.
66. Corrigan, C.J., Kay, A.B., 1992. T cells and eosinophils in the pathogenesis of asthma. Immunology Today 13, 501–507.
67. Dawson, T.M., Dawson, V.L., 1995. Nitric oxide: actions and pathological roles. Neuroscientist 1, 7–18.
68. de Lamirande, E., Jiang, H., Zini, A., Kodama, H., Gagnon, C., 1997. Reactive oxygen species and sperm physiology. Reviews in Reproduction 2, 48–54.
69. Deaton, C.M., 2006. The role of oxidative stress in an equine model of human asthma. Redox Report 11, 46–52.
70. Dennery, P.A., 2004. Role of redox in fetal development and neonatal diseases. Antioxidants & Redox Signaling 6, 147–153.
71. Descalzo, A.M., Insani, E.M., Biolatto, A., Sancho, A.M., García, P.T., Pensel, N.A., Josifovich, J.A., 2005. Influence of pasture or grain-based diets supplemented with vitamin E on antioxidant/oxidative balance of Argentine beef. Meat Science 70, 35–44.
72. Di Giuseppe, D., Di Simplicio, P., Capecchi, P.L., Lazzerini, P.E., Pasini, F.l., 2003. Alteration in the redox state of plasma in heart-transplant patients with moderate hyperhomocysteinemia. Journal of Laboratory and Clinical Medicine 142, 21–28.
73. Di Giuseppe, D., Frosali, S., Priora, R., Di Simplicio, F.C., Buonocore, G., Cellesi, C., Capecchi, P.L., Pasini, F.L., Lazzerini, P.E., Jakubowski, H., Di Simplicio, P., 2004. The effects of age and hyperhomocysteinemia on the redox forms of plasma thiols. Journal of Laboratory and Clinical Medicine 144, 235–245.
74. Di Trana, A., Celi, P., Claps, S., Fedele, V., Rubino, R., 2006. The effect of hot season and nutrition on the oxidative status and metabolic profile in dairy goats during mid lactation. Animal Science 82, 717–722.
75. Dobryszycka, W., 1997. Biological functions of haptoglobin – new pieces to an old puzzle. European Journal of Clinical Chemistry and Clinical Biochemistry 35, 647–654.
76. Drackley, J.K., 1999. Biology of dairy cows during the transition period: the final frontier. Journal of Dairy Science 82, 2259–2273.
77. Dröge, W., 2002. Free radicals in the physiological control of cell function. Physiological Reviews 82, 47–95.
78. Dröge, W., 2006. Redox regulation in anabolic and catabolic processes. Current Opinion in Clinical Nutrition & Metabolic Care 9, 190–195.
79. Dröge, W., Schulze-Osthoff, K., Mihm, S., Galter, D., Schenk, H., Eck, H.P., Roth, S., Gmunder, H., 1994. Functions of glutathione and glutathione disulfide in immunology and immunopathology. FASEB Journal 8, 1131–1138.
80. Duff, G.C., Galyean, M.L., 2007. Board-invited review: recent advances in management of highly stressed, newly received feedlot cattle. Journal of Animal Science 85, 823–840.
81. Ealy, A.D., Arechiga, C.F., Bray, D.R., Risco, C.A., Hansen, P.J., 1994. Effectiveness of short-term cooling and vitamin E for alleviation of infertility induced by heat stress in dairy cows. Journal of Dairy Science 77, 3601–3607.
82. Eger, S., Drori, D., Kadoori, I., Miller, N., Schindler, H., 1985. Effects of selenium and vitamin E on incidence of retained placenta. Journal of Dairy Science 68, 2119–2122.
83. Eiler, H., Hopkins, F.M., 1992. Bovine retained placenta: effects of collagenase and hyaluronidase on detachment of placenta. Biology of Reproduction 46, 580–585.
84. Ekstrand-Hammarström, B., Österlund, C., Lilliehook, B., Bucht, A., 2006. Vitamin E down-modulates mitogen-activated protein kinases, nuclear factor-kB and inflammatory responses in lung epithelial cells. Clinical Experimental Immunology 147, 359–369.

85. Erdogan, H., Karapehlivan, M., Citil, M., Atakisi, O., Uzlu, E., Unver, A., 2008. Serum sialic acid and oxidative stress parameters changes in cattle with leptospirosis. Veterinary Research Communications 32, 333–339.
86. Erskine, R.J., 1993. Nutrition and mastitis. Update on bovine mastitis. Veterinary Clinics of North America Food Animal Practice 9, 551–556.
87. Erskine, R.J., Eberhart, R.J., Hutchimon, L.J., 1987. Herd management and mastitis prevalence in high and low cell count dairy herds. Journal of American Veterinary Medical Association 190, 1411–1416.
88. Fan, J., Shek, P.N., Suntres, Z.E., Li, Y.H., Oreopoulos, G.D., Rotstein, O.D., 2000. Liposomal antioxidants provide prolonged protection against acute respiratory distress syndrome. Surgery 128, 332–338.
89. Fang, Y.-Z., Yang, S., Wu, G., 2002. Free radicals, antioxidants, and nutrition. Nutrition 18, 872–879.
90. Fialovà, L., Malbohan, I., Kalousova, M., Soukupova, J., Krofta, L., Štipek, S., Zima, T., 2006. Oxidative stress and inflammation in pregnancy. Scandinavian Journal of Clinical and Laboratory Investigtion 66, 121–128.
91. Fivaz, B.H., Hobson, C., Wellington, A.C., Williams, E.J., 1993. Influence of selenium supplementation on helminth burdens of marginally deficient suckling Angora goat kids. Small Ruminant Research 10, 75–79.
92. Flohe, L., Gunzler, W.A., Schock, H.H., 1973. Glutathione peroxidase: a selenoenzyme. FEBS Letters 32, 132–134.
93. Frei, B., 1994. Natural Antioxidants in Human Health and Disease, Academic Press, San Diego, CA.
94. Frei, B., Stocker, R., Ames, B.N., 1988. Antioxidant defenses and lipid peroxidation in human blood plasma. Proceedings of the National Academy of Science USA 85, 9748–9752.
95. Fürll, M., Dabbagh, M.N., Fürll, B., Sattler,T., Spielmann C., 1999. Ätiologie und Prophylaxe von Reperfusionsschäden. Dtsch Tierärztl Wochenschr 106, 389–393.
96. Fürll, M., Dabbagh, M.N., Kirbach, H., Nauruschat, C., Sattler, T., Wilken, H., 2003. The anti-oxidative state before and after a diagnosed abomasal displacement in cows. Acta Veterinaria Scandinavica 44(Suppl 1), P37.
97. Gaál, T., Ribiczeyné-Szabó, P., Stadler, K., Jakus, J., Reiczigel, J., Kövér, P., Mézes, M., Sümeghy, L., 2006. Free radicals, lipid peroxidation and the antioxidant system in the blood of cows and newborn calves around calving. Comparative Biochemistry and Physiology Part B: Biochemistry and Molecular Biology 143, 391–396.
98. Gabai, G., Testoni, S., Piccinini, R., Marinelli, L., Stradaioli, G., 2004. Oxidative stress in primiparous cows in relation to dietary starch and the progress of lactation. Animal Science 79, 99–108.
99. Galyean, M.L., Perino, L.J., Duff, G.C., 1999. Interaction of cattle health/immunity and nutrition. Journal of Animal Science 77, 1120–1134.
100. Gatellier, P., Mercier, Y., Renerre, M., 2004. Effect of diet finishing mode (pasture or mixed diet) on antioxidant status of Charolais bovine meat. Meat Science 67, 385–394.
101. Gazi, M.R., Yokota, M., Tanaka, Y., Kanda, S., Itabashis, H., 2007. Effects of protozoa on the antioxidant activity in the ruminal fluid and blood plasma of cattle. Animal Science Journal 78, 34–40.
102. Georgieva, N.V., 2005. Oxidative stress as a factor of disrupted ecological oxidative balance in biological systems – a review. Bulgarian Journal of Veterinary Medicine 8, 1–11.
103. Ghiselli, A., Serafini, M., Maiani, G., Azzini, E., Ferro-Luzzi, A., 1995. A fluorescence-based method for measuring total plasma antioxidant capability. Free Radical Biology and Medicine 18, 29–36.
104. Ghiselli, A., Serafini, M., Natella, F., Scaccini, C., 2000. Total antioxidant capacity as a tool to assess redox status: critical view and experimental data. Free Radical Biology and Medicine 29, 1106–1114.
105. Glantzounis, G.K., Tsimoyiannis, E.C., Kappas, A.M., Galaris, D.A., 2005. Uric acid and oxidative stress. Current Pharmaceutical Design 11, 4145–4151.

106. Glass, E.J., Jensen, K., 2007. Resistance and susceptibility to a protozoan parasite of cattle – gene expression differences in macrophages from different breeds of cattle. Veterinary Immunology and Immunopathology. Disease Resistance and Immunity 120, 20–30.
107. Goff, J.P., 2006. Major advances in our understanding of nutritional influences on bovine health. Journal of Dairy Science 89, 1292–1301.
108. Goff, J.P., Stabel, J.R., 1990. Decreased plasma retinol, alpha-tocopherol, and zinc concentration during the periparturient period: effect of milk fever. Journal of Dairy Science 73, 3195–3199.
109. Goff, J.P., Horst, R.L., Mueller, F.J., Miller, J.K., Kiess, G.A., Dowlen, H.H., 1991. Addition of chloride to a prepartal diet high in cations increases 1,25-dihydroxyvitamin D response to hypocalcemia preventing milk fever. Journal of Dairy Science 74, 3863–3871.
110. Goff, W.L., Carl Johnson, W., Wyatt, C.R., Cluff, C.W., 1996. Assessment of bovine mononuclear phagocytes and neutrophils for induced L-arginine-dependent nitric oxide production. Veterinary Immunology and Immunopathology 55, 45–62.
111. Grandin, T., 1997. Assessment of stress during handling and transport. Journal of Animal Science 75, 249–257.
112. Grewal, A., Ahuja, C.S., Singha, S.P.S., Chaudhary, K.C., 2005. Status of lipid peroxidation, some antioxidant enzymes and erythrocytic fragility of crossbred cattle naturally infected with *Theileria annulata*. Veterinary Research Communications 29, 387–394.
113. Griffiths, H.R., Møller, L., Bartosz, G., Bast, A., Bertoni-Freddari, C., Collins, A., Cooke, M., Coolen, S., Haenen, G., Hoberg, A.-M., Loft, S., Lunec, J., Olinski, R., Parry, J., Pompella, A., Poulsen, H., Verhagen, H., Astley, S.B., 2002. Biomarkers. Molecular Aspects of Medicine 23, 101–208.
114. Grunert, E., Ahlers, D., Heuwieser, W., 1989. The role of endogenous estrogens in the maturation process of the bovine placenta. Theriogenology 31, 1081–1091.
115. Gu, B., Zhu, Y., Zhu, W., Miao, J., Deng, Y.E., Zou, S., 2009. Retinoid protects rats against neutrophil-induced oxidative stress in acute experimental mastitis. International Immunopharmacology 9, 223–229.
116. Guerin, P., El Mouatassim, S., Menezo, Y., 2001. Oxidative stress and protection against reactive oxygen species in the pre-implantation embryo and its surroundings. Human Reproduction Update 7, 175–189.
117. Haider, M.J., 1992. Hematological study of water buffalo (*Bubalus bubalis*) during theileriosis (*Theileria annulata*). Annals of the New York Academy of Sciences 653, 191–193.
118. Halliwell, B., 1988. Albumin – an important extracellular antioxidant. Biochemical Pharmacology 37, 569–571.
119. Halliwell, B., Chirico, S., 1993. Lipid peroxidation: its mechanism, measurement, and significance. American Journal of Clinical Nutrition 57, 715S–725S.
120. Halliwell, B., Gutteridge, J.M.C., 1989. Free Radicals in Biology and Medicine, Oxford University Press, Oxford, UK.
121. Hansen, P.J., Soto, P., Natzke, R.P., 2004. Mastitis and fertility in cattle – possible involvement of inflammation or immune activation in embryonic mortality. American Journal of Reproductive Immunology 51, 294–301.
122. Harrison, J.H., Hancock, D.D., Conrad, H.R., 1984. Vitamin E and selenium for reproduction of the dairy cow. Journal of Dairy Science 67, 123–132.
123. Hartung, J., 2003. Effects of transport on health of farm animals. Veterinary Research Communications 27, 525–527.
124. Herdt,T.H., 2000. Ruminant adaptation to negative energy balance: influences on the etiology of ketosis and fatty liver. Veterinary Clinics of North America Food Animal Practice 16, 215–231.
125. Hermes-Lima, M., Storey, J.M., Storey, K.B., 1988. Antioxidants defenses and metabolic depression. The hypothesis of preparation for oxidative stress in land snails. Comparative Biochemistry and Physiology B, Biochemistry and Molecular Biology 120, 437–448.
126. Hogan, J.S., Smith, K.L., Weiss, W.P., Todhunter, D.A., Schockey, W.L., 1990. Relationships among vitamin E, selenium, and bovine blood neutrophils. Journal of Dairy Science 73, 2372–2378.

127. Hogan, J.S., Weiss, W.P., Smith, K.L., 1993. Role of vitamin E and selenium in host defense against mastitis. Journal of Dairy Science 76, 2795–2803.
128. Hopkins, G.J., Roberts, J., Evans, J.V., 1975. Red blood cell concentrations of reduced glutathione and potassium as biochemical markers of wool growth in Merino sheep. Journal of Agricultural Science 84, 481–486.
129. Hoste, H., Torres-Acosta, J.F., Aguilar-Caballero, A.J., 2008. Nutrition–parasite interactions in goats: is immunoregulation involved in the control of gastrointestinal nematodes? Parasite Immunology 30, 79–88.
130. Hostetler, C.E., Kincaid, R.L., Mirando, M.A., 2003. The role of essential trace elements in embryonic and fetal development in livestock. Veterinary Journal 166, 125–139.
131. Huie, R.E., Padmaja, S., 1993. The reaction of no with superoxide. Free Radical Research 18, 195–199.
132. Huntley, J.F., Patterson, M., Mackellar, A., Jackson, F., Stevenson, L.M., Coop, R.L., 1995. A comparison of the mast cell and eosinophil responses of sheep and goats to gastrointestinal nematode infections. Research in Veterinary Science 58, 5–10.
133. Inanami, O., Shiga, A., Okada, K., 1999. Lipid peroxides and antioxidants in serum of neonatal calves. American Journal of Veterinary Research 60, 452–457.
134. Ingvartsen, K.L., Andersen, J.B., 2000. Integration of metabolism and intake regulation: a review focusing on periparturient animals. Journal of Dairy Science 83, 1573–1597.
135. Jain, S.K., McVie, R., 1999. Hyperketonemia can increase lipid peroxidation and lower glutathione levels in human erythrocytes in vitro and in type 1 diabetic patients. Diabetes 48, 1850–1855.
136. Janero, D.R., 1990. Malondialdehyde and thiobarbituric acid-reactivity as diagnostic indices of lipid peroxidation and peroxidative tissue injury. Free Radical Biology and Medicine 9, 515–540.
137. Jelinek, P.D., Ellis, T., Wroth, R.H., Sutherland, S.S., Masters, H.G., Petterson, D.S., 1988. The effect of selenium supplementation on immunity, and the establishment of an experimental Haemonchus contortus infection, in weaner Merino sheep fed a low selenium diet. Australian Veterinary Journal 65, 214–217.
138. Jobgen, W.S., Fried, S.K., Fu, W.J., Meininger, C.J., Wu, G., 2006. Regulatory role for the arginine-nitric oxide pathway in metabolism of energy substrates. The Journal of Nutritional Biochemistry 17, 571–588.
139. Julien, W.E., Conrad, H.R., Jones, J.E., Moxon, A.L., 1976. Selenium and vitamin E and incidence of retained placenta in parturient dairy cows. Journal of Dairy Science 59, 1954–1959.
140. Kalousová, M., Skrha, J., Zima, T., 2002. Advanced glycation end products and advanced oxidation protein products in patients with diabetes mellitus. Physiological Research 51, 597–604.
141. Kalousová, M., Zima, T., Tesar, V., Dusilová-Sulková, S., Skrha, J., 2005. Advanced glycoxidation end products in chronic diseases – clinical chemistry and genetic background. Mutation Research/Fundamental and Molecular Mechanisms of Mutagenesis Inflammation, cellular and redox signalling mechanisms in cancer and degenerative diseases 579, 37–46.
142. Kaneda, H., Taguchi, J., Ogasawara, K., Aizawa, T., Ohno, M., 2002. Increased level of advanced oxidation protein products in patients with coronary artery disease. Atherosclerosis 162, 221–225.
143. Kankofer, M., 1999. The enzymes responsible for the metabolism of prostaglandins in bovine placenta. Prostaglandins, Leukotrienes and Essential Fatty Acids 61, 359–362.
144. Kankofer, M., 2001. Antioxidative defence mechanisms against reactive oxygen species in bovine retained and not-retained placenta: activity of glutathione peroxidase, glutathione transferase, catalase and superoxide dismutase. Placenta 22, 466–472.
145. Kankofer, M., 2001. The levels of lipid peroxidation products in bovine retained and not retained placenta. Prostaglandins, Leukotrienes and Essential Fatty Acids 64, 33–36.
146. Kankofer, M., 2002. 8-iso-prostaglandin F2[alpha] as a marker of tissue oxidative damage in bovine retained placenta. Prostaglandins & Other Lipid Mediators 70, 51–59.

147. Kankofer, M., 2002. Placental release/retention in cows and its relation to peroxidative damage of macromolecules. Reproduction in Domestic Animals 37, 27–30.
148. Kankofer, M., Wiercinski, J., 1999. Prostaglandin E29-keto reductase from bovine term placenta. Prostaglandins, Leukotrienes and Essential Fatty Acids 61, 29–32.
149. Kankofer, M., Wierci, Å., Ski, J., Fidecki, M., 1998. Activity of hyaluronidase in placental tissues from cows with and without retained fetal membranes. Journal of Veterinary Medicine Series A: Physiology Pathology Clinical Medicine 45, 337–341.
150. Kankofer, M., Wiercinski, J., Zerbe, H., 2000. Activity of placental ß-N-acetyl-glucosaminidase in cows with and without retained foetal membranes. Reproduction in Domestic Animals 35, 97–100.
151. Kankofer, M., Lipko, J., Zdunczyk, S., 2005. Total antioxidant capacity of bovine spontaneously released and retained placenta. Pathophysiology 11, 215–219.
152. Kappel, L.C., Ingraham, R.H., Morgan, E.B., Dixon, J.M., Zeringue, L., Wilson, D., Babcock, D.K., 1984. Selenium concentrations in feeds and effects of treating pregnant Holstein cows with selenium and vitamin E on blood selenium values and reproductive performance. American Journal of Veterinary Research 45, 691–694.
153. Kato, H., Sugino, N., Takiguchi, S., Kashida, S., Nakamura, Y., 1997. Roles of reactive oxygen species in the regulation of luteal function. Reviews of Reproduction 2, 81–83.
154. Kawabata, K., Hagio, T., Matsuoka, S., 2002. The role of neutrophil elastase in acute lung injury. European Journal of Pharmacology 451, 1–10.
155. Kehrer, J.P., Lund, L.G., 1994. Cellular reducing equivalents and oxidative stress. Free Radical Biology and Medicine 17, 65–75.
156. Kehrer, J.P., Smith., C.V., 1994. Free radicals in biology: sources, reactivities, and roles in the etiology of human dieseases. In: B. Frei (Ed.), Natural Antioxidants in Human Health and Diseases, Academic Press, San Diego, CA, pp. 25–62.
157. Kim, Y.G., Jeon, D.Y., Yang, M.K., 1997. Superoxide dismutase activity and lipid peroxidation in the liver of guinea pig infected with Leptospira interrogans. Free Radical Research 26, 1–6.
158. Kleczkowski, M., Klucinski, W., Shaktur, A., Sikora, J., 2005. Concentration of ascorbic acid in the blood of cows with subclinical mastitis. Polish Journal of Veterinary Sciences 8, 121–125.
159. Knaapen, A.M., Seiler, F., Schilderman, P.A.E.L., Nehls, P., Bruch, J., Schins, R.P.F., Borm, P.J.A., 1999. Neutrophils cause oxidative DNA damage in alveolar epithelial cells. Free Radical Biology and Medicine 27, 234–240.
160. Knowles, T.G., 1998. A review of the road transport of slaughter sheep. Veterinary Record 143, 212–219.
161. Knowles, T.G., 1999. A review of the road transport of cattle. Veterinary Record 144, 197–201.
162. Komine, K.-i., Kuroishi, T., Komine, Y., Watanabe, K., Kobayashi, J., Yamaguchi, T., Kamata, S.-i., Kumagai, K., 2004. Induction of nitric oxide production mediated by tumor necrosis factor alpha on staphylococcal enterotoxin c-stimulated bovine mammary gland cells. Clinical and Diagnostic Laboratory Immunology 11, 203–210.
163. Laven, R.A., Drew, S.B., 1999. Dietary protein and the reproductive performance of cows. Veterinary Record 145, 687–695.
164. LeBlanc, S.J., 2008. Postpartum uterine disease and dairy herd reproductive performance: a review. The Veterinary Journal. Production Diseases of the Transition Cow 176, 102–114.
165. LeBlanc, S.J., Herdt, T.H., Seymour, W.M., Duffield, T.F., Leslie, K.E., 2004. Peripartum serum vitamin E, retinol, and beta-carotene in dairy cattle and their associations with disease. Journal of Dairy Science 87, 609–619.
166. Leidl, W., Hegner, D., Rockel, P., 1980. Investigations on the PGF2a concentration in maternal and foetal cotyledons of cows with and without retained foetal membranes. Journal of Veterinary Medicine A 27, 691.
167. Levett, P.N., 2001. Leptospirosis. Clinical Microbiology Reviews 14, 296–326.
168. Liebler, D.C., 1993. The role of metabolism in the antioxidant function of vitamin E. Critical Reviews in Toxicology 23, 147–169.

169. Lightbody, J.H., Stevenson, L.M., Jackson, F., Donaldson, K., Jones, D.G., 2001. Comparative aspects of plasma antioxidant status in sheep and goats, and the influence of experimental abomasal nematode infection. Journal of Comparative Pathology 124, 192–199.
170. Liu, S.M., Eady, S.J., 2005. Glutathione: its implications for animal health, meat quality, and health benefits of consumers. Australian Journal of Agricultural Research 56, 775–780.
171. Liu, S.M., Masters, D.G., 2000. Quantitative analysis of methionine and cysteine requirements for wool production of sheep. Animal Science 71, 175–185.
172. Lohrke, B., Viergutz, T., Kanitz, W., Gollnitz, K., Becker, F., Hurtienne, A., Schweigert, F.J., 2004. High milk yield in dairy cows associated with oxidant stress. Online Journal of Veterinary Research 8, 70–78.
173. Lykkesfeldt, J., 2001. Determination of malondialdehyde as dithiobarbituric acid adduct in biological samples by HPLC with fluorescence detection: comparison with ultraviolet-visible spectrophotometry. Clinical Chemistry 47, 1725–1727.
174. Lykkesfeldt, J., Svendsen, O., 2007. Oxidants and antioxidants in disease: oxidative stress in farm animals. The Veterinary Journal 173, 502–511.
175. Maciel, E.A.P., Athanazio, D.A., Reis, E.A.G., Cunha, F.Q., Queiroz, A., Almeida, D., McBride, A.J.A., Ko, A.I., Reis, M.G., 2006. High serum nitric oxide levels in patients with severe leptospirosis. Acta Tropica 100, 256–260.
176. MacMicking, J., Xie, Q.-w., Nathan, C., 1997. Nitric oxide and macrophage function. Annual Review of Immunology 15, 323–350.
177. Makkar, H.P.S., Francis, G., Becker, K., 2007. Bioactivity of phytochemicals in some lesser-known plants and their effects and potential applications in livestock and aquaculture production systems. Animal 1, 1371–1391.
178. Martin, G.B., Kadokawa, H., 2006. "Clean, green and ethical" animal production. Case study: reproductive efficiency in small ruminants. The Journal of Reproduction and Development 52, 145–152.
179. McClelland, G.B., 2004. Fat to the fire: the regulation of lipid oxidation with exercise and environmental stress. Comparative Biochemistry and Physiology Part B: Biochemistry and Molecular Biology. Special Issue Dedicated to the Memory of Peter W. Hochachka (1937–2002) 139, 443–460.
180. McEvoy, T.G., Robinson, J.J., Aitken, R.P., Findlay, P.A., Robertson, I.S., 1997. Dietary excesses of urea influence the viability and metabolism of preimplantation sheep embryos and may affect fetal growth among survivors. Animal Reproduction Science 47, 71–90.
181. McNaughton, A.P., Murray, R.D., 2009. Structure and function of the bovine fetomaternal unit in relation to the causes of retained fetal membranes. Vetetinary Record 165, 615–622.
182. Merlo, M., Barbato, O., Stefani, A., Celi, P., Da Dalt, L., Gabai, G., 2008. Embryonic mortality and plasma advanced oxidation protein products (AOPP) increase in dairy cows. Current Topics in Dairy Production, Proceedings of the Dairy Research Foundation Symposium, Camden, Australia, pp. 49–53.
183. Merlo, M., Celi, P., Barbato, O., Gabai, G., 2008. Relationships between oxidative status and pregnancy outcome in dairy cows. Animal Production in Australia 27, 84.
184. Miller, J.K., Brzezinska-Slebodzinska, E., Madsen, F.C., 1993. Oxidative stress, antioxidants, and animal function. Journal of Dairy Science 76, 2812–2823.
185. Miller, N.J., Rice-Evans, C., Davies, M.J., Gopinathan, V., Milner, A., 1993. A novel method for measuring antioxidant capacity and its application to monitoring the antioxidant status in premature neonates. Clinical Science 84, 407–412.
186. Miller, J.K., Brzezinska-Slebodzinska, E., Campbell, M.H., Mosley, S.A., Holt, J.A., 1999. Vitamin and mineral involvement in antioxidants. Proceedings of the 50th California Animal Nutrition Conference, Fresno, 79–85.
187. Milne, G.L., Musiek, E.S., Morrow, J.D., 2005. F2-Isoprostanes as markers of oxidative stress in vivo: an overview. Biomarkers 10, 10–23.
188. Moilanen, E., Vapaatalo, H., 1995. Nitric oxide in inflammation and immune response. Annals of Medicine 27, 359–367.

189. Moran, L.K., Gutteridge, J., Quinlan, G.J., 2001. Thiols in cellular redox signalling and control. Current Medicinal Chemistry 8, 763.
190. Morrison, I.W., 2009. Progress towards understanding the immunobiology of Theileria parasites. Parasitology 136, 1415–1426.
191. Mudron, P., Rehage, J., Qualmann, K., Sallmann, H.P., Scholz, H., 1999. A study of lipid peroxidation and vitamin E in dairy cows with hepatic insufficiency. Journal of Veterinary Medicine A 46, 219–224.
192. Mukherjee, R., 2008. Selenium and vitamin E increases polymorphonuclear cell phagocytosis and antioxidant levels during acute mastitis in riverine buffaloes. Veterinary Research Communications 32, 305–313.
193. Mulligan, F.J., Doherty, M.L., 2008. Production diseases of the transition cow. The Veterinary Journal 176, 3–9.
194. Musicki, B., Aten, R., Behrman, H., 1994. Inhibition of protein synthesis and hormone-sensitive steroidogenesis in response to hydrogen peroxide in rat luteal cells. Endocrinology 134, 588–595.
195. Naresh, R., Dwivedi, S.K., Swarup, D., Patra, R.C., 2002. Evaluation of ascorbic acid treatment in clinical and subclinical mastitis of Indian dairy cows. Asian-Australasian Journal of Animal Science 15, 905–911.
196. Naziroglu, M., Åžaki, C.E., Sevgili, M., 1999. The effect of buparvaquone treatment on the levels of some antioxidant vitamins, lipid peroxidation and glutathione peroxidase in cattle with theileriosis. Journal of Veterinary Medicine, Series B 46, 233–239.
197. Ness, R.B., 2004. The consequences for human reproduction of a robust inflammatory response. The Quarterly Review of Biology 79, 383–393.
198. Nockels, C.F., Odde, K.G., Craig, A.M., 1996. Vitamin E supplementation and stress affect tissue alpha-tocopherol content of beef heifers. Journal of Animal Science 74, 672–677.
199. Norman, J.E., Thomson, A.J., Telfer, J.F., Young, A., Greer, I.A., Cameron, I.T., 1999. Myometrial constitutive nitric oxide synthase expression is increased during human pregnancy. Molecular Human Reproduction 5, 175–181.
200. Notebaert, S., Demon, D., Vanden Berghe, T., Vandenabeele, P., Meyer, E., 2008. Inflammatory mediators in *Escherichia coli*-induced mastitis in mice. Comparative Immunology, Microbiology and Infectious Diseases 31, 551–565.
201. Noyan, T., Güler, A., Sekeroglu, M.R., Kamaci, M., 2006. Serum advanced oxidation protein products, myeloperoxidase and ascorbic acid in pre-eclampsia and eclampsia. Australian and New Zealand Journal of Obstetrics and Gynaecology 46, 486–491.
202. O'Flaherty, C.M., Beorlegui, N.B., Beconi, M.T., 1999. Reactive oxygen species requirements for bovine sperm capacitation and acrosome reaction. Theriogenology 52, 289–301.
203. O'Flaherty, C., Beorlegui, N., Beconi, M.T., 2003. Participation of superoxide anion in the capacitation of cryopreserved bovine sperm. International Journal of Andrology 26, 109–114.
204. Okamoto, T., Akaike, T., Nagano, T., Miyajima, S., Suga, M., Ando, M., Ichimori, K., Maeda, H., 1997. Activation of human neutrophil procollagenase by nitrogen dioxide and peroxynitrite: a novel mechanism for procollagenase activation involving nitric oxide. Archives of Biochemistry and Biophysics 342, 261–274.
205. Owens, F.N., Secrist, D.S., Hill, W.J., Gill, D.R., 1998. Acidosis in cattle: a review. Journal of Animal Science 76, 275–286.
206. Ozawa, M., Nagai, T., Fahrudin, M., Karja, N.W.K., Kaneko, H., Noguchi, J., Ohnuma, K., Kikuchi, K., 2006. Addition of glutathione or thioredoxin to culture medium reduces intracellular redox status of porcine IVM/IVF embryos, resulting in improved development to the blastocyst stage. Molecular Reproduction and Development 73, 998–1007.
207. Paula-Lopes, F.F., Al-Katanani, Y.M., Majewski, A.C., McDowell, L.R., Hansen, P.J., 2003. Manipulation of antioxidant status fails to improve fertility of lactating cows or survival of heat-shocked embryos. Journal of Dairy Science 86, 2343–2351.

208. Pedernera, M., Celi, P., García, S.C., Salvin, H.E., Barchia, I., Fulkerson, W.J., 2010. Effect of diet, energy balance and milk production on oxidative stress in early-lactating dairy cows grazing pasture. The Veterinary Journal 186, 352–357.
209. Perl, A., Gergely, P., Puskas, F., Banki, K., 2002. Metabolic switches of T-cell activation and apoptosis. Antioxidants & Redox Signaling 4, 427–443.
210. Petreccia, D.C., Nauseef, W.M., Clark, R.A., 1987. Respiratory burst of normal human eosinophils. Journal of Leukocyte Biology 41, 283–288.
211. Pregel, P., Bollo, E., Cannizzo, F., Biolatti, B., Contato, E., Biolatti, P., 2005. Antioxidant capacity as a reliable marker of stress in dairy calves transported by road. Veterinary Record 156, 53–54.
212. Radi, R., Beckman, J.S., Bush, K.M., Freeman, B.A., 1991. Peroxynitrite-induced membrane lipid peroxidation: the cytotoxic potential of superoxide and nitric oxide. Archives of Biochemistry and Biophysics 288: 481–487.
213. Ranjan, R., Swarup, D., Naresh, R., Patra, R.C., 2005. Enhanced erythrocytic lipid peroxides and reduced plasma ascorbic acid, and alteration in blood trace elements level in dairy cows with mastitis. Veterinary Research Communication 29, 27–34.
214. Reilly, P.M., Schiller, H.J., Bulkley, G.B., 1991. Pharmacologic approach to tissue injury mediated by free radicals and other reactive oxygen metabolites. The American Journal of Surgery 161, 488–503.
215. Reynolds, L.P., Borowicz, P.P., Vonnahme, K.A., Johnson, M.L., Grazul-Bilska, A.T., Redmer, D.A., Caton, J.S., 2005. Placental angiogenesis in sheep models of compromised pregnancy. Journal of Physiology 565, 43–58.
216. Riollet, C., Rainard, P., Poutrel, B., 2000. Differential induction of complement fragment C5a and inflammatory cytokines during intramammary infections with *Escherichia coli* and *Staphylococcus aureus*. Clincal and Diagnostic Laboratory Immunology 7, 161–167.
217. Rivera, J.D., Duff, G.C., Galyean, M.L., Walker, D.A., Nunnery, G.A., 2002. Effects of supplemental vitamin E on performance, health, and humoral immune response of beef cattle. Journal of Animal Science 80, 933–941.
218. Rivera, M.T., Souza, A.P.D., Araujo-Jorge, T.C., De Castro, S.L., Vanderpas, J., 2003. Trace elements, innate immune response and parasites. Clinical Chemistry and Laboratory Medicine 41, 1020–1025.
219. Rizzo, A., Mutinati, M., Spedicato, M., Minoia, G., Trisolini, C., Jirillo, F., Sciorsci, R.L., 2008. First demonstration of an increased serum level of reactive oxygen species during the peripartal period in the Ewes. Immunopharmacology and Immunotoxicology 30, 741–746.
220. Robson, S., Plant, J.W., 2007. Selenium deficiency in sheep. Primefact 471 (NSW Department of Primary Industries), http://www.dpi.nsw.gov.au/agriculture/livestock/sheep/health/selenium-deficiency-sheep.
221. Rocksen, D., Ekstrand-Hammarstrom, B., Johansson, L., Bucht, A., 2003. Vitamin E reduces transendothelial migration of neutrophils and prevents lung injury in endotoxin-induced airway inflammation. American Journal of Respiratory Cell Molecular Biology 28, 199–207.
222. Rosselli, M., Dubey, R., Imthurn, B., Macas, E., Keller, P., 1995. Effects of nitric oxide on human spermatozoa: evidence that nitric oxide decreases sperm motility and induces sperm toxicity. Human Reproduction 10, 1786–1790.
223. Rosselli, M., Dubey, R., Rosselli, M., Macas, E., Fink, D., Lauper, U., Keller, P., Imthurn, B., 1996. Identification of nitric oxide synthase in human and bovine oviduct. Molecular Human Reproduction 2, 607–612.
224. Rosselli, M., Keller, P., Dubey, R., 1998. Role of nitric oxide in the biology, physiology and pathophysiology of reproduction. Human Reproduction Update 4, 3–24.
225. Rueda, B., Tilly, K., Hansen, T., Hoyer, P., Tilly, J., 1995. Expression of superoxide dismutase, catalase and glutathione peroxidase in the bovine *corpus luteum*: evidence supporting a role for oxidative stress in luteolysis. Endocrine 3, 227–232.

226. Sahoo, S., Patra, R., Behera, P., Swarup, D., 2009. Oxidative stress indices in the erythrocytes from lactating cows after treatment for subclinical ketosis with antioxidant incorporated in the therapeutic regime. Veterinary Research Communications 33, 281–290.
227. Sakatani, M., Yamanaka, K., Kobayashi, S., Takahashi, M., 2008. Heat shock-derived reactive oxygen species induce embryonic mortality in in vitro early stage bovine embryos. Journal of Reproduction and Development 54, 496–501.
228. Saluja, P.S., Gupta, S.L., Malhotra, D.V., Ambawat, H.K., 1999. Status of plasma malondialdehyde in experimental *theileria annulata* infection in crossbred bovine calves. Indian Veterinary Journal 76, 379–381.
229. Sauberlich, H.E., 1994. Pharmacology of vitamin C. Annual Reviews in Nutrition 14, 371–391.
230. Sawada, M., Carlson, J.C., 1991. Rapid plasma membrane changes in superoxide radical formation, fluidity, and phospholipase A2 activity in the *Corpus luteum* of the rat during induction of luteolysis. Endocrinology 128, 2992–2998.
231. Sawada, M., Carlson, J.C., 1994. Studies on the mechanism controlling generation of superoxide radical in luteinised rat ovaries during regression. Endocrinology 135, 1645–1650.
232. Scaletti, R.W., Trammell, D.S., Smith, B.A., Harmon, R.J., 2003. Role of dietary copper in enhancing resistance to *Escherichia coli* Mastitis. Journal of Dairy Science 86, 1240–1249.
233. Schingoethe, D.J., Kirkbride, C.A., Palmer, I.S., Owens, M.J., Tucker, W.L., 1982. Response of cows consuming adequate selenium to vitamin E and selenium supplementation prepartum. Journal of Dairy Science 65, 2338–2344.
234. Schwarz, K.B., 1996. Oxidative stress during viral infection: a review. Free Radical Biology and Medicine 21, 641–649.
235. Sconberg, S., Nockels, C.F., Bennet, D.W., Bruyninclese, W., Blancquaret, A.M.B., Craig, A.M., 1993. Effect of shipping handling, adrenocortiocotropic hormone and epinephrine on L-tocopherol content of bovine blood. American Journal of Veterinary Research 54, 1287–1293.
236. Segerson, E.C., Riviere, G.J., Dalton, H.L., Whitacre, M.D., 1981. Retained placenta of holstein cows treated with selenium and vitamin E. Journal of Dairy Science 64, 1833–1836.
237. Selkirk, M.E., Smith, V.P., Thomas, G.R., Gounaris, K., 1998. Resistance of filarial nematode parasites to oxidative stress. International Journal for Parasitology 28, 1315–1332.
238. Sengoku, K., Tamate, K., Yoshida, T., Takaoka, Y., Miyamoto, T., Ishikawa, M., 1998. Effects of low concentrations of nitric oxide on the zona pellucida binding ability of human spermatozoa. Fertility and Sterility 69, 522–527.
239. Sharma, R., Agarwal, A., 2004. Role of reactive oxygen species in gynecologic diseases. Reproductive Medicine and Biology 3, 177–199.
240. Shi, W., Meininger, C., Haynes, T., Hatakeyama, K., Wu, G., 2004. Regulation of tetrahydrobiopterin synthesis and bioavailability in endothelial cells. Cell Biochemistry and Biophysics 41, 415–433.
241. Shi, X.Y., Hou, F.F., Niu, H.X., Wang, G.B., Xie, D., Guo, Z.J., Zhou, Z.M., Yang, F., Tian, J.W., Zhang, X., 2008. Advanced oxidation protein products promote inflammation in diabetic kidney through activation of renal nicotinamide adenine dinucleotide phosphate oxidase. Endocrinology 149, 1829–1839.
242. Shimamura, K., Sugino, N., Yoshida, Y., Nakamura, Y., Ogino, K., Kato, H., 1995. Changes in lipid peroxide and antioxidant enzyme activities in corpora lutea during pseudopregnancy in rats. Journal of Reproduction and Fertility 105, 253–257.
243. Shiono, H., Yagi, Y., Thongnoon, P., Kurabayashi, N., Chikayama, Y., Miyazaki, S., Nakamura, I., 2001. Acquired methemoglobinemia in anemic cattle infected with Theileria sergenti. Veterinary Parasitology 102, 45–51.
244. Shiono, H., Yagi, Y., Chikayama, Y., Miyazaki, S., Nakamura, I., 2003. The influence of oxidative bursts of phagocytes on red blood cell oxidation in anemic cattle infected with Theileria sergenti. Free Radical Research 37, 1181–1189.
245. Shiono, H., Yagi, Y., Chikayama, Y., Miyazaki, S., Nakamura, I., 2003. Oxidative damage and phosphatidylserine expression of red blood cells in cattle experimentally infected with Theileria sergenti. Parasitology Research 89, 228–234.

246. Silanikove, N., 2000. Effects of heat stress on the welfare of extensively managed domestic ruminants. Livestock Production Science 67, 1–18.
247. Slama, H., Vaillancourt, D., Goff, A.K., 1994. Control of in vitro prostaglandin F2[alpha] and E2 synthesis by caruncular and allantochorionic tissues from cows that calved normally and those with retained fetal membranes. Domestic Animal Endocrinology 11, 175–185.
248. Smith, R., 1978. S-methylcysteine sulphoxide, the brassica anaemia factor (a valuable dietary factor for man?). Veterinary Research Communications 2, 47–61.
249. Smith, J.A., 1994. Neutrophils, host defense, and inflammation: a double-edged sword. Journal of Leukocyte Biology 56, 672–686.
250. Smith, R.H., Watt, W.B., Lawson, W.J., Rice-Evans, C., 1982. Red cell membrane changes in a Heinz body anaemia. Protides Biological Fluid 29, 125–128.
251. Smith, K.L., Harrison, J.H., Hancock, D.D., Todhunter, D.A., Conrad, H.R., 1984. Effect of vitamin E and selenium supplementation on incidence of clinical mastitis and duration of clinical symptoms. Journal of Dairy Science 67, 1293–1300.
252. Smith, K.L., Hogan, J.S., Weiss, W.P., 1997. Dietary vitamin E and selenium affect mastitis and milk quality. Journal of Animal Science 75, 1659–1665.
253. Smith, A., Madden, K.B., Yeung, K.J.A., Zhao, A., Elfrey, J., Finkelman, F., Levander, O., Shea-Donohue, T., Urban, J.F., Jr., 2005. Deficiencies in selenium and/or vitamin E lower the resistance of mice to *Heligmosomoides polygyrus* infections. Journal of Nutrition 135, 830–836.
254. Sordillo, L.M. and Aitken, S.L., 2009. Impact of oxidative stress on the health and immune function of dairy cattle. Veterinary Immunology and Immunopathology 128(1–3), 104–109.
255. Soriani, M., Pietraforte, D., Minetti, M., 1994. Antioxidant potential of anaerobic human plasma: role of serum albumin and thiols as scavengers of carbon radicals. Archives of Biochemistry and Biophysics 312, 180–188.
256. Spears, J.W., 2000. Micronutrients and immune function in cattle. Proceedings of the Nutrition Society 59, 587–594.
257. Spears, J.W., Weiss, W.P., 2008. Role of antioxidants and trace elements in health and immunity of transition dairy cows. The Veterinary Journal. Production Diseases of the Transition Cow 176, 70–76.
258. Staats, D.A., Lohr, D.P., Colby, H.D., 1988. Effects of tocopherol depletion on the regional differences in adrenal microsomal lipid peroxidation and steroid metabolism. Endocrinology 123, 975.
259. Stabel, J.R., Spears, J.W., Brown Jr., T.T., Brake, J., 1989. Selenium effects on glutathione peroxidase and the immune response of stressed calves challenged with Pasteurella hemolytica. Journal of Animal Science 67, 557–564.
260. Sugino, N., 2006. Roles of reactive oxygen species in the *corpus luteum*. Animal Science Journal 77, 556–565.
261. Takahashi, M., Keicho, K., Takahashi, H., Ogawa, H., Schulte, R.M., Okano, A., 2000. Effect of oxidative stress on development and DNA damage in in-vitro cultured bovine embryos by comet assay. Theriogenology 54, 137–145.
262. Tanaka, M., Miyazaki, T., Tanigaki, S., Kasai, K., Minegishi, K., Miyakoshi, K., Ishimoto, H., Yoshimura, Y., 2000. Participation of reactive oxygen species in PGF2alpha-induced apoptosis in rat luteal cells. Journal of Reproduction & Fertility 120, 239–245.
263. Torres-Acosta, J.F.J., Hoste, H., 2008. Alternative or improved methods to limit gastro-intestinal parasitism in grazing sheep and goats. Small Ruminant Research. Sheep and Goat Farming: grazing systems of production and development 77, 159–173.
264. Trout, J.P., McDowell, L.R., Hansen, P.J., 1998. Characteristics of the estrous cycle and antioxidant status of lactating holstein cows exposed to heat stress. Journal of Dairy Science 81, 1244–1250.
265. Tucker, E.M., Kilgour, L., 1973. The effect of anaemia on sheep with inherited differences in red cell reduced glutathione (GSH) concentrations. Research in Veterinary Science 14, 306–311.

266. Turk, R., Juretic, D., Geres, D., Turk, N., Rekic, B., Simeon-Rudolf, V., Svetina, A., 2004. Serum paraoxonase activity and lipid parameters in the early postpartum period of dairy cows. Research in Veterinary Science 76, 57–61.
267. Turrens, J.F., 2004. Oxidative stress and antioxidant defenses: a target for the treatment of diseases caused by parasitic protozoa. Molecular Aspects of Medicine Oxidative Stress in Aging and Disease - Mitochondrial Aging, Neuronal Fuction and Neurodegeneration, and Oxidative Metabolic Disorders and Diseases 25, 211–220.
268. Tüzün, A., Erdil, A., Inal, V., Aydin, A., Bagci, S., Yesilova, Z., Sayal, A., Karaeren, N., Dagalp, K., 2002. Oxidative stress and antioxidant capacity in patients with inflammatory bowel disease. Clinical Biochemistry 35, 569–572.
269. Ueland, P.M., Mansoor, M.A., Guttormsen, A.B., Muller, F., Aukrust, P., Refsum, H., Svardal, A.M., 1996. Reduced, oxidized and protein-bound forms of homocysteine and other aminothiols in plasma comprise the redox thiol status – a possible element of the extracellular antioxidant defense system. Journal of Nutrition 126, 1281S–1284S.
270. Upcroft, P., Upcroft, J.A., 2001. Drug targets and mechanisms of resistance in the anaerobic protozoa. Clinical Microbiology Reviews 14, 150–164.
271. Urban-Chmiel, R., 2006. The influence of transport stress on oxidative stress parameters in bovine leukocytes. Slovakian Veterinary Research 43, 243–246.
272. Urban-Chmiel, R., Kankofer, M., Wernicki, A., Albera, E., Puchalski, A., 2009. The influence of different doses of [alpha]-tocopherol and ascorbic acid on selected oxidative stress parameters in in vitro culture of leukocytes isolated from transported calves. Livestock Science 124, 89–92.
273. Van Houtert, M.F.J., Sykes, A.R., 1996. Implications of nutrition for the ability of ruminants to withstand gastrointestinal nematode infections. International Journal for Parasitology 26, 1151–1167.
274. Vega, M., Castillo, T., Retamales, I., Las Heras, J., Devoto, L., Videla, L.A., 1994. Steroidogenic capacity and oxidative stress-related parameters in human luteal cell regression. Free Radical Biology and Medicine 17, 493–499.
275. Velasquez-Pereira, J., McDowell, L.R., Risco, C.A., Prichard, D., Martin, F.G., Calhoun, M.C., Williams, S.N., Wilkinson, N.S., Ogebe, P., 1998. Effects on performance, tissue integrity, and metabolism of vitamin E supplementation for beef heifers fed a diet that contains gossypol. Journal of Animal Science 76, 2871–2884.
276. Velez-Pardo, C., Morales, A.T., Rio, M.J.D., Olivera-Angel, M., 2007. Endogenously generated hydrogen peroxide induces apoptosis via mitochondrial damage independent of NF-[kappa]B and p53 activation in bovine embryos. Theriogenology 67, 1285–1296.
277. Venditti, P., Meo, S., 2006. Thyroid hormone-induced oxidative stress. Cellular and Molecular Life Sciences (CMLS) 63, 414–434.
278. Vincent, H.K., Innes, K.E., Vincent, K.R., 2007. Oxidative stress and potential interventions to reduce oxidative stress in overweight and obesity. Diabetes, Obesity and Metabolism 9, 813–839.
279. Wang, P.G., Xian, M., Tang, X., Wu, X., Wen, Z., Cai, T., Janczuk, A.J., 2002. Nitric oxide donors: chemical activities and biological applications. Chemical Reviews 102, 1091–1134.
280. Warriss, P., Brown, S., Knowles, T., Kestin, S., Edwards, J., Dolan, S., Phillips, A., 1995. Effects on cattle of transport by road for up to 15 hours. Veterinary Record 136, 319–323.
281. Waterman, W.R., John, M.E., Simpson, E.R., 1986. Regulation of synthesis and activity of cytochrome P-450 enzymes in physiological pathways. In: P.R. Ortiz de Montellano (Ed.), Cytochrome P-450: Structure, Mechanism, and Biochemistry, Plenum Press, New York, NY, 345 pp.
282. Weiss, W.P., Smith, K.L., Hogan, J.S., Steiner, T.E., 1995. Effect of forage to concentrate ratio on disappearance of vitamins A and E during in vitro ruminal fermentation. Journal of Dairy Science 78, 1837–1842.

283. Weiss, W.P., Hogan, J.S., Todhunter, D.A., Smith, K.L., 1997. Effect of vitamin E supplementation in diets with a low concentration of selenium on mammary gland health of dairy cows. Journal of Dairy Science 80, 1728–1737.
284. Weiss, W.P., Hogan, J.S., Smith, K.L., 2004. Changes in vitamin C concentrations in plasma and milk from dairy cows after an intramammary infusion of *Escherichia coli*. Journal of Dairy Science 87, 32–37.
285. Wernicki, A., Urban-Chmiel, R., Kankofer, M., Mikucki, P., Puchalski, A., Tokarzewski, S., 2006. Evaluation of plasma cortisol and TBARS levels in calves after short-term transportation. Revue de Médicine Vétérinaire 157, 30–34.
286. Wessely-Szponder, J., Urban-Chmiel, R., Wernicki, A., Bobowiec, R., 2005. Effect of leukotoxin of *M. haemolytica* and LPS of *E. coli* on secretory response of bovine neutrophils in vitro. Polish Journal of Veterinary Science 8, 99–105.
287. Whelan, B.R., Barrow, N.J., Peter, D.W., 1994. Selenium fertilizers for pastures grazed by sheep. 2. Wool and liveweight responses to selenium. Australian Journal of Agricultural Research 45, 877–887.
288. White, C.L., Rewell, L., 2007. Vitamin E and selenium status of sheep during autumn in Western Australia and its relationship to the incidence of apparent white muscle disease. Australian Journal of Experimental Agriculture 47, 535–543.
289. Whittaker, D., 1999. Care of dairy cows during turnout to grass. UKVet 4, 24–28.
290. Williams, A.J., 1976. Metabolism of cysteine by Merino sheep genetically different in wool production. 4. Rates if entry of cystine in to plasma, measured with a single intravenous infection of L-[34S] cystine, and the subsequent incorporation of 35S into wool fibres. Australian Journal of Biological Sciences 29, 513–524.
291. Witko-Sarsat, V., Friedlander, M., Khoa, T.N., Capeillere-Blandin, C., Nguyen, A.T., Canteloup, S., Dayer, J.-M., Jungers, P., Drueke, T., Descamps-Latscha, B., 1998. Advanced oxidation protein products as novel mediators of inflammation and monocyte activation in chronic renal failure1, 2. Journal of Immunology 161, 2524–2532.
292. Witko-Sarsat, V., Nguyen-Khoa, T., Jungers, P., Drueke, T., Descamps-Latscha, B., 1999. Advanced oxidation protein products as a novel molecular basis of oxidative stress in uraemia. Nephrology Dialysis Transplantation 14, 76–78.
293. Woolaston, R.R., Baker, R.L., 1996. Prospects of breeding small ruminants for resistance to internal parasites. International Journal for Parasitology. Novel Approaches to the Control of Helminth Parasites of Livestock 26, 845–855.
294. Wu, G., Fang, Y.-Z., Yang, S., Lupton, J.R., Turner, N.D., 2004. Glutathione metabolism and its implications for health. Journal of Nutrition 134, 489–492.
295. Yagi, K., 1998. Simple assay for the level of total lipid peroxides in serum or plasma. Methods in Molecular Biology (Clifton, N.J.) 108, 101–106.
296. Yagi, Y., Thongnoon, P., Shiono, H., Chikayama, Y., 2002. Increase in oxidized proteins in Theileria sergenti-infected erythrocyte membrane. Journal of Veterinary Medical Science 64, 623–625.
297. Yamashita, T., Someya, A., 1987. Comparative study on the stimulation of superoxide production in guinea-pig eosinophils by the calcium ionophore A23187. Biochimica et Biophysica Acta (BBA) – Molecular Cell Research 927, 359–365.

Antioxidant and Anti-inflammatory Effects of Common Pharmaceuticals and Nutraceuticals

Lester Mandelker and Peter Vajdovich

Abstract These charts list the cellular effects of various herbs, antioxidants and nutrients and their proposed mechanism of action and traditional uses and applications. Much of the information pertains to proposed cellular mechanisms and cellular effects. The information was gathered from scientific sources and experimental research from Medline. The information is new and may be considered controversial and is intended for educational and scientific usage. It is suggested that practitioners use their best judgment and sound reasoning when applying these supplements in clinical situations.

Keywords Glucose autoxidation • Advanced glycation end-products (AGEs) • Polyol pathway • Protein kinase C • Diabetic neuropathy • Diabetic retinopathy • Ketoacidosis • Sorbitol

P. Vajdovich (✉)
Faculty of Veterinary Sciences, Department of Internal Medicine and Clinics,
Szent Istvan University, Istvan u. 2, 1143, Budapest, Hungary
e-mail: Vajdovich.Peter@aotk.szie.hu

L. Mandelker and P. Vajdovich (eds.), *Studies on Veterinary Medicine*,
Oxidative Stress in Applied Basic Research and Clinical Practice 5,
DOI 10.1007/978-1-61779-071-3_14, © Springer Science+Business Media, LLC 2011

The following information consists of proposed cellular effects regarding the use of various antioxidants. The information was gathered from scientific sources and experimental research from Medline. It is suggested that practitioners use their best available judgment and sound reasoning when applying these supplements in clinical situations.

Represented		Proposed mechanism of action	Proposed cellular effects	Traditional and potential uses	Adverse effects	Dose
Common name	Chemicals					
Adaptogens	Schisandra, gingseng	Antistress	Immune stimulant	Health tonic	None	Proportional to human dose
(Prime one[a])	Goldenseal, molasses	Modulates blood sugar	Modulates eicosanoid production	Cardiotonic		
Numerous brands	Rhodiola Ashwagandra	Reduces oxygen Free radicals	Modulates cortiosteroid production	Mood elevator		
Allopurinol	Allopurinol	Reduces vascular oxidative stress Decreases oxidative stress during ischemia–reperfusion	Xanthine oxidase inhibitor, which reduces formulation of free radicals	Antigout Antifibrosis Heart disease	Vomiting, diarrhea, rash, bone marrow, suppression, hepatoxicity	10–20 mg/ kg/day
Avarol	Marine sesquiterpenoid	Anti-inflammatory	Reduces TNF alpha	Anti-arthritis	None noted	10–40 mg/kg
	Hydroquinone	Antiarthritic	Inhibits binding of NF-kB, blocks Cox II	Psoriasis		
		Antiseborrheic	Prostaglandins II			
Alpha-lipoic acid	Dihydrolipoic acid (DL)	Antioxidant Anti-inflammatory effects	Modulates apoptosis	Antioxidant	Adverse effects may be seen in cats at 30 mg	1–5 mgkg/day in dogs Unpublished data suggest that supplementing >30 mg/ day in cats may not be safe

		Metal chelator potentiates levels of vitamin C and vitamin E				
		Reduces oxidative stress	Modulates NF-kB activation Increases cellular glutathione levels and improves cellular redox state	Diabetic neuropathy Ischemia–reperfusion injury		
Bioflavonoids	Bilberry	Antioxidants scavenges	Upregulates antioxidant	Antiaging	None	25–1,000 mg/day
Polyphenols	Pycnogenol Rutin Quercetin Resveratrol	Free radicals Chelates iron	Enzymes catalase and glutathione perioxidase	Pomotes cardiovascular health		
(grape seed extract)	Anthocyanidins Proanthocyanidins	Anti-inflammatory, antimicrobial, hepatoprotective, antithrombotic, antiviral, and anticancer effects	Protects SOD levels Reduce mitochondrial lipid perioxidation Reduce DNA damage May reduce apoptosis May inhibit stress-induced transcription Factor NF-kB May inhibit leukotriene synthesis May reduce adhesion molecules	Immune stimulation retinopathy, lymph edema; antiallergies, cancer, improves wound healing		

(continued)

(continued)

Represented		Proposed mechanism of action	Proposed cellular effects	Traditional and potential uses	Adverse effects	Dose
Common name	Chemicals					
Vitamin B3	Niacinamide	Precursor for NAD+ and improves NAD and NADH content of cells	Reduce reactive oxygen species and oxidative damage to cell, diseases	Antiaging	Unknown	100–500 mg/day
		Essential B vitamin	Decreases calcium influx into cell	Effects immune diseases		
			May decrease telomere shortening			
			May suppress leukocyte chemotaxis			
			Inhibits release of proteases			
Carnitine	L-Carnitine	Reduces oxidative stress	Improve mitochondrial oxidation	Neurodegeneration	Unknown	20–150 mg/kg tid
		Cytoprotective	Transports fatty acids from cytoplasm to mitochondria	Heart disease		
		May modulate cellular stress	Reduces damage to mitochondrial electron chain	Hepatic disease		
		Essential for mitochondrial energy production and vitamin C-dependant synthesis	Improves cellular redox state	Lipidosis		

		Essential factor in fatty acid metabolism		Weight loss		
Carnosine	L-Carnosine	Antioxident, detoxifies free radicals and acts as metal ion scavenger	Scavenger of free radical aldehyde	Antiaging	None reported	Proportional to human dose
			Supresses heme oxygenase activity			
			May reduce glycation of proteins			
Chondroitin	Chondroitin sulfate	Improves synthesis of GAGs, antioxidant effects, antifibrotic effects	Restores SOD and catalase glutathione perioxidase activity in cartilage cells	Osteoarthritis	Unknown	50–1,000 mg/day
		Reduces proteoglycan degradation	May reduce lipid perioxidation		May alter glycemic control	
			May reduce DNA fragmentation			
Coenzyme Q10	Ubiquinone	Has antioxidant; prooxidant effects	Modulates intracellular oxidation reduction reactions	Heart disease	Rare	1–2 mg/kg/day
		Scavenges free radicals	Transfer of reducing equivalents in the mitochondrial electron transport chain	Hepatic disease		
		Essential for mitochondrial energy production and catalyst for ATP production	Protects against lipid perioxidation	Energy production		
			Improves extracellular superoxide dismutase activity	Neuroprotective		
			Improves mitochondrial function and reduces oxidative stress	Cardiomyopathy Hypertension		

(continued)

(continued)

Represented		Proposed mechanism of action	Proposed cellular effects	Traditional and potential uses	Adverse effects	Dose
Common name	Chemicals					
Creatine	Creatine monohydrate	S phosphocreatine is the ready energy source for muscle function	Modulates creatine mitochondrial transport	May enhance athletic performance	Unknown	Label dose on veterinary products
		May promote protein muscle synthesis	Improves ATP synthesis	May ameliorate progress of neuromuscular diseases		
			Reduces cell death	May improve geriatric muscle wasting		
			Alters mitochondrial permeability transition pores			
Carotenoids	B-Carotene	Antioxidant, pro-oxidant	Reduce oxidative stress	Cancer prevention	None	Proportional to human dose
	Lutein	Immune stimulant	May reduce DNA damage			
	Lycopene	Scavenges free radicals	May inhibit oxidant-induced cytokine production			
	Zeaxanathin	Reduces lipid peroxidation	Can form free radicals and inhibit oxidative phosphorylation	Retinal disorders		
	Crytpxanthin		Upregulate connexin genes			
	Astaxanthin		May improve GJIC			
	A-Carotene					

Catechins	Green Tea extract	Anti-inflammatory	Protects SOD levels	Immune stimulation	None reported	50–1,000 mg/day
	Proanthocyanidins	Antimicrobial	Reduce mitochondrial lipid perioxidation			
		Hepatoprotective	Reduce DNA damage	Antiallergy		
		Antithrombotic	May reduce apoptosis			
		Antiviral	May inhibit stress-induced transcription factor NF-κB	Anticancer		
		Anticancer effects	May inhibit leukotriene synthesis			
			May reduce adhesion molecules	Improves wound healing		
			Binds iron			
Choline	Choline chloride	Promotes lipid transport	Precursor to acetylcholine	Heart disease	None	Proportional to human dose
	Phosphatidylcholine (lecithin)	Is a structural component of biologic membranes, neurotransmitters, and transmethylation reactions	Reduces perioxidative damage to cell membranes	Fatigue Senility Performance enhancer		
Deferoxamine	(DFO)	Iron chelator Free radical Scavenger (OH)	Prevents Fenton reaction by chelating iron	Hemochromatosis Volvulus (GDV)	Unknown	10–40 mg/kg IV

(continued)

(continued)

Represented						
Common name	Chemicals	Proposed mechanism of action	Proposed cellular effects	Traditional and potential uses	Adverse effects	Dose
DHA (from fish oil)	Docosahexaenoic acid	Reduces oxidative stress	Modulates cell membrane synthesis and mitochondrial membrane phospholipids	Cancer Immune-mediated disease	Unknown	60 mg/kg/day (day/day) or proportional to human dose
		Immune modulator, omega-3 fatty acid with anti-inflammatory properties	May facilitate apoptosis	Kidney disease		
DHEA	Dehydroepiandro-sterone	CNS antioxidant effects	Improves thioredoxin antioxidant system and cell redox	Lupus	May increase cancer growth abnormalities	5–50 mg/day
		Immune modulator	May stimulate Th2 response	Antiaging		
		Precursor to sex hormones	May modulate NF-jB activation	Neurodegeneration		
		Modulates plasma membrane proteins		Anabolic effects Hypercholes Terolemia		
		Cytoprotective of cells		Mood Elevator Obesity		
DMSO	DMSO	Reduce oxidative stress	Hydroxy radical	Arthritis	Garlic odor (DMSO)	250–1,000 mg (MSM)
MSM	MSM	Anti-inflammatory	Scavenger	Inflammation		

		Natural solvent	May reduce cell membrane injury	Interstitial cystitis		
			Promote GJIC	Spinal cord injury		
DMG	N, N-dimethylglycine (pangamic acid)	Antioxidant	May modulate homocysteine metabolism (by methionine pump)	Athletic endurance	Unknown	1–3 mg/kg; doses of 50–400 mg/day have been recommended
		May improve oxygen use and promote liver detoxification	May enhance both humoral and cell-mediated immune response	Circulatory stimulant		
		May enhance immune function		Hyperlipidemias		
				Metabolic enhancer		
DMAE	Dimethylamino-thanol	Intermediate phospholipid metabolite	May stimulate DNA synthesis in fibroblasts	Antiaging	Unknown	Proportional to human dose
		May enhance insulin activity		Stimulates fibrous tissue		Rarely used clinically

(continued)

(continued)

Represented						
Common name	Chemicals	Proposed mechanism of action	Proposed cellular effects	Traditional and potential uses	Adverse effects	Dose
EPA	Eicoapentaenoic acid	Antioxidant, stabilizes cell membranes	Reduces cellular reactive oxygen species	Arthritis	May reduce blood clotting	60 mg/kg
		Modulates immune functions	Reduces inflammatory cytokines	Cancer		
		Modulates eicosanoid production	Depresses nitric oxide production	Immune modulator		
			May alter cell membrane phospholipids			
		May reduce cancer cachexia	Downregulates NF-kB	Cardiovascular disease		
			May induce apoptosis in cancer cells	Kidney disease		
			May inhibit lipoxygenase			
			May reduce TNF-s ynthesis and IL-2 production			
Glutamine	L-Gutamine	Antioxidant effects	Regulates cell redox	Gastrointestinal supplement for bowel disease	None	500 mg/kg divided daily
		Preferred fuel for enterocytes	Provides fuel for mitochondrial	Conditionally essential in chronic debilitating illness		
		May enhance cellular immunity	May stimulate DNA synthesis			

		Modulate tumor cell metabolism and improve clinical outcome in stress situations	May block apoptosis essential donor for nucleotide precursor synthesis			
			May modulate inflammatory activities of IL-8 and TNF-a precursor to GABA and L-glutamic acid			
GAGs and Precursors	Glucosamine HcI	Reduce oxidative stress to chondrocytes	May inhibit degradative enzymes	Osteoarthritis	May increase bleeding and partial thromboplastin time	50–2,000 mg/day
	Gucosamine sulfate	Building block of cartilage (GAGs)	May inhibit nitric oxide activity	Rheumatoid arthritis		
	Chondroitin sulfate	Modify joint damage	May modulate cytokine activity	Wound healing	Studies done on one brand (cosequin) show no hematologic adverse effects	
		May stimulate synthesis of proteo-glycans in vascular endothelium and bladder mucosa		Feline lower tract disease		

(continued)

(continued)

Represented						
Common name	Chemicals	Proposed mechanism of action	Proposed cellular effects	Traditional and potential uses	Adverse effects	Dose
Ginkgo biloba	Ginkgo flavone	Antioxidant	Improves mitochondrial membrane	May improve memory	Unknown	Proportional to human dose
	Glycosides	Scavengers	Stimulates complex I & III of the mitochondrial electron chain			
		Free radicals	Protects against mitrochondrial oxidative uncoupling	May promote better circulation		
		Improves mitochondrial functions	Forms cell redox ratio of reduced glutathione to oxidized glutathione			
Glutathione (reduced form)	L-glutathione	Primary intracellular antioxidant	GSH/GSSH measures oxidative stress	Beneficial in all pathologic states	Unknown	Proportional to human dose
		Primary agent for detoxification of drugs	Modulates apoptosis	Antiaging		
		Reduces oxidative stress	Primes DNA synthesis	Immune enhancer Liver disease		Absorption is questionable when dosed orally
Lactoferrin	Siderophillin	Anti-inflammatory, antioxidant effects, binds metal ions, antiviral effects	Stimulates glycolysis and mitochondrial ATP	Feline stomatitis	Unknown	40 mg/kg topically

		May enhance immune function	Enhances phagocytic activity Regulates iron activity	Immune disorders		
Mannitol	D-mannitol	Exogenous antioxidant	Hydroxyl radical (OH) Scavenger	Renal failure Osmatic agent Ischemia injuries Cerebral edema Glaucoma	Fluid and electrolyte abnormalities	0.25–1 g/kg dose IV slowly repeated as needed
Melatonin	Melatonin	Antioxidant May enhance T (helper) cell response to antigen	May protect DNA and cellular membranes from oxidative stress	Insomnia Anticancer effects alopecia in certain breeds	Mild hypothermia May suppress male infertility	0.3–5 mg total, depending on animal size
		Regulates circadian cycles May upregulate cell lineage-specific stem cell activity	May stimulate IL-2 release by T helper cells and lymphocytes	Regulates circadian rhythm May be used for episodes of thumder-phobia	May increase adverse lung effects in asthmatics	6 mg tid for alopecia in boxers
Methionine	DL-methionine	Antioxidant, source of sulfur	Component of glutathione redox system	Controversial use in liver support	High doses may increase iron levels	Human dose is 800–1,000 mg/day
		Essential for energy production and muscle building Methionine also has pro-oxidant effects	Precursor to S-adenosyl-methionine		Cause pancreatic damage, or induce neurologic change	

(continued)

(continued)

Represented						
Common name	Chemicals	Proposed mechanism of action	Proposed cellular effects	Traditional and potential uses	Adverse effects	Dose
NAC	*N*-acetyl-L-cysteine	Increases glutatahione levels	Improves cell redox state	Neurodegenerative conditions	Unknown	25 mg/kg tid
		Radical oxygen scavenger	Modulates apoptosis	Bronchitis		
		Improves mitochondrial function	May inhibit TNF-a and stress-mediated NF-jB activation	Liver Toxicity Acetaminophen toxicity		
Phospholipids	Phosphatidylserine	Essential components of cell membranes	Membrane phospholipid that facilitates signal transduction	Cognitive dysfunction	None reported	25–100 mg/kg bid
	Phosphatidylcholine (lecithin)	Regulates CNS neurotransmitters	May reduce proinflammatory effects of lipid peroxidation	Mood enhancer		
		Antioxidant Anti-inflammatory Antifiibrobitic	Decreases leukocyte Infiltration	Depression Kidney damage		
SAM-e	S-adenosyl-methionine	Involved in three major biochemical pathways (transmethylation, transsulfuration, and ainopropylation)	Component of glutathione perioxidase	Liver damage		20 mg/kg/day
		Upregulates genes for proteoglycan synthesis	Inhibits oxidation of lipids	Osteoarthritis		

		May facilitate DNA production and brain neurotransmitters	Scavenges free radicals Protects cellular DNA; hepatic microsomal oxidation and detoxification	Depression in humans Antiaging Acetaminophen toxicity		
Selenium	Selenium sodium	Antioxidant	Increases cellular glutathione levels	Muscular disease	High levels are toxic	5–50 μg/day IV 0.3 mg/kg for acute pancreatitis
	Selenite	Modulates thyroid hormone	Cofactor in many antioxidant enzymes	May help prevent cancer in humans		Lethal dose in dogs is 2 mg/kg
	Sodium selenate	Synergy with vitamin E	Major component of GSH-perioxidase	Useful in managing pancreatitis in dogs		
		Binds to some toxins Improves hepatic function Increases bile flow Aids liver detoxification enzyme activity		Very high levels are hepatoxic		

(continued)

(continued)

Represented		Proposed mechanism of action	Proposed cellular effects	Traditional and potential uses	Adverse effects	Dose
Common name	Chemicals					
Silymarin	Silibinin (milk thistle)	Antioxidant	Regulator of intracellular glutathione	Hepatitis	Rare allergic reaction	50–150 mg/kg orally, bid for hepatotoxicosis
		Anticancer	Reduces mitocondrial oxidation	Hepatic		Lower doses for chronic use
		Hypocholesterolemic effects	Protects against lipid peroxidation	Fibrosis		7–15 mg/kg/day
		Antihepatotoxic agent	Cell membrane stabilizers may increase DNA synthesis	Hyperlipidemias		
SOD	Superoxide dismutase	Anti-inflammatory	Free radical scavenger	Liver detoxicant	Unknown	5–20 IU/kg
		Endogenous antioxidant enzyme	Protects mitochondrial membranes from oxidation	Osteoarthritis		
		Neutralizes superoxide radicals		Inflammatory bowel disease		
				Heart disease		
				Chronic diseases		
Taurine	L-taurine	Antioxidant reduces ischemic damage	Modulates intracellular calcium levels	Congestive heart failure, diabetes	None	250–500 mg bid
		Improves endogenous and exogenous antioxidant defense	Maintains cell membrane	Liver disease		

		Modulates CNS activity (osmoregulation, neuroprotection, and neuromodulation)	Restores glutathione levels in cells	Hypertension		
			Vital for the proper use of sodium potassium and calcium	Anxiety		
			Bile acid conjugation	CNS disturbances		
				Dilated cardiomyopathy of acts and cocker spaniels		
Vitamin C	Ascorbic acid	Antioxidant, pro-oxidant under certain circumstances	Reduces TNFa-induced activation of NF-jB	Allergies	Increases oxalate crystalluriax	50 mg/kg, up to 1,000 mg/day in large dogs
	Sodium ascorbate	Aids synthesis of collagen, catecholamines, steroids, carnitine, iron absorbption	May modulate apoptosis	Chronic inflammation	May interfere with activity of glycosides in urine	Doses of 3–5 g/day have been used without adverse effect in large dogs
	Magnesium	Improves immune function	May protect DNA by reduction of reactive oxygen species	Immune function	Depleted by tetracyclines salicylates	
	Ascorbate	Antihistamine activity	May reduce oxidation of LDLs	Macular degeneration		
	Esterified ascorbate		May improve endothelial GJIC	Cataracts		
				Cancer prevention		
				May reduce effectiveness of chemo-therapeutic agents		

(continued)

(continued)

Represented						
Common name	Chemicals	Proposed mechanism of action	Proposed cellular effects	Traditional and potential uses	Adverse effects	Dose
Vitamin E	D-alpha-tocopherol (active form)	Antioxidant, neuroprotective and antiatherogenic effects	Fat-soluble membrane stabilizer	Cancer prevention	Rare	10–20 IU/kg, up to 800 IU/day for large dogs
	DL-alpha-tocopherol	Suppresses lipid peroxidation	Modulates apoptosis	Cholestatic liver disease	Potential effects of anticoagulants and digoxin	
	Mixed tocopherol	Modulates synthesis of coenzyme A and ATP	Modulates growth factors	Cardio-vascular disease		
	Tocotrienols	Modulates immune response	May decrease androgen concentrations	Diabetes		
		Improves oxygen use	May decrease LDL oxidation	Inflammatory disorders		
			May reduce adhesion molecules	Immune function enhancement		
			May decrease transcription factor NF-kB	Senility		

Zinc	Zinc gluconate	Antioxidant enzyme system factor	Oxygen scavenger through induction of metallothionein	Zinc responsive dermatitis	High doses cause	Zinc methionine, 4 mg/kg/day po
	Zinc sulfate	Antifibrotic stabilizes cell membranes	Zinc-dependent transcription factor	Improves immunity		
	Zinc methionine	Aids protein synthesis	Modulates apoptosis NF-kB	Wound repair	Gastritis	Zinc sulfate, 10 mg/kg/day po
		Component of SOD	Modulates genetic expression of cytokines (IL-2)	Skin disorders	Hemolytic anemia	
			Modulates T-cell function and ratio	Viral infections	Icterus	Zinc gluconate, 5 mg/kg/day po
				Mental health		
				Wilson's disease (binds copper)	Hypotention	
				Sexual maturation		

Index

L. Mandelker and P. Vajdovich (eds.), *Studies on Veterinary Medicine*, Oxidative Stress in Applied Basic Research and Clinical Practice 5,
DOI 10.1007/978-1-61779-071-3, © Springer Science+Business Media, LLC 2011

FSC
www.fsc.org

MIX
Papier aus verantwortungsvollen Quellen
Paper from responsible sources
FSC® C105338

If you have any concerns about our products,
you can contact us on
ProductSafety@springernature.com

In case Publisher is established outside the EU,
the EU authorized representative is:
Springer Nature Customer Service Center GmbH
Europaplatz 3, 69115 Heidelberg, Germany

Printed by Libri Plureos GmbH
in Hamburg, Germany